Fundamental
Concepts
in the Design
of Experiments

Fourth Edition

Fundamental Concepts in the Design of Experiments

Charles R. Hicks

Professor Emeritus of Statistics and Education
Purdue University

Saunders College Publishing
New York   Chicago   San Francisco
Philadelphia   Montreal   Toronto   London
Sydney   Tokyo   Mexico City
Rio de Janeiro   Madrid

Copyright ©1993, 1982, 1973, 1964 by Holt, Rinehart, and Winston, Inc.

Text Typeface: Times Roman
Compositor: Technique Typesetting
Acquisitions Editor: Emily Barrosse
Developmental Editor: Laura Shur
Managing Editor: Carol Field
Project Editor: Kimberly A. LoDico
Copy Editor: Technical Editing Services
Manager of Art and Design: Carol Bleistine
Art Director: Anne Muldrow
Art Assistant: Caroline McGowan
Text Designer: Anne O'Donnell
Cover Designer: Lawrence Didona
Text Artwork: Grafacon, Inc.
Director of EDP: Tim Frelick
Production Manager: Joanne Cassetti
Marketing Manager: Monica Wilson

Printed in the United States of America

Fundamental Concepts in the Design of Experiments, 4th Edition

0-03-097710-X

Library of Congress Catalog Card Number: 93-083224

345 169 987654321

Contents

Preface xi

1 The Experiment, the Design, and the Analysis 1

1.1 Introduction to Experimental Design 1
1.2 The Experiment 3
1.3 The Design 4
1.4 The Analysis 6
1.5 Summary in Outline 6
1.6 Examples 7
Problems 14

2 Review of Statistical Inference 16

2.1 Introduction 16
2.2 Estimation 18
2.3 Tests of Hypotheses 20
2.4 The Operating Characteristic Curve 22
2.5 How Large a Sample? 25
2.6 Application to Tests on Variances 26
2.7 Application to Tests on Means 28
2.8 Application to Tests on Proportions 32
2.9 Analysis of Experiments with SAS 33
Problems 45

3 Single-Factor Experiments with No Restrictions on Randomization 49

3.1 Introduction 49
3.2 Analysis of Variance Rationale 51
3.3 After ANOVA—What? 56

3.4 Tests on Means 56
3.5 Confidence Limits on Means 64
3.6 Components of Variance 64
3.7 Summary and Assumptions 69
3.8 Analysis of Residuals 70
3.9 SAS Programs for ANOVA and Tests after ANOVA 74
Problems 87

4 Single-Factor Experiments—Randomized Block and Latin Square Designs 93

4.1 Introduction 93
4.2 Randomized Complete Block Design 95
4.3 ANOVA Rationale 99
4.4 Missing Values 101
4.5 Latin Squares 103
4.6 Interpretations 105
4.7 Graeco–Latin Squares 106
4.8 Extensions 107
4.9 Summary 107
4.10 SAS Programs for Randomized Blocks and Latin Squares 108
Problems 115

5 Factorial Experiments 121

5.1 Introduction 121
5.2 Factorial Experiments 126
5.3 Interpretations 129
5.4 ANOVA Rationale 131
5.5 One Observation per Treatment 136
5.6 Summary 138
5.7 SAS Programs for Factorial Experiments 138
Problems 148

6 Fixed, Random, and Mixed Models 156

6.1 Introduction 156
6.2 Single-Factor Models 157
6.3 Two-Factor Models 158
6.4 EMS Rules 160
6.5 EMS Derivations 165
6.6 The Pseudo-F Test 167
6.7 Remarks 169
Problems 170

7 Nested and Nested-Factorial Experiments 173

7.1 Introduction 173
7.2 Nested Experiments 174
7.3 ANOVA Rationale 178
7.4 Nested-Factorial Experiments 179
7.5 Repeated-Measures Design and Nested-Factorial Experiments 184
7.6 Summary 189
7.7 SAS Programs for Nested and Nested-Factorial Experiments 189
Problems 193

8 Experiments of Two or More Factors—Restrictions on Randomization 201

8.1 Introduction 201
8.2 Factorial Experiment in a Randomized Block Design 201
8.3 Factorial Experiment in a Latin Square Design 207
8.4 Remarks 208
8.5 Summary 209
8.6 SAS Programs 210
Problems 213

9 2^f Factorial Experiments 217

9.1 Introduction 217
9.2 2^2 Factorial 217
9.3 2^3 Factorial 225
9.4 2^f Remarks 228
9.5 Summary 229
Problems 230

10 3^f Factorial Experiments 235

10.1 Introduction 235
10.2 3^2 Factorial 236
10.3 3^3 Factorial 246
10.4 Summary 254
10.5 SAS Programs 255
Problems 255

11 Factorial Experiment—Split-Plot Design 258

11.1 Introduction 258
11.2 A Split-Split-Plot Design 266

11.3 Summary 270
11.4 SAS Programs 270
Problems 275

12 Factorial Experiment—Confounding in Blocks 278

12.1 Introduction 278
12.2 Confounding Systems 279
12.3 Block Confounding—No Replication 282
12.4 Block Confounding with Replication 287
12.5 Summary 296
12.6 SAS Programs 297
Problems 300

13 Fractional Replication 303

13.1 Introduction 303
13.2 Aliases 304
13.3 2^f Fractional Replications 306
13.4 Design Resolution 310
13.5 3^f Fractional Factorials 312
13.6 Plackett–Burman Designs 317
13.7 Summary 318
13.8 SAS Programs 319
Problems 321

14 Taguchi Approach to the Design of Experiments 325

14.1 Introduction 325
14.2 $L_4(2^3)$ 325
14.3 $L_8(2^7)$ 330
14.4 $L_{16}(2^{15})$ 337
14.5 $L_9(3^4)$ 339
14.6 Some Other Taguchi Designs 344
14.7 Summary 345
Problems 346

15 Regression 347

15.1 Introduction 347
15.2 Linear Regression 347
15.3 Curvilinear Regression 354
15.4 Orthogonal Polynomials 357

15.5 SAS Programs 358
15.6 Multiple Regression 365
15.7 SAS Programs 367
15.8 Summary 376
Problems 377

16 Miscellaneous Topics 387

16.1 Introduction 387
16.2 Covariance Analysis 387
16.3 Response-Surface Experimentation 402
16.4 Evolutionary Operation (EVOP) 413
16.5 Analysis of Attribute Data 423
16.6 Randomized Incomplete Blocks—Restriction on Experimentation 427
16.7 Youden Squares 434
Problems 436

Summary and Special Problems 441

Glossary of Terms 448

References 453

Statistical Tables 455

Table A Areas under the Normal Curve 455
Table B Student's t Distribution 457
Table C Cumulative Chi-Square Distribution 458
Table D Cumulative F Distribution 459
Table E.1 Upper 5 Percent of Studentized Range q^* 469
Table E.2 Upper 1 Percent of Studentized Range q 470
Table F Coefficients of Orthogonal Polynomials 471

Answers to Odd-Numbered Problems 473

Index 503

Preface

It is the primary purpose of this book to present the fundamental concepts in the design of experiments using simple numerical problems, many from actual research work. These problems are selected to emphasize the basic philosophy of design. Another purpose is to present a logical sequence of designs that fit into a consistent outline; for every type of experiment, the distinctions among the experiment, the design, and the analyses are emphasized. Since this theme of experiment–design–analysis is to be highlighted throughout the text, the first chapter presents and explain this theme. The second chapter reviews statistical inference, and the remaining chapters follow a general outline, with an experiment–design–analysis summary at the end of each chapter.

The book is written for anyone engaged in experimental work who has a good background in statistical inference. It will be most profitable reading to those with a background in statistical methods including analysis of variance.

This text, now in its fourth edition, introduces the Statistical Analysis System (SAS). Computer programs from SAS have been added to most of the examples. Additionally, several new topics have been introduced, including the analysis of residuals, the concept of resolution in fractional replications, the Plackett–Burman designs, and Taguchi design procedures.

New problems have been added to most chapters, all of which are real problems that have arisen during my consulting experiences.

The order of chapter presentations has been changed to better conform to the way courses are usually taught. In this vein, the new Chapter 14, "Taguchi Approach to the Design of Experiments," follows the Chapter 13 presentation of fractional factorials.

I am grateful to McElrath and Associates, Inc., and feel particularly indebted to Mr. P. Duane Saylor, a colleague with McElrath and Associates, for developing all of the SAS programs that appear in the examples in the text. These programs have been used intensively in my consulting work and have proved to be very helpful in analyzing statistical data.

I am also grateful to Purdue University and to my wife, Helen, for her patience and help as we worked through this edition.

Charles R. Hicks
Lafayette, Indiana

The Experiment, the Design, and the Analysis

1

1.1 Introduction to Experimental Design

One of the most abused words in the English language is probably the word "research." Research means many things to many people. The high school student who visits the library to gather information on the Crusades is doing research. A man who checks the monthly sales of his company for the year 1980 is doing research. Many such studies do make use of bibliothecal or statistical tools or similar research tools but are a far cry from a scientist's conception of the term. Let us narrow the concept a bit and define *research* as "a systematic quest for undiscovered truth." (See Leedy [17].[1])

In this definition we note that the quest must be "systematic," using sound scientific procedures—not a haphazard, seat-of-the-pants approach—to solving a problem. The truth sought should be something that is not already known, which implies a thorough literature search to ascertain that the answer is not available in previous studies.

Not all studies are necessarily research, nor is all research experimental. A *true experiment* may be defined as a study in which certain independent variables are manipulated, their effect on one or more dependent variables is determined, and the levels of these independent variables are assigned at random to the experimental units in the study.

An *experimental unit* must be carefully defined. It may be a part, a person, a class, a lot of parts, a month, or some other unit depending on the problem at hand. The experimental units make up a *universe* of all present or future such units, and a *population* is defined as measurements that might be taken on all experimental units in the universe. There may be several populations defined on the same experimental units. For example, if the universe is made up of all persons in a given school, we may conceive of two populations: one made up

---

[1] Numbers in brackets refer to the references at the end of the book.

of the heights of all these persons and another of the weights of these same persons.

Manipulation and randomization are essential for a true experiment from which one may be able to infer cause and effect. It is not always physically possible to assign some variables to the experimental units at random but it may be possible to run a quasi-experiment in which groups of units are assigned to various levels of the independent variable at random. For example, to study the effects of two methods of in-plant instruction on worker output, it may not be possible to assign workers at random to the two methods, but only to decide by a random flip of a coin which two of four intact classes will receive method I and which two will receive method II. In such quasi-experiments the experimenter is obligated to demonstrate that the four classes were similar at the start of the study. Since they can be dissimilar in many different ways, this is no easy task. Randomizing all individuals to the four groups will assure a high probability of group equality on many variables.

There are other types of research that are not experimental. One of the most common is *ex-post-facto* research. The variables have already acted and the research only measures what has occurred. An example of some note would be the studies showing a high incidence of cancer in people who smoke heavily. There is no manipulation here as the researchers do not assign X number of cigarettes per day to some people and none to another group, and then wait to see whether or not cancer develops. Instead they study what they can find from people's historical records, and then attempt to make an inference. However, such inferences are dangerous, as the researchers must be able to rule out other competing hypotheses as to what may have caused cancer in the people studied.

In this category are survey research, which is also ex post facto, historical research, and developmental research. All of these may use statistical methods to analyze data collected but none are really experimental in nature.

An illustration familiar to statisticians is found in the techniques of regression and correlation. In a regression problem levels of the independent variable X are set and observations are made on the dependent variable Y, and the given level of X is assigned at random to each of the experimental units in the study. The purpose of such a study is to find an equation for predicting Y from X over the range of specified X's. In a correlation problem each experimental unit is measured on two variables, the n pairs (X, Y) are plotted to look for a relationship, and certain statistics are computed to indicate the strength of the relationship. In this case a strong correlation (high r) does not imply that X caused Y or that Y caused X. Hence regression analysis can often be used in a true experiment whereas correlational techniques are more likely to be used in ex-post-facto studies. This distinction between regression and correlation is often misunderstood.

Assuming that true experiments are our basic concern, what steps should be taken to make a meaningful study?

To attack this problem one may easily recognize three important phases of every project: the experimental or planning phase, the design phase, and the analysis phase. The theme of this book might be called experiment, design, and analysis.

1.2 The Experiment

The experiment includes a statement of the problem to be solved. This sounds rather obvious, but in practice it often takes quite a while to get general agreement as to the statement of a problem. It is important to bring out all points of view to establish just what the experiment is intended to do. A careful statement of the problem goes a long way toward its solution.

Too often problems are tossed about in very general terms, a sort of "idea burping" (see [17]). As an example, note the difference in these two statements of a problem: What can Salk vaccine do for polio? Is there a difference in the percentage of first, second, and third grade children in the United States who contract polio within a year after being inoculated with Salk vaccine and those who are not inoculated? Much care must be taken to spell out the problem in terms that are well understood and that point the way in which the research might be conducted.

The statement of the problem must include reference to one or more criteria to be used in assessing the results of the study. The criterion is also called the dependent variable or the response variable Y; and there may be more than one Y of interest in a given study. The dependent variable may be qualitative, such as the number or percentage contracting polio, or quantitative, such as yield in grams of penicillin per kilogram of corn steep. Knowledge of this dependent variable is essential because the shape of its distribution often dictates what statistical tests can be used in the subsequent data analysis.

One must also ask: Is the criterion measurable, and if so, how accurately can it be measured? What instruments are necessary to measure it?

It is also necessary to define the independent variables or factors that may affect the dependent or response variable. Are these factors to be held constant, to be manipulated at certain specified levels, or to be averaged out by a process of randomization? Are levels of the factors to be set at certain fixed values, such as temperature at 70°F, 90°F, and 110°F, or are such levels to be chosen at random from among all possible levels? Factors whose levels are set at specified values are called *fixed effects,* and those set at random levels *random effects*. Are the factors to be varied quantitative (such as temperature) or qualitative (operators)? And how are the various factor levels to be combined?

If one factor A is set at three levels and a second factor B is set at two levels, can one set all six combinations as follows?

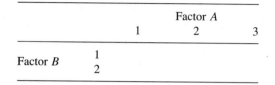

| | | Factor A | |
	1	2	3
Factor B 1			
2			

This is called a factorial experiment as all levels of *A* are combined with all levels of *B*. Sometimes one chooses three suppliers and then randomly selects two batches from each supplier to be analyzed chemically. Here the two batches are unique to a given supplier and the resulting arrangement is a nested or hierarchical arrangement. A *nested or hierarchical experiment* is one in which the levels of one factor are chosen within the levels of another factor.

| | Supplier | |
1	2	3
Batch	Batch	Batch
1 2	3 4	5 6

This involves six sets of data as did the above factorial, but now the factors are in a decidedly different arrangement.

All of these points should be considered as part of the experiment phase of any research.

1.3 The Design

Of primary importance, since the remainder of this book will be devoted to it, is the design phase of a project. Many times an experiment is agreed upon, data are collected, and conclusions are drawn with little or no consideration given to *how* the data were collected. First, how many observations are to be taken? Considerations of how large a difference is to be detected, how much variation is present, and what size risks are to be tolerated are all important in deciding on the size of the sample to be taken for a given experiment. Without this information the best alternative is to take as large a sample as possible. In practice this sample size is often quite arbitrary. However, as more and more tables become available, it should be possible to determine the sample size in a much more objective fashion (see Section 2.5).

Also of prime importance is the order in which the experiment is to be run, which should be random order. Once a decision has been made to control certain variables at specified levels, there are always a number of other variables that cannot be controlled. Randomization of the order of experimentation will tend to average out the effect of these uncontrolled variables.

For example, if an experimenter wishes to compare the average current flow through two types of computers and five of each type are to be tested, in what order are all ten to be tested? If the five of type I are tested, followed by the five of type II, and any general "drift" in line voltage occurs during the testing, it may appear that the current flow is greater on the first five (type I) than on the second five (type II), yet the real cause is the "drift" in line voltage. A random order for testing allows any time trends to average out. It is desirable to have the average current flow in the type I and type II computers equal, if the computer types do not differ in this respect. Randomization will help accomplish this. Randomization will also permit the experimenter to proceed as if the errors of measurement were independent, a common assumption in most statistical analyses.

What is meant by random order? Is the whole experiment to be completely randomized, with each observation made only after consulting a table of random numbers, or tossing dice, or flipping a coin? Or, possibly, once a temperature bath is prepared, is the randomization made only within this particular temperature and another randomization made at another temperature? In other words, what is the randomization procedure and how are the units arranged for testing? Once this step has been agreed upon, it is recommended that the experimenter keep a watchful eye on the experiment to see that it is actually conducted in the order prescribed.

Having agreed upon the experiment and the randomization procedure, a mathematical model can now be set up, which should describe the experiment. This model will show the response variable as a function of all factors to be studied and any restrictions imposed on the experiment as a result of the method of randomization.

Since the objective of the research project is to shed light on a stated problem, one should now express the problem in terms of a testable hypothesis or hypotheses. A research hypothesis states what the experimenter expects to find in the data. For example: "Inoculation with Salk vaccine will lower the incidence of polio." Or: "The four treatments will produce different average yields."

To test hypotheses, a statistician usually states the hypothesis in null form, since this is the only testable hypothesis. For example: "There will be no difference in the percentage of youngsters contracting polio between those inoculated and those not inoculated." Or: "The treatment means will have the same average yield." This null hypothesis form can usually be expressed in terms of the mathematical model set up in the last step of the design phase.

In this discussion of variables that influence the criterion either by being purposefully introduced into the study, representing restrictions on randomization, or being somewhat extraneous, one notes that there are basically three ways to handle independent variables in an experiment:

1. Rigidly controlled, the variables remain fixed throughout the experiment, which, of course, means that any inferences drawn from the experiment are

valid only for these fixed conditions. Here, for example, one might decide to study only children from the first three grades.

2. Manipulated or set at levels of interest. These would be the variables whose effect on Y are to be studied. They could be either qualitative or quantitative and their levels either fixed or random. To be effective, every effort should be made to include the extreme levels of such variables to provide opportunity to maximize their effect, if present.

3. Randomized. The order of experimentation is randomized to average out the effects of variables that cannot be controlled. Such averaging does not remove their effect completely as they still increase the variance of the observed data. Proper design of the experiment can also reduce or minimize the experimental error when such factors are anticipated.

A basic principle in such experimentation is referred to as the *max-min-con principle* (see Kerlinger [15]), where we seek to maximize the effect of the independent variables of interest, minimize the error variance, and control some of the variables at specified levels.

1.4 The Analysis

The final step, analysis, includes the procedure for data collection, data reduction, and the computation of certain test statistics to be used in making decisions about various aspects of an experiment. Analysis involves the computation of test statistics such as t, F, χ^2, and their corresponding decision rules for testing hypotheses about the mathematical model. Once the test statistics have been computed, decisions must be made. These decisions should be made in terms that are meaningful to the experimenter. They should not be couched in statistical jargon such as "the third-order $A \times B \times E$ interaction is significant at the 1 percent level," but instead should be expressed in graphical or tabular form, in order that they be clearly understood by the experimenter and by those persons who are to be "sold" by the experiment. The actual statistical tests should probably be included only in the appendix of a report on the experiment. These results should also be used as "feedback" to design a better experiment, once certain hypotheses seem tenable.

1.5 Summary in Outline

I. *Experiment*
   A. Statement of problem
   B. Choice of response or dependent variable
   C. Selection of factors to be varied

D. Choice of levels of these factors
1. Quantitative or qualitative
2. Fixed or random
E. How factor levels are to be combined

II. *Design*
A. Number of observations to be taken
B. Order of experimentation
C. Method of randomization to be used
D. Mathematical model to describe the experiment
E. Hypotheses to be tested

III. *Analysis*
A. Data collection and processing
B. Computation of test statistics
C. Interpretation of results for the experimenter

*F) How other Variables in the Exp-
are to be Handled*
1) Blocking - Background Variable
2) Randomize - Uncontrolled Var.
3) Variables Held Constant
4) Co Variables

*G) Obtain Preliminary Estimates
& Repeatability*

1.6 Examples

■ **EXAMPLE 1.1 Salk Vaccine Experiment.**   The experiment conducted in the United States in 1954 concerning the use of Salk vaccine in the control of polio has become a classic in its employment of good design principles. A more detailed description of the study can be found in *Statistics: A Guide to the Unknown* (see [28]).

Before the statement of the problem was agreed upon as given in Section 1.2, there was much discussion as to what universe was to be sampled. Clearly the experimental unit is a child and the universe was considered to be made up of all children in the United States in grades 1, 2, and 3. The population consists of 0's and 1's, meaning the child did not (0) contract polio or the child did (1) contract polio. Children in grades 1, 2, and 3 were chosen as most appropriate, as it was among children of these ages that the disease was most prevalent.

Implied in the statement of the problem "Is there a difference in the percentage of first, second, and third grade children in the United States contracting polio within a year after being inoculated with Salk vaccine and those not inoculated?" is the concept of a criterion. The stated criterion is the percentage of children contracting polio. This dependent variable could be either the percent contracting the disease or the number contracting it. In discussing this variable it was noted that the overall percentage of children contracting polio was quite small—about 8 or 10 cases in 10,000 children. It was therefore necessary to take a fairly large sample in order to get any percentages large enough to discriminate between those who were inoculated and those who were not. The basic variable here is a qualitative one—a binomial variable—as children fall into just two categories: contracting polio or not contracting polio. One can safely assume an approximately normal distribution of this Y if the average number in the sample is expected to be four or more.

The main factor of interest in affecting Y is whether or not the child is inoculated, so it was planned to have an approximately equal number of children receive the vaccine and be given a placebo (a salt solution). This latter group would act as a control.

Other factors that might affect the contraction of polio included socioeconomic status, grade level, doctors who make the diagnosis, and geographic area of residence. Grade level was considered and they agreed to sample from the first, second, and third grades. The other factors, or nuisance variables, were handled in the design procedures.

Since there is only one qualitative, fixed factor of interest—inoculated or not inoculated—there is no basic concern about combining factors. However, because three grade levels were to be considered, it was decided to stratify their sample by grades; that is, take approximately equal sample sizes of inoculated and noninoculated children from each grade level.

Much planning went into the design phase. At first it was proposed that children in grades 1 and 3 receive the vaccine and those in grade 2 be given the placebo treatment. This plan was later abandoned because polio is a contagious disease, and if children in a given grade get polio, others in that grade might contract it as well. Another early idea was to use children of parents who would consent to the inoculation as "experimental" and children of parents not consenting as a "control." This too was abandoned because parents of a higher socioeconomic status are more likely to consent than parents with a lower socioeconomic status and polio is a socially related disease—it affects more children from the upper echelons of society than from the lower!

The final decision was to do a complete randomization countrywide involving about 1 million randomly selected children from grades 1 to 3 in areas of the United States where polio had been quite prevalent in the past. Of these 1 million, more than 400,000 parents consented to have their children participate in the study. These children were then divided into two groups at random, with group 1 getting the Salk vaccine and group 2 getting the placebo. To handle other concerns, the experiment was run as a "double-blind" experiment. The children did not know whether they were receiving the vaccine or the salt solution and the doctors who diagnosed the children did not know which children were inoculated and which were not.

The samples of approximately 200,000 each seemed adequate to permit the detection of differences in the percent contracting the disease, if it did show up.

A mathematical model could be written as

$$Y_{ij} = \mu + \tau_j + \varepsilon_{ij}$$

where Y_{ij} is equal to 0 or 1: 0 if no polio is diagnosed and 1 if polio is found. The subscript i represents the child, a number from 1 to approximately 200,000, and j is 1 or 2: 1 if treated with Salk vaccine and 2 if not so treated. μ is a constant or general average of the 0's and 1's for the whole population. τ_j is the treatment effect: $j = 1$ if treated, $j = 2$ if not treated. ε_{ij} is a random error associated with a child receiving treatment j.

To see the hypothesis to be tested, note that

$$p_j = \frac{\sum_{i=1}^{n_j} Y_{ij}}{n_j}$$

is the proportion of children contracting polio who received treatment j. n_j is the number who received treatment j.

The statistical hypothesis to be tested is then

$$H_0: p_1' = p_2'$$

with alternative

$$H_1: p_1' < p_2'$$

where the sign $<$ indicates that we are interested in showing that the true proportion contracting polio will be less for treatment 1 (vaccine) than for treatment 2 (placebo). The primes on the p values are to show that the hypotheses are statements about population proportions whereas the samples will give p_1 and p_2—observed sample proportions. The data are shown in Table 1.1.

Table 1.1 Polio Experiment Results

Treatment	Sample Size	Number Contracting Polio	Proportion Contracting Polio
Salk vaccine	$n_1 = 200,745$	56	$p_1 = 28 \times 10^{-5}$
Placebo	$n_2 = 201,229$	142	$p_2 = 71 \times 10^{-5}$
Totals	401,974	198	$\hat{p} = 49 \times 10^{-5}$

The test statistic for testing the hypothesis stated is

$$z = \frac{p_1 - p_2}{\sqrt{\hat{p}\hat{q}(1/n_1 + 1/n_2)}}$$

where $\hat{p}$ is the proportion contracting polio in the total sample of $n_1 + n_2$ children. Here $\hat{q} = 1 - \hat{p}$.

If one now assumes an α risk of 0.001—taking one chance in 1000 of rejecting H_0 when it is true—the rejection region for z is given by $z < -3.09$ (from the normal distribution table). Substituting the values above:

$$z = \frac{28 \times 10^{-5} - 71 \times 10^{-5}}{\sqrt{49 \times 10^{-5}(1 - 49 \times 10^{-5})(1/200{,}745 + 1/201{,}229)}}$$

$$z = -6.14$$

which gives a very, very strong indication that we should reject H_0 and conclude that the vaccine was indeed effective.

In fact, the probability of getting a z as low as or lower than -6.14 when there is really no difference in the two groups is less than one chance in a billion. These amazing results, of course, have been strongly substantiated by the almost complete eradication of polio in the United States. ■

■ **EXAMPLE 1.2**   The following example [19] is presented to show the three phases of the design of an experiment. It is not assumed that the reader is familiar with the design principles or analysis techniques in this problem. The remainder of the book is devoted to a discussion of many such principles and techniques, including those used in this problem.

An experiment was to be designed to study the effect of several factors on the power requirements for cutting metal with ceramic tools. The metal was cut on a lathe and the y or vertical component of a dynamometer reading was recorded. As this y component is proportional to the horsepower requirements in making the cut, it was taken as the measured variable. The y component is measured in millimeters of deflection on a recording instrument. Some of the factors that might affect this deflection are tool types, angle of tool edge bevel, type of cut, depth of cut, feed rate, and spindle speed. After much discussion, it was agreed to hold depth of cut constant at 0.100 in., feed rate constant at 0.012 in./min, and spindle speed constant at 1000 rpm. These levels were felt to represent typical operating conditions. The main objective of the study was to determine the effect of the other three factors (tool type, angle of edge bevel, and type of cut) on the power requirements. As only two ceramic tool types were available, this factor was considered at two levels. The angle of tool edge bevel was also set at two levels, 15° and 30°, representing the extremes for normal operation. The type of cut was either continuous or interrupted—again, two levels.

There are therefore two fixed levels for each of three factors, or eight experimental conditions (2^3), which may be set and which may affect the power requirements or y deflection on the dynamometer. This is called a 2^3 factorial experiment, since both levels of each of the three factors are to be combined with both levels of all other factors. The levels of two factors (type of tool and

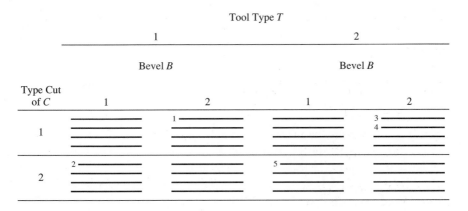

Figure 1.1 Data layout of power consumption for ceramic tools.

type of cut) are qualitative, whereas the angle of edge bevel (15° and 30°) is a quantitative factor.

The question of design for this experiment involves the number of tests to be made under each of the eight experimental conditions. After some preliminary discussion of expected variability under the same set of conditions and the costs of wrong decisions, it was decided to take four observations under each of the eight conditions, making a total of 32 runs. The order in which these 32 units were to be put in a lathe and cut was to be completely randomized.

In order to completely randomize the 32 readings, the experimenter decided on the order of experimentation from the results of three coin tossings. A penny was used to represent the tool type T: heads for one type, tails for the other; a nickel represented the angle of bevel B: heads 30°, tails 15°; and a dime represented the type of cut C: heads interrupted, tails continuous.

Thus if the first set of tosses came up *THT*, it would mean that the first tool to be used in the lathe would be of tool type 1, bevel 2, and subjected to the continuous type of cut. A data layout is given in Figure 1.1 that shows each of the 32 experimental conditions. The numbers 1, 2, 3, 4, and 5 indicate the first five conditions to be run on the lathe, assuming the coins came up *THT, TTH, HHT, HHT, HTH*.

In this layout note that the same set of conditions may be repeated (for example, runs 3 and 4) before all eight conditions are run once. The only restriction on complete randomization here is that once four repeated measures have occurred in the same cell, no more will be run using those same conditions.

The coin flipping continues until the order of all 32 runs has been decided upon. This is a 2^3 factorial experiment with four observations per cell, run in a completely randomized manner. Complete randomization ensures the averaging out of any effects that might be correlated with the time of the experiment. If the lathe-spindle speed should vary and all of the type 1 tools are run through first and the type 2 tools follow, this extraneous effect of lathe-spindle speed might appear as a difference between tool types if the speed were faster at first and slower near the end of the experiment.

The mathematical model for this experiment and design would be

$$Y_{ijkm} = \mu + \underbrace{T_i + B_j + TB_{ij} + C_k + TC_{ik} + BC_{jk} + TBC_{ijk}}_{model} + \underbrace{\varepsilon_{m(ijk)}}_{Error}$$

where Y_{ijkm} represents the measured variable, μ a common effect in all observations (the true mean of the population from which all the data came), T_i the tool type effect where $i = 1, 2$, B_j the angle of bevel where $j = 1, 2$, and C_k the type of cut where $k = 1, 2$. $\varepsilon_{m(ijk)}$ represents the random error in the experiment where $m = 1, 2, 3, 4$. The other terms stand for interactions between the main factors T, B, and C.

The analysis of this experiment consists of collecting 32 items of data in the spaces indicated in Figure 1.1 in a completely randomized manner. The results in millimeter deflection are given in Table 1.2.

Table 1.2 Data for Power Requirement Example

	Tool Type T			
	1		2	
	Bevel Angle B		Bevel Angle B	
Type of Cut C	15°	30°	15°	30°
Continuous	29.0	28.5	28.0	29.5
	26.5	28.5	28.5	32.0
	30.5	30.0	28.0	29.0
	27.0	32.5	25.0	28.0
Interrupted	28.0	27.0	24.5	27.5
	25.0	29.0	25.0	28.0
	26.5	27.5	28.0	27.0
	26.5	27.5	26.0	26.0

This experiment and the mathematical model suggest a three-way analysis of variance (ANOVA), which yields the results in Table 1.3.

In testing the hypotheses that there is no type of tool effect, no bevel effect, no type of cut effect, and no interactions if all mean squares are tested against the error mean square of 2.23 with 24 degrees of freedom (df), the proper test statistic is the F statistic (Statistical Table D) with 1 and 24 df. At the 5 percent significance level ($\alpha = 0.05$) the critical region of F if $F \geq 4.26$. Comparing each mean square with the error mean square indicates that only two hypotheses can be rejected: bevel has no effect on deflection, and type of cut has no effect on

Table 1.3 ANOVA for Power Requirement Example

Source of Variation	Degrees of Freedom (df)	Sum of Squares (SS)	Mean Square (MS)	F	PR > F
Tool type	1	2.82	2.82	1.27	0.2715
Bevel	1	20.32	20.32	9.13*	0.0059
T, B interaction	1	0.20	0.20	0.09	0.7696
Type of cut	1	31.01	31.01	13.93*	0.0010
T, C interaction	1	0.01	0.01	0.00	0.9533
B, C interaction	1	0.94	0.94	0.42	0.5209
T, B, C interaction	1	0.19	0.19	0.09	0.7696
Error	24	53.44	2.23		
Totals	31	108.93			

* One asterisk indicates significance at the 5 percent level; two, the 1 percent level; and three, the 0.1 percent level. This notation will be used throughout this book.

$Total = 32$ $df\ 1 - Total = 31$

deflection. None of the other hypotheses can be rejected, and it is concluded that only the angle of bevel and type of cut affect power consumption as measured by the y deflection on the dynamometer. Tool type appears to have little effect on the y deflection, and all interactions are negligible. Calculations on the original data of Table 1.2 show the average y deflections given in Table 1.4.

Table 1.4 Average y Deflections

Tool type T	1	28.1	⎤ .6
	2	27.5	⎦
Bevel angle B	15°	27.0	⎤ 1.6
	30°	28.6	⎦
Type of cut C	1	28.8	⎤ 2.0
	2	26.8	⎦

These averages seem to bear out the conclusions that bevel affects y deflection, with a 30° bevel requiring more power than the 15° bevel (note the difference in average y deflection of 1.6 mm), and that type of cut affects y deflection, with a continuous cut averaging 2.0 mm more deflection than an interrupted cut. The difference of 0.6 mm in average deflection due to tool type is not significant at the 5 percent level. Graphing all four B–C combinations in Figure 1.2 shows the meaning of no significant interaction.

A brief examination of this graph indicates that the y deflection is increased by an increase in degree of bevel. The fact that the line for the continuous cut

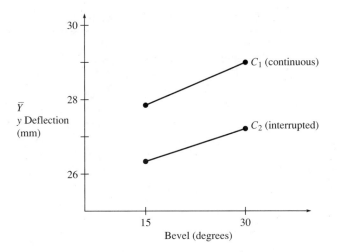

Figure 1.2 $B \times C$ interaction for power requirement example.

C_1 is above the line for the interrupted cut C_2 shows that the continuous cut requires more power. The fact that the lines are nearly parallel is characteristic of no interaction between two factors. Or it can easily be seen that an increase in the degree of bevel produced about the same average increase in y deflection regardless of which type of cut was made. This is another way to interpret the presence of no interaction.

The experiment here was a three-factor experiment with two levels for each factor. The design was a completely randomized design and the analysis was a three-way ANOVA with four observations per cell. From the results of this experiment the experimenter not only found that two factors (angle of bevel and type of cut) affect the power requirements, but also determined that within the range of the experiment it makes little difference which ceramic tool type is used and that there are no significant interactions among the three factors. ∎

These are but two examples that might have been used to illustrate the three phases of the project: experiment, design, and analysis.

Problems

1.1 In your own area of interest, write out the statement of a problem that you believe is researchable.

1.2 For the problem stated in (1), name the experimental unit.

1.3 For the problem stated in (1), list the dependent variable or variables of interest and tell how you propose to measure them.

1.4 For the problem stated in (1), list the independent variables that might affect your dependent variable and explain how you would handle each variable.

1.5 Prepare a data layout for collecting data on your problem.

1.6 From your experience name some experimental unit and name at least two populations that might be generated on these same experimental units.

Review of Statistical Inference

2

2.1 Introduction

In any experiment the experimenter is attempting to draw certain inferences or make a decision about some hypothesis or "hunch" concerning the situation being studied. Life consists of a series of decision-making situations. As you taste your morning cup of coffee, almost unconsciously you decide that it is better than, the same as, or worse than the coffee you have been drinking in the past. If, based on your idea of a "standard" cup of coffee, the new cup is sufficiently "bad," you may pour it out and make a new batch. If you believe it superior to your "standard," you may try to determine whether the brand is different than the one you are used to or the brewing is different, and so forth. In any event, it is the extreme differences that may cause you to decide to take action. Otherwise, you behave as if nothing has changed. In any decision you run the risk that the decision is wrong since it is based on a small amount of data.

In deciding whether or not to carry an umbrella on a given morning, you "collect" certain data: you tap the barometer, look at the sky, read the newspaper forecast, listen to the radio, and so on. After quickly assimilating all available data—including such predictions as "a 30 percent probability of rain today"—you make a decision. Somehow a compromise is made between the inconvenience of carrying the umbrella and the possibility of having to spend money for a cleaning bill for your clothes.

In these instances, and in most everyday events, decisions are made in the light of uncertainty. *Statistics* may be defined as a tool for decision making in the light of uncertainty. Uncertainty does not imply no knowledge, but only that the exact outcome is not completely predictable. If 10 coins are tossed, one knows that the number of heads will be some integer between 0 and 10. However, each specific integer has a certain chance of occurring and various results may be predicted in terms of their chance or probability of occurrence. When the results of an experiment that are observed could have occurred only 5 times in 100 by chance alone, most experimenters consider that this is a rare event and will state that the results are statistically significant at the 5 percent significance level. In such cases the hypothesis being tested is usually rejected as untenable. When statistical methods are used in experimentation, one can assess the magnitude of the risks taken in making a particular decision.

Statistical inference refers to the process of inferring something about a population from a sample drawn from that population. The population consists of all possible values of some random variable Y, where Y represents the response or dependent variable being measured. The response Y may represent tensile strength, weight, score, reaction time, or whatever criterion is being used to evaluate the experimental results. Characteristics of the population of this random variable are called *parameters*. Typically θ is used to designate any given population parameter. The average or expected value of the random variable is designated as $E(Y) = \mu$. If the probability function defining the random variable is known,

$$E(Y) = \sum Y_i p(Y_i)$$

where $p(Y_i)$ is a discrete probability function, and

$$E(Y) = \int Y f(Y) \, dY$$

where $f(Y)$ is a continuous probability density function.

The long-range average of squared deviations from the mean of a population is called the *population variance* σ_Y^2. Or

$$E\left[(Y - \mu_Y)^2\right] = \sigma_Y^2$$

The square root of this parameter is called *population standard deviation* σ_Y. Quantities computed from the sample values drawn from the population are called *sample statistics* or, simply, statistics. Examples include the sample mean

$$\bar{Y} = \sum_i^n Y_i / n$$

where n is the number of observations in the sample, and the sample variance

$$s^2 = \sum_{i=1}^n (Y_i - \bar{Y})^2 / (n - 1)$$

Italic letters will be used to designate sample statistics. The symbol u will be used to designate a general statistic corresponding to the population parameter θ.

Most statistical theory is based on the assumption that samples drawn are *random samples,* that is, that each member of the population has an equal chance of being included in the sample and that the pattern of variation in the population is not changed by this deletion of the n members for the sample.

The notion of statistical inference may be divided into two parts: (1) estimation and (2) tests of hypotheses.

2.2 Estimation

The objective of statistical estimation is to make an estimate of a population parameter based on a sample statistic drawn from this population. Two types of estimates are usually needed, point estimates and interval estimates.

A *point estimate* is a single statistic used to estimate a parameter. For example, the sample mean $\bar{Y}$ is a point estimate of the population mean μ. Point estimates are usually expected to have certain desirable characteristics. They should be unbiased, consistent, and have minimum variance.

An *unbiased statistic* is one whose expected or average value taken over an infinite number of similar samples equals the population parameter being estimated. Symbolically, $E(u) = \theta$, where $E(\cdot)$ means the expected value of the statistic in $(\cdot)$. The sample mean is an unbiased statistic, since it can be proved that

$$E(\bar{Y}) = \mu_Y \qquad (\text{or simply, } \mu)$$

Likewise, the sample variance as defined above is unbiased, since

$$E(s^2) = E\left[\sum_i (Y_i - \bar{Y})^2 / (n-1)\right] = \sigma_Y^2 \qquad (\text{or } \sigma^2)$$

Note that the sum of squares $\sum_{i=1}^{n}(Y_i - \bar{Y})^2$ must be divided by $n-1$ and not by n if s^2 is to be unbiased. The sample standard deviation

$$s = \sqrt{\sum_{i=1}^{n} (Y_i - \bar{Y})^2 / (n-1)}$$

is not unbiased, since it can be shown that

$$E(s) \neq \sigma$$

This somewhat subtle point is proved in Burr (see [8], pp. 101–104).

A few basic theorems involving expected values that will be useful later in this book include

$$E(k) = k \qquad (\text{where } k \text{ is a constant})$$
$$E(kY) = kE(Y)$$
$$E(Y_1 + Y_2) = E(Y_1) + E(Y_2)$$
$$E(Y - \mu_Y)^2 = E(Y^2) - \mu_Y^2 = \sigma_y^2$$

The statistic $\sum_i (Y_i - \bar{Y})^2$ is used so frequently that it is often referred to as the *sum of squares*, or SS_Y. It is actually the sum of the squares of deviations of the sample values from their sample mean.

It is now easily seen that

$$E\left[\sum_i (Y_i - \bar{Y})^2\right] = (n-1)\sigma_Y^2$$

or

$$E(SS_Y) = (n - 1)\sigma_Y^2$$

where $n-1$ represents the degrees of freedom associated with the sum of squares on the left. In general, for any statistic u,

$$E\left[\sum_i (u_i - \bar{u})^2\right] = (df \text{ on } SS_u)\sigma_u^2$$

or

$$E(SS_u) = (df \text{ on } SS_u)\sigma_u^2$$

When a sum of squares is divided by its degrees of freedom, it indicates an averaging of the squared deviations from the mean, and this statistic, SS/df, is called a *mean square,* or MS. Hence

$$E(MS) = E(SS / df) = \sigma^2$$

of the variable or statistic being studied.

A *consistent statistic* is one whose value comes closer and closer to the parameter as the sample size is increased. Symbolically,

$$\lim_{n \to \infty} \Pr(|u_n - \theta| < \varepsilon) \to 1$$

which expresses the notion that as n increases, the probability approaches certainty (or 1) that the statistic u_n, which depends on n, will be within a distance ε, however small, of the true parameter θ.

Minimum variance applies where two or more statistics are being compared. If u_1 and u_2 are two estimates of the same parameter θ, the estimate having the smaller standard deviation is called the minimum-variance estimate. This is shown diagrammatically in Figure 2.1. In Figure 2.1, $\sigma_{u_1} < \sigma_{u_2}$, and u_1 is then said to be the *minimum-variance estimate* as variance is the square of the standard deviation.

An *interval estimate* consists of the interval between two values of a sample statistic that is asserted to include the parameter in question. The band of values between these two limits is called a *confidence* interval for the parameter, since

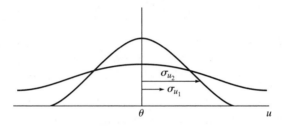

Figure 2.1 Minimum-variance estimates.

its width may be determined from the degree of confidence that is assumed when we say that the parameter lies in this band. Confidence limits (the end points of the confidence interval) are determined from the observed sample statistic, the sample size, and the degree of confidence desired. A 95 percent confidence interval on μ is given by

$$\bar{Y} \pm 1.96 \frac{\sigma}{\sqrt{n}}$$

where 1.96 is taken from normal distribution values (see Statistical Table A), n is the sample size, and σ is the population standard deviation. If 100 sample means based on n observations each are computed, and 100 confidence intervals are set up using the above formula, we expect that about 95 of the 100 intervals will include μ. If only one interval is set up based on one sample of n, as is usually the case, we can state that we have 95 percent confidence that this interval includes μ. If σ is unknown and the population is normal, Student's t distribution (Statistical Table B) is used, and confidence intervals are given by

$$\bar{Y} \pm t_{1-\alpha/2} \frac{s}{\sqrt{n}}$$

where s is the sample standard deviation and t has $n - 1$ df. Then $100(1 - \alpha)$ percent gives the degree of confidence desired.

2.3 Tests of Hypotheses

A *statistical hypothesis* is an assumption about the population being sampled. It usually consists of assigning a value to one or more parameters of the population. For example, it may be hypothesized that the average number of miles per gallon obtained with a certain carburetor is 19.5. This is expressed as $H_0: \mu = 19.5$ mi/gal. The basis for the assignment of this value to μ usually rests on past experience with similar carburetors. Another example would be to hypothesize that the variance in weight of filled vials for the week is 40 grams^2 or $H_0: \sigma^2 = 40$ g^2. When such hypotheses are to be tested, the other parameters of the population are either assumed or estimated from data taken on a random sample from this population.

A *test of a hypothesis* is simply a rule by which a hypothesis is either accepted or rejected. Such a rule is usually based on sample statistics, called *test statistics,* when they are used to test hypotheses. For example, the rule might be to reject $H_0: \mu = 19.5$ mi/gal if a sample of 25 carburetors averaged 18.0 mi/gal ($\bar{Y}$) or less when tested. The *critical region* of a test statistic consists of all values of the test statistic where the decision is made to reject H_0. In the preceding example, the critical region for the test statistic $\bar{Y}$ is where $\bar{Y} \leq 18.0$ mi/gal.

Since hypothesis testing is based on observed sample statistics computed on n observations, the decision is always subject to possible errors. If the hypothesis is really true and it is rejected by the sample, a *type I error* is committed. The probability of a type I error is designated as α. If the hypothesis is accepted when it is not true, that is, some alternative hypothesis is true, a *type II error* has been made and its probability is designated as β. These α and β error probabilities are often referred to as the risks of making incorrect decisions, and one of the objectives in hypothesis testing is to design a test whose α and β risks are both small. In most such test procedures α is set at some predetermined level, and the decision rule is then formulated in such a way as to minimize the other risk, β. In quality control work, α is the producer's risk and β the consumer's risk.

To review hypothesis testing, a series of steps can be taken that will apply to most types of hypotheses and test statistics. To help clarify these steps and to illustrate the procedure, a simple example is given parallel to the steps.

Pop $U = 19.5$

Assume

Steps in Hypothesis Testing

1. Set up the hypothesis and its alternative.

2. Set the significance level of the test α and the sample size n.

3. Choose a test statistic to test H_0, noting any assumptions necessary when applying this statistic.

4. Determine the sampling distribution of this test statistic when H_0 is true.

5. Set up a critical region on this test statistic where H_0 will be rejected in $(100)\alpha$ percent of the samples when H_0 is true.

6. Choose a random sample of n observations, compute the test statistic, and make a decision on H_0.

Examples

1. H_0: $\mu = 19.5$ mi/gal.
   H_1: $\mu < 19.5$ mi/gal.

2. $\alpha = 0.05$, $n = 25$.

   Test on mean

3. Test statistic: $\bar{Y}$ or standardized $\bar{Y}$
   $$Z = \frac{\bar{Y} - \mu}{\sigma / \sqrt{n}}$$
   (Assume σ known and $= 2$.)

4. $\bar{Y}$ is normally distributed with mean μ and standard deviation $\sigma / \sqrt{n}$, or Z is $N(0, 1)$.

5.

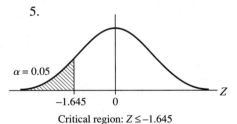

$\alpha = 0.05$

$-1.645 \quad 0$

Critical region: $Z \leq -1.645$

6. If $n = 25$ and $\bar{Y} = 18.9$ mi/gal,
   $$Z = \frac{18.9 - 19.5}{2 / \sqrt{25}} = -1.5.$$
   As $-1.5 > -1.645$, do not reject H_0.

In this example a one-sided or one-tailed test was used. This is dictated by the alternative hypothesis, since we only wish to reject H_0 when low values of $\bar{Y}$ are observed. The size of the significance level α is often set in an arbitrary fashion such as 0.05 or 0.01. It should reflect the seriousness of rejecting many carburetors when they are really satisfactory, or when the actual mean of the lot (population) is 19.5 mi/gal or better. In using the normal variate Z, σ is assumed known; a different test statistic would be used if σ were unknown, namely, Student's t. The critical region may also be expressed in terms of $\bar{Y}$ using the critical Z value of -1.645:

$$-1.645 = \frac{\bar{Y}_c - 19.5}{2 / \sqrt{25}}$$

or $\bar{Y}_c = 18.8$, and the decision rule can be expressed as: Reject H_0 if $\bar{Y} \le 18.8$. Here $\bar{Y} = 18.9$ and the hypothesis is not rejected.

The procedure outlined may be used to test many different hypotheses. The nature of the problem will indicate what test statistic is to be used, and proper tables can be found to set up the required critical region. Well known are tests such as those on a single mean, two means with various assumptions about the corresponding variances, one variance, and two variance. These tests are reviewed in Sections 2.6–2.8.

2.4 The Operating Characteristic Curve

In the previous example no mention was made of the type II error of size β. β is the probability of accepting the original hypothesis H_0 when it's not true or when some alternative hypothesis, H_1, is true. Now H_1 in this example states that $\mu < 19.5$ mi/gal, and there are many possible means that satisfy this alternative hypothesis. Thus β is a function of the μ that is less than the hypothesized μ value. To see how β varies with μ, let us consider several possible μ values and sketch the distributions of sample means $\bar{Y}$ that these various μ values would generate. Also note the critical region that already has been established as $\bar{Y} = 18.8$ based on H_0.

From Figure 2.2 it can be seen that as one considers μ farther and farther to the left of 19.5, the β error decreases. The size of any β error can be determined for any given μ from Figure 2.2 and a normal distribution table. For example, if μ is assumed to be 19.0, the critical mean 18.8 is standardized with respect to this assumed mean and

$$Z = \frac{18.8 - 19.0}{2 / \sqrt{25}} = \frac{-0.2}{0.4} = -0.5$$

indicating that the rejection point is one half of a standard deviation below 19.0. For a normal distribution (see Statistical Table A), a $Z = -0.5$ has an area above it of $1 - 0.3085$ or 0.6915. Thus, the β error for $\mu = 19.0$ is 0.69, or 69 percent of the samples would still accept the original hypothesis that $\mu = 19.5$ when it has really dropped to 19.0.

If one considers the mean μ at 18.0 as in the bottom sketch of Figure 2.2,

$$Z = \frac{18.8 - 18.0}{0.4} = 2.00$$

and the β error is only 0.0228.

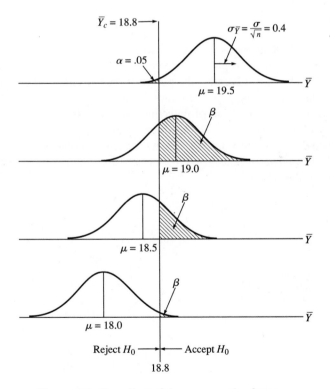

Figure 2.2 The effect of the mean on the β error.

If β is plotted for various values of μ, the resulting curve is called the *operating characteristic* (or OC) *curve* for this test. Table 2.1 summarizes the computations for several assumed values of μ.

Table 2.1 Data for Operating Characteristic Curve

μ	$Z = \dfrac{18.8 - \mu}{2/\sqrt{25}}$	β	$1 - \beta$
18.0	+2.00	0.02	0.98
18.3	+1.25	0.11	0.89
18.5	+0.75	0.23	0.77
19.0	−0.50	0.69	0.31
19.2	−1.00	0.84	0.16
19.5	−1.75	0.96	0.04[*]

[*] (approximately $= \alpha$)

The plotted OC curve is then as shown in Figure 2.3.

Some people prefer to plot $1 - \beta$ versus μ. This curve is called the *power curve* of the test as $1 - \beta$ is called the power of the test against the alternative μ.

Note that when μ is quite a distance from the hypothesized value of 19.5, the test does a fairly good job of detecting this shift; that is, if $\mu = 18.5$, the probability of rejecting 19.5 is 0.77, which is fairly high. On the other hand, if μ has shifted only slightly from 19.5, say 19.2, the probability of detection is only 0.16. The power of a test may be increased by increasing the sample size or increasing the risk α.

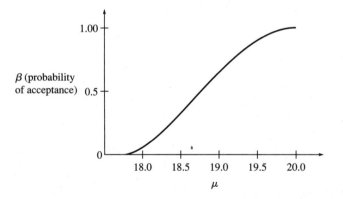

Figure 2.3 Operating characteristic curve.

2.5 How Large a Sample?

The question of how large a sample to take from a population for making a test is one often asked of a statistician. This question can be answered provided the experimenter can answer each of the following questions:

1. How large a shift (from 19.5 to 19.0) in a parameter do you wish to detect?
2. How much variability is present in the population? (Based on past experience, $\sigma = 2$ mi/gal.)
3. What size risks are you willing to take? ($\alpha = 0.05$ and $\beta = 0.10$.)

If numerical values can be at least estimated in answering these questions, the sample size may be determined.

Set up two sampling distributions of $\bar{Y}$, one when H_0 is true, $\mu = 19.5$, and the other when the alternative to be detected is true, $\mu = 19.0$ (Fig. 2.4). Indicate by $\bar{Y}_c$ a value between the two μ's that will become a critical point, rejecting H_0 for observed values of $\bar{Y}$ below it and accepting H_0 for $\bar{Y}$ values above it. Indicate the α and β risks on the diagram. Set up two simultaneous equations, standardizing $\bar{Y}_c$ first with respect to a μ of 19.5 (α equation) and second with respect to a μ of 19.0 (β equation). Solve these equations for n and $\bar{Y}_c$:

z Table

$$\alpha \text{ equation:} \quad \frac{\bar{Y}_c - 19.5}{2/\sqrt{n}} = -1.645 \quad \text{(based on } \alpha = 0.05\text{)} \qquad 1.645$$

z table

$$\beta \text{ equation:} \quad \frac{\bar{Y}_c - 19.0}{2/\sqrt{n}} = +1.282 \quad \text{(based on } \beta = 0.10\text{)} \qquad -2.085$$

Subtracting the second equation from the first and multiplying both sides of

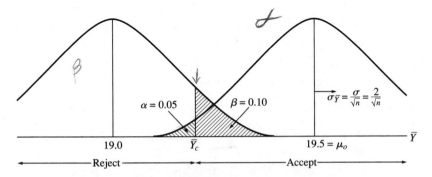

Figure 2.4 Determining sample size.

each equation by $2 / \sqrt{n}$ gives

$$-0.5 = -2.927 \left(\frac{2}{\sqrt{n}} \right)$$

$$\sqrt{n} = \frac{5.854}{0.5} = 11.71$$

$$n = (11.71)^2 = 136.9 \text{ or } 137$$

Keeping $\alpha = 0.05$ gives

$$\bar{Y}_c = 19.5 - 1.645 \left(\frac{2}{\sqrt{137}} \right) = 19.22 \text{ mi} / \text{gal}$$

The decision rule is then: Choose a random sample of 137 carburetors, and if the mean mi/gal of these is less than 19.22, reject H_0; otherwise, accept H_0.

Excellent tables are available for determining n for several tests of hypotheses, such as those in Davies [10, pp. 606–615] or Owen [23, pp. 19, 23, 36, 41 ff.].

2.6 Application to Tests on Variances

To review how the concepts of confidence limits and hypothesis testing apply to variance tests consider the following.

I. *Tests on a single variance.*
    $H_0: \sigma^2 = \sigma_0^2$
    $H_1: \sigma^2 \neq \sigma_0^2 \ \left[\text{or } \sigma^2 > \sigma_0^2 \text{ or } \sigma^2 < \sigma_0^2 \right]$
    Choose α and n.
    Test statistic is

$$\frac{(n-1)s^2}{\sigma_0^2}$$

which follows the chi-square distribution with $(n-1)$ degrees of freedom, if one can assume that Y is normally distributed.

If $\chi^2 \geq \chi_{1-\alpha/2}^2$, or $\chi^2 \leq \chi_\alpha^2$, reject H_0.
[If $\chi^2 \geq \chi_{1-\alpha}^2$, reject, or if $\chi^2 \leq \chi_\alpha^2$, reject.]

To set confidence limits on σ^2 one computes

$$\frac{(n-1)s^2}{\chi_{1-\alpha/2}^2} \leq \sigma^2 \leq \frac{(n-1)s^2}{\chi_{\alpha/2}^2}$$

using the chi-square table with $(n-1)$ df to give $100(1-\alpha)$ percent confidence limits.

As an example consider that one might wish to determine whether a process standard deviation is greater than a specified value of 5 oz. If a sample of 10 weights is found to have an average of 35.2 oz and a standard deviation of 7 oz, what can we conclude about the standard deviation of this process?

We hypothesize $H_0: \sigma^2 = 25$ oz^2 and take as alternative $H_1: \sigma^2 > 25$ oz^2. Selecting $\alpha = 0.05$ and $n = 10$ gives $n - 1 = 9$ df. One should reject H_0 if

$$\chi^2 \geq \chi^2_{0.95}(9 \text{ df}) = 16.9$$

based on Statistical Table C.

From the sample data

$$\chi^2 = \frac{(n-1)s^2}{\sigma_0^2} = \frac{9(49)}{25} = 17.64$$

so we reject the hypothesis and conclude that our process standard deviation is greater than 5 oz.

We might wish to set 90 percent confidence limits on σ^2 (or σ) based on our observed variance of 49:

$$\frac{9(49)}{\chi^2_{0.95}} \leq \sigma^2 \leq \frac{9(49)}{\chi^2_{0.05}}$$

$$\frac{9(49)}{16.9} \leq \sigma^2 \leq \frac{9(49)}{3.33}$$

$$26.09 \leq \sigma^2 \leq 132.43$$

or

$$5.1 \leq \sigma \leq 11.5 \text{ oz}$$

and we feel 90 percent confident that the process standard deviation is between 5.1 and 11.5. We note that this band does not include the hypothesized value of 5 oz. We also note that this is a two-sided confidence band whereas the hypothesis test was one-sided. By making the α for confidence limits twice the α in the test of H_0, the results are comparable.

II. *Tests on two independent variances.*

As another example of a test of a hypothesis, consider testing whether or not the variances of two normal populations are equal. This example is included here, as the test statistic involved will have many applications in later chapters.

In accordance with the steps outlined in Section 2.3, we have the following.

$H_0: \sigma_1^2 = \sigma_2^2 \qquad H_1: \sigma_1^2 > \sigma_2^2.$
$\alpha = 0.05.$
Test statistic $F = s_1^2 / s_2^2$ (often called the variance ratio, where s_1^2 is based on $n_1 - 1$ df and s_2^2 is based on $n_2 - 1$ df).
If the two samples are independently chosen from normal populations and H_0 is true, the F statistic follows a skewed distribution, formed as the ratio of two independent chi-square distributions. A table giving a few percentiles

of this F distribution appears as Statistical Table D. This table is entered with $n_1 - 1$ df for the numerator and $n_2 - 1$ df for the denominator.

The critical region is set at $F \geq F_{0.95}$ in this example ($\alpha = 0.05$) with the degrees of freedom dependent on the size of the two samples.

If the first sample results are

$$n_1 = 8 \qquad s_1^2 = 156$$

and the second are

$$n_2 = 10 \qquad s_2^2 = 100$$

then

$$F = \frac{156}{100} = 1.56$$

and the critical region is $F \geq 3.29$ for 7 and 9 df. Hence the hypothesis is not rejected.

In this example only a one-sided test has been considered as one is usually concerned with showing that one variance is greater or less than another. If one is testing $H_0: \sigma_1^2 = \sigma_2^2$ versus $H_1: \sigma_1^2 \neq \sigma_2^2$, the critical region is set as follows: If $F \leq F_{\alpha/2}$ or $F \geq F_{1-\alpha/2}$, reject H_0.

2.7 Application to Tests on Means

I. *Tests on a single mean.*
   $H_0: \mu = \mu_0$

   $H_1: \mu \neq \mu_0$ [or $\mu > \mu_0$ or $\mu < \mu_0$]
   Choose n and α.

   Test statistic when:

   A. σ is known.

$$Z = \frac{\bar{Y} - \mu_0}{\sigma / \sqrt{n}}$$

   Z is $N(0, 1)$
   Reject H_0 if $|Z| \geq Z_{1-\alpha/2}$ [or $Z \geq Z_{1-\alpha}$, or $Z \leq Z_\alpha$].

The problem of the carburetors in Section 2.3 illustrates this test.

   B. σ is unknown, normal population.

$$t = \frac{\bar{Y} - \mu_0}{s / \sqrt{n}}$$

   t follows a t distribution (Statistical Table B) with $n - 1$ degrees of freedom.
   Reject H_0 if $|t| \geq t_{1-\alpha/2}$ [or $t \geq t_{1-\alpha}$ or $t < -t_{1-\alpha}$].

Consider a sample of six cylinder blocks whose cope hardness values are 70, 75, 60, 75, 65, and 80. Is there evidence here that the average cope hardness has changed from its specified value of 75?

$$H_0: \mu = 75$$
$$H_1: \mu \neq 75$$

If $\alpha = 0.05$, $n = 6$, one should reject H_0 if $|t| \geq t_{0.975}$ with 5 df or if $|t| \geq 2.57$. From the given sample of six: $\bar{Y} = 70.8$, $s = 7.4$. So,

$$t = \frac{70.8 - 75}{7.4 / \sqrt{6}} = -1.39$$

Since $|t| = 1.39 < 2.57$, one cannot reject H_0. We might have used our data to set 95 percent confidence limits on μ. From Section 2.2, $100(1 - \alpha)$ percent confidence limits are

$$\bar{Y} \pm t_{1-\alpha/2} s / \sqrt{n}$$
$$70.8 \pm 2.57(7.4) / \sqrt{6}$$
$$70.8 \pm 7.8$$

or from 63.0 to 78.6, which includes our hypothesized value of 75.

II. **Tests on two means.**

A. If independent samples:

1. If variances are known: σ_1^2 and σ_2^2.

$H_0: \mu_1 = \mu_2$
$H_1: \mu_1 \neq \mu_2$ [or $\mu_1 > \mu_2$]
Choose α, n_1, and n_2.
Test statistic is

$$Z = \frac{\bar{Y}_1 - \bar{Y}_2}{\sqrt{\sigma_1^2 / n_1 + \sigma_2^2 / n_2}}$$

Reject H_0 if $|Z| \geq Z_{1-\alpha/2}$ [or $Z \geq Z_{1-\alpha}$].

2. If variances are unknown but equal and populations are normal: Test statistic is

$$t = \frac{\bar{Y}_1 - \bar{Y}_2}{\sqrt{\dfrac{(n_1 - 1)s_1^2 + (n_2 - 1)s_2^2}{n_1 + n_2 - 2} \left(\dfrac{1}{n_1} + \dfrac{1}{n_2} \right)}}$$

with $n_1 + n_2 - 2$ df. One rejects if $|t| \geq t_{1-\alpha/2}$ [or $t \geq t_{1-\alpha}$].

3. If variances are unknown and unequal: If a preliminary F test shows the variances to be unequal, no test on means may be necessary as the two samples are so heterogeneous with respect to their variances. If,

however, a test on means is desired, some tests are given in statistics books. This situation is often referred to as the Behrens–Fisher problem. One method is to take as the test statistic

$$t' = \frac{\bar{Y}_1 - \bar{Y}_2}{\sqrt{s_1^2 / n_1 + s_2^2 / n_2}}$$

and check its significance using a special formula for its degrees of freedom:

$$df = \frac{\left(s_1^2 / n_1 + s_2^2 / n_2\right)^2}{\dfrac{\left(s_1^2 / n_1\right)^2}{n_1 + 1} + \dfrac{\left(s_2^2 / n_2\right)^2}{n_2 + 1}} - 2$$

B. If dependent or correlated samples:
Here one often has the same sample "before and after" some treatment has been applied. The usual procedure is to take differences between the first and second observations on the same piece, person, or whatever, and test the hypothesis that the mean difference μ_D is zero. This reduces the problem to a test on a single mean:

$$H_0: \mu_D = 0$$
$$H_1: \mu_D \neq 0 \quad (\mu_D > 0)$$

n differences, α
Test statistic is

$$t = \frac{\bar{d}}{s_d / \sqrt{n}} \qquad \text{with } n - 1 \text{ df}$$

Reject if $|t| \geq t_{1-\alpha/2}$ [or $t \geq t_{1-\alpha}$].
Choose a random sample of n, take two sets of measurements, one on each unit in the sample, and compute the differences. Using these differences compute $\bar{d}$ and s_d, then t, and render a decision on H_0.

Some examples illustrating the foregoing tests are given in the following.

■ **EXAMPLE 2.1** Measurements were made at each of two plants on the tensile strength of samples of the same type of steel. The data are given below in thousands of pounds per square inch (psi). Can it be stated that the steels made in the two plants have the same mean tensile strength?

Plant 1	Plant 2
219	207
228	195
222	209
197	218
225	196
184	217
218	
211	

From these data:

$$n_1 = 8 \qquad \bar{Y}_1 = 213 \qquad s_1 = 15.2$$
$$n_2 = 6 \qquad \bar{Y}_2 = 207 \qquad s_2 = 9.9$$

To test $H_0: \sigma_1^2 = \sigma_2^2$ versus $H_1: \sigma_1^2 \neq \sigma_2^2$ with $\alpha = 0.05$, the test statistic is

$$F = \frac{s_1^2}{s_2^2} = \frac{(15.2)^2}{(9.9)^2} = 2.35$$

with df 7 and 5. Reject if $F \leq F_{0.025}$ or $F \geq F_{0.975}$ or reject if $F < 0.189$ or $F > 6.85$. Since our F falls between these two values, we assume variances are homogeneous.

To test the means $H_0: \mu_1 = \mu_2$ versus $\mu_1 \neq \mu_2$, reject if $|t| \geq t_{0.975, 12\text{ df}} = 2.18$.

$$t = \frac{213 - 207}{\sqrt{[(7(15.2)^2 + 5(9.9)^2)/12][1/8 + 1/6]}}$$

$$t = 0.84$$

Since this t is less than 2.18, we do not reject $H_0: \mu_1 = \mu_2$, and we assume means are also homogeneous. ∎

■ **EXAMPLE 2.2** Reflection light-box readings before and after dichromating the interior of a metal cone were

Test Number	1	2	3	4	5	6	7	8
Before	6.5	6.0	7.0	6.8	6.5	6.8	6.2	6.5
After	4.4	4.2	5.0	5.0	4.8	4.6	5.2	4.9
Differences	2.1	1.8	2.0	1.8	1.7	2.2	1.0	1.6

Test for a significant difference in mean light-box readings. The hypotheses are $H_0: \mu_d = 0$, $H_1: \mu_d \neq 0$, $\alpha = 0.05$. Here $\bar{d} = 1.775$, and $s_d = 0.373$.

$$t = \frac{1.775}{\sqrt{0.373 / 8}} = 13.45$$

Since $t > t_{0.975, 7} = 2.35$, we reject H_0 and conclude that there is a difference in mean readings. ■

2.8 Application to Tests on Proportions

I. *Tests on a single proportion* p.

$H_0: p = p_0$

$H_1: p \neq p_0$ [or $p > p_0$ or $p < p_0$]

Choose α and n (n had best be large enough so that $np \geq 4$), then the test statistic is

$$Z = \frac{\hat{p} - p_0}{\sqrt{p_0 q_0 / n}}$$

where $q_0 = 1 - p_0$ and $\hat{p}$ is the observed proportion in a sample of n. Here Z is $N(0, 1)$ and one rejects if $|Z| > Z_{1-\alpha/2}$ [or $Z > Z_{1-\alpha}$ or $Z < -Z_{1-\alpha}$].

As an example consider whether or not a class of 25 students has an excessive number of left-handed students if we find 10 left-handed students and the national average is 20 percent.

$H_0: p = 0.20$

$H_1: p > 0.20$

$n = 25$, $\alpha = 0.05$

Reject H_0 if $Z \geq 1.65$. Here

$$Z = \frac{10 / 25 - 0.20}{\sqrt{(0.2)(0.8) / 25}} = 2.5$$

so we reject H_0 and conclude that our class does have an excessive number of left-handers.

II. *Tests on two independent proportions.*

$H_0: p_1 = p_2$

$H_1: p_1 \neq p_2$ [or $p_1 > p_2$]

Choose α, n_1, and n_2.

Test statistic is

$$Z = \frac{\hat{p}_1 - \hat{p}_2}{\sqrt{\hat{p}\hat{q} (1 / n_1 + 1 / n_2)}}$$

where $\hat{p}$ is the proportion of defectives in both samples together:

$$\hat{p} = \frac{n_1\hat{p}_1 + n_2\hat{p}_2}{n_1 + n_2}$$

Reject H_0 if $|Z| \geq Z_{1-\alpha/2}$ [or $Z \geq Z_{1-\alpha}$]

If the number of defectives found in a sample of 100 drawn from a production process was 12 on Monday and a sample of 200 from the same process showed 16 defectives on Tuesday, has the process improved?

$H_0: p_1 = p_2$
$H_1: p_1 > p_2$ where p_1 is proportion on Monday.
$\alpha = 0.05$, $n_1 = 100$, $n_2 = 200$.
Reject if $Z \geq 1.64$.
Here

$$Z = \frac{12/100 - 16/200}{\sqrt{(28/300)(272/300)[1/100 + 1/200]}}$$

$$Z = 1.12$$

so one cannot say that the process has improved significantly.

2.9 Analysis of Experiments with SAS

The calculations necessary to summarize a relatively small amount of data can be tedious and time-consuming. To simplify these calculations for various types of studies and experiments, SAS (Statistical Analysis System) [24], a preprogrammed package of statistical techniques, is very useful. We present a brief introduction to SAS programming to give an idea of the power of this language and the ease with which we obtain results.

SAS is a special-purpose compiler composed of many preprogrammed modules for manipulating data and providing solutions through the use of various statistical procedures. The SAS language is almost the same on all types of computer systems, but due to the many variations at a given computer site there are unique requirements for getting on and off any given system.

SAS has expanded since 1966 from a statistical package to a general-purpose language of its own. It is this total capability (flexibility) that makes it difficult to learn. Fortunately, approximately 10 percent of the SAS commands will handle 90 percent of the applications. Reports are preformatted. Special report formats are programmable, but for analysis purposes a great amount of time is saved by accepting the results as is and concentrating on the interpretation of the output.

This introduction gives examples using some of the procedures and options available to solve some of the text examples and problems.

Consider Problem 2.1 at the end of this chapter.

For the following data on tensile strength in psi, determine the mean, variance, and other statistics of this sample and produce a variety of charts with each output page titled.

Tensile Strength	Frequency
18,461	2
18,466	12
18,471	15
18,476	10
18,481	8
18,486	3
Total	50

We will write the SAS program assuming that the data are entered with the program.

The first step in using SAS is to create a SAS data set. This is done by a series of statements that name the data set (the **DATA** statement), describe the arrangement of the data lines (the **INPUT** statement), and signal the beginning of the data itself (the **CARDS** statement). SAS statements may begin in any column on your screen that is available and must always end with a semicolon (;).

The name chosen for the data set must be a one-word name of eight or fewer characters starting with an alphabetic that usually relates to the data. We will name our data set "TENSILE" as we type our first line.

1. DATA TENSILE;

The next statement is the **INPUT** statement. This statement names the variables and describes the order in which their values appear on the data lines. Variable names also follow the rules for naming the data set. We have two variables, PSI and COUNT. We type on the second line.

2. INPUT PSI COUNT;

Note that spacing is required between the keyword **INPUT** and each of the variable names. We avoid using **FREQ** as a variable name since it is a keyword used in many SAS procedures. Also, if any of the data were nonnumeric, we would so indicate with a $ following the variable name.

The computer now knows that each line contains the value of two variables PSI and COUNT and the order in which the data occurs. The PSI is the first value, and COUNT is the second.

The input statement is followed by a statement that signals that the data will follow. This statement is typed on line 3 as

3. CARDS;

The data are to follow immediately with only two observations per typed line. When all the data have been entered, a line containing only a semicolon is typed. This signals the end of the data. Thus far we have

1.	DATA TENSILE;		names the data set
2.	INPUT PSI COUNT;		names the variables in order
3.	CARDS;		signals that data follows
4.	18461	2	data (two entries per line)
5.	18466	12	
6.	18471	15	
7.	18476	10	
8.	18481	8	
9.	18486	3	
10.	;		signals the end of data

The data are now in a SAS data set called TENSILE. We now tell the computer what to do with the data. This is done by using one or more procedure statements. Each procedure statement begins with the keyword **PROC** followed by the name of the procedure. Obviously the programmer must be aware of the procedure, what it will do, and some of the keywords required to obtain some of the options available with the procedure.

We will first use the **UNIVARIATE** procedure to obtain the desired sample statistics followed by the **CHART** procedure asking for a variety of charts for the variable PSI. We now type

11. PROC UNIVARIATE PLOT FREQ;

The mean, variance, and many other values will be calculated for each of the numeric variables in our data set. A stem-and-leaf plot, a box plot, and a normal probability plot have been requested with the option **PLOT**. The **FREQ** option requests additional tables of frequencies and percentages. Since we want results for only the variable PSI, we will identify the procedure by use of a keyword VAR followed by the variable name PSI. We next type

12. VAR PSI;
13. FREQ COUNT;

Since each value of PSI has a frequency (COUNT) associated with it, we add the statement FREQ COUNT. If we had not used this last statement the mean and variance of just the variable PSI would have been calculated with no consideration for frequency of occurrence.

The histogram is obtained by using the **CHART** procedure. We specify a vertical bar chart for PSI with VBAR PSI, the midpoints of our histogram, and tell the computer to use the variable COUNT as the frequency by so stating the values desired and using the statement FREQ=COUNT. All of this follows the slash (/).

The statements required are

```
13. PROC CHART;
14. VBAR PSI/MIDPOINTS=18461 18466 18471 18476 18481 18486
    FREQ=COUNT;
```

Note that more than one line may be used for a SAS statement if a semicolon has not been used until the end of the statement.

As many charts as desired can be requested following the **PROC CHART** command. As an example, we could request a cumulative bar chart by adding TYPE = CFREQ or TYPE = CPERCENT to an additional **VBAR** statement such as

```
15. VBAR PSI/FREQ=COUNT TYPE=CPERCENT;
```

Since we did not specify the midpoints, SAS will automatically create some that may or may not be what we want.

If horizontal charts are required, the **VBAR** of each of the two previous statements would be changed to **HBAR**. Output for these two charts is included.

A pie chart is also available along with many other variations to horizontal and vertical charts.

```
16. PIE PSI/FREQ=COUNT;
```

For our title we use the keyword **TITLE (n)** where n is the line number for the title to be printed. For example if a two-line title is desired, we type

```
17. TITLE1'TENSILE STRENGTH';
18. TITLE2'STUDY';
```

The **TITLE** statements should be placed after the **PROC** statement with which it is to be associated and be enclosed within single quotes.

Our last command tells the system to run the program, and we do this with the keyword **RUN**.

Our entire program is as follows:

```
DATA TENSILE;
INPUT PSI COUNT;
CARDS;
(Data Lines)
;
PROC UNIVARIATE PLOT FREQ;
VAR PSI;
FREQ=COUNT;
PROC CHART;
VBAR PSI/MIDPOINTS=18461 18466 18471 18476 18481 18486
```

```
FREQ=COUNT;
VBAR PSI/FREQ=COUNT TYPE=CPERCENT;
PIE PSI/FREQ=COUNT;
TITLE1'TENSILE STRENGTH';
TITLE2'STUDY';
RUN;
```

Our output is on pages 38–39.

UNIVARIATE PRINTED OUTPUT (copied from SAS)

For each variable, **UNIVARIATE** prints:

1. VARIABLE=, the name of the variable
2. The variable label
3. N, the number of observations on which the calculations are based
4. SUM WGTS, the sum of the weights of these observations
5. The MEAN
6. The SUM
7. STD DEV, the standard deviation
8. The VARIANCE
9. The measure of SKEWNESS
10. The measure of KURTOSIS
11. USS, the uncorrected sum of squares
12. CSS, the corrected sum of squares
13. CV, the coefficient of variation
14. STD MEAN, the standard error of the mean
15. T: MEAN = 0, Student's t value for testing the hypothesis that the population mean is 0
16. PROB> $|T|$, the probability of a greater absolute value for this t value
17. SGN RANK, the centered signed rank statistic for testing the hypothesis that the population mean is 0
18. PROB> $|S|$, an approximation to the probability of a greater absolute value for this statistic
19. NUM, the number of nonzero observations
20. MAX, the largest value
21. Q3, Q1, and MED, the upper and lower quartiles, and the median
22. MIN, the smallest value
23. The RANGE
24. Q3−Q1, the difference between the upper and lower quartiles

SAS

UNIVARIATE

VARIABLE=PSI

MOMENTS

N	50	SUM WGTS	50
MEAN	18472.9	SUM	923645
STD DEV	6.45945	VARIANCE	41.7245
SKEWNESS	0.305914	KURTOSIS	-0.659197
USS	1.706E+10	CSS	2044.5
CV	0.0349672	STD MEAN	0.913504
T:MEAN=0	20222	PROB>|T|	0.0001
SGN RANK	637.5	PROB>|S|	0.0001
NUM	50		

QUANTILES(DEF=4)

100%	MAX	18486	99%		18486
75%	Q3	18476	95%		18481
50%	MED	18471	90%		18471
25%	Q1	18466	10%		18466
0%	MIN	18461	5%		18461
			1%		18461
RANGE	25				
Q3-Q1	10				
MODE	18471				

	EXTREMES	
	LOWEST	HIGHEST
	18461	18466
	18466	18471
	18471	18476
	18476	18481
	18481	18486

18463.7
18461

Stem Leaf   #
18E3 000   3
18E3
18E3
18E3 00000000   8
18E3
18E3 0000000000   10
18E3
18E3 000000000000000   15
18E3
18E3 000000000000   12
18E3
18E3
18E3 00   2

Box plot

Normal Probability Plot

FREQUENCY TABLE

		PERCENTS	
VALUE	COUNT	CELL	CUM
18461	2	4.0	4.0
18466	12	24.0	28.0
18471	15	30.0	58.0
18476	10	20.0	78.0
18481	8	16.0	94.0
18486	3	6.0	100.0

38

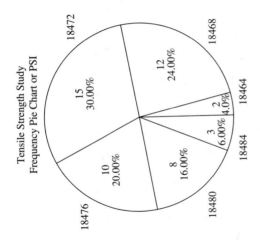

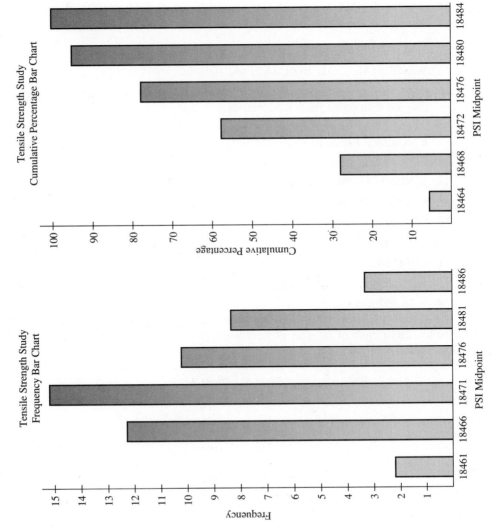

25. The MODE
26. The first, fifth, tenth, ninetieth, ninety-fifth, and ninety-ninth percentiles (1 percent, 5 percent, 10 percent, 90 percent, 95 percent, 99 percent)
27. The five largest, HIGHEST, and five smallest, LOWEST, values.

    If missing values occur for a variable, the following are printed:

28. The MISSING VALUE
29. COUNT, the number of occurrences
30. %COUNT/NOBS, the count as a percentage of the total number of observations
31. %COUNT/NMV, the count as a percentage of the total number of missing values (not shown)

    UNIVARIATE can also print

32. W:NORMAL . . . D:NORMAL, test statistic
33. Associated probabilities, PROB<W or PROB>D, for testing the hypothesis that the data come from a normal distribution
34. STEM LEAF, a stem-and-leaf plot (if any value's count is greater than 48, a horizontal bar chart is printed instead)
35. BOX PLOT, a box plot
36. A NORMAL PROBABILITY PLOT
37. VALUE, a frequency table of variable values
38. COUNT, frequencies
39. CELL, percentages
40. CUM, cumulative percentages

The procedure **PROC TTEST** can be used to test for equality of both variances and means.

Consider Example 2.1. Measurements were made at each of two plants on the tensile strength of samples of the same type of steel. The data are given here in thousands of psi.

Plant 1	Plant 2
207	219
195	228
209	222
218	197
196	225
217	184
	218
	211

Here Plant 1 and Plant 2 data have been switched from Example 2.1, which will not change the *t* test on two means. Can it be stated that the steels made in the two plants have the same mean tensile strength?

Since we are testing two populations, we need what is called a **CLASS** statement, which identifies a variable that has two populations (plant 1 and plant 2). We begin, as before, by naming our data set and variables. However, our data will be classified as Plant 1 or Plant 2 with the corresponding test reading as PSI.

```
DATA STLTST;   (STeeL Tensile STrength)
INPUT PLANT PSI @@;
```

Something new has been added. The symbols @@ tell the computer that more than one set of variables will be entered on a line instead of a single set as before. Our data entry will be:

```
CARDS;
1 207 2 219 1 195 2 228 1 209 2 222
1 218 2 197 1 196 2 225 1 217 2 184
2 218 2 211
;
```

Note that our sample sizes are not equal. However, since we have identified to the computer what each pair of entries represent we have no problem. Continuing with our program, we write

```
PROC TTEST;
CLASS PLANT;
TITLE1'TESTING OF STEEL EQUALITY';
TITLE2'OF';
TITLE3'MEANS AND VARIANCES';
RUN;
```

Our output is on page 42.

PROC TTEST PRINTED OUTPUT (copied from SAS)

For each variable included in the analysis, **TTEST** prints the following statistics for each group:

1. The name of the variable
2. The levels of the classification variable
3. N, the number of nonmissing values
4. The MEAN or average
5. STD DEV, or the standard deviation
6. STD ERROR, or the standard error
7. The MINIMUM value
8. The MAXIMUM value

SAS

OBS	_PLANT	PSI
1	1	207
2	2	219
3	1	195
4	2	228
5	1	209
6	2	222
7	1	218
8	2	197
9	1	196
10	2	225
11	1	217
12	2	184
13	2	218
14	2	211

TESTING OF STEEL EQUALITY
OF
MEANS AND VARIANCES

TTEST PROCEDURE

VARIABLE: PSI

PLANT	N	MEAN	STD DEV	STD ERROR	MINIMUM	MAXIMUM	VARIANCES	T	DF	PROB > |T|
1	6	207.00000000	9.89949494	4.04145188	195.00000000	218.00000000	UNEQUAL	-0.8932	11.9	0.3895
2	8	213.00000000	15.17516769	5.36523199	184.00000000	228.00000000	EQUAL	-0.8394	12.0	0.4176

FOR H0: VARIANCES ARE EQUAL, F'= 2.35 WITH 7 AND 5 DF   PROB > F'= 0.3640

Handwritten annotations:

$H_0: M_1 = M_2$

Since P Value = .4176 > .05

No difference between plants

Cannot Reject H_0 @ .05 level

2 = 6v

G = Populations

Assume the same

Have the same Tailed

Variance Two Tailed

T-Value

0.3895 > .05

0.4176 > .05

Under the assumption of UNEQUAL variances, **TTEST** prints

9. T, and approximate t statistic
10. DF, Satterthwaite's approximation for the degrees of freedom
11. PROB> |T|, the probability of a greater absolute value of t

TTEST then gives the results of the test of equality of variances:

12. The F′ (folded) statistic
13. The degrees of freedom, DF, in each group
14. PROB>F′, the probability of a greater F value

The result agrees, except for sign, with the t value in Section 2.7.

PROC MEANS can be used to test a hypothesis on μ (Mu) when σ^2 is unknown. If requested, a T and the probability of a T occurring with that magnitude or larger will be calculated. Unless otherwise constructed the T is for H_0: $\mu = 0$.

Consider Example 2.2. Test for a significant difference in mean light-box readings of the following sample.

Test No.	1	2	3	4	5	6	7	8
Before	6.5	6.0	7.0	6.8	6.5	6.8	6.2	6.5
After	4.4	4.2	5.0	5.0	4.8	4.6	5.2	4.9

This particular problem is a typical "before–after" experiment performed on the same items. Since the same sample is used under both conditions, we have dependent or correlated samples. The usual procedure is to take the difference between "before" and "after" and test for μ difference = 0.

As before we must name our data set (LIGHTBOX) and our variables (BEFORE) and (AFTER). We type

```
1. DATA LIGHTBOX;
2. INPUT BEFORE AFTER @@;
     6.5   4.4   6.0   4.2  7.0  5.0  6.8  5.0
     6.5   4.8   6.8   4.6  6.2  5.2  6.5  4.9
     ;
```

Since the **INPUT** statement identified two variables, the computer expects to read them in pairs as listed, with spaces between to identify not only the variables but also the following sets.

We must also create a new variable (DIFF). This is easily done by using what is called an operator (=, +, −, ÷). We will create our new variable with the statement

```
DIFF=BEFORE-AFTER;
```

Our program for testing the H_0: $\mu_{DIFF} = 0$ is as follows:

```
DATA LIGHTBOX;
INPUT BEFORE AFTER @@;
DIFF=BEFORE-AFTER
CARDS;
 (Data Lines)
 ;
PROC MEANS T PRT MEAN VAR STDERR;
VAR DIFF;
```

We specify the variable on which we want the test made. If we did not use this command the test would be made on all variables (BEFORE, AFTER, DIFF).

```
TITLE1'TESTING LIGHT READINGS';
```

Our output is as follows:

TESTING LIGHT READINGS

VARIABLE	T	PR >|T|	MEAN	VARIANCE	STD ERROR OF MEAN
DIFF	13.45	0.0000	1.77500000	0.13928571	0.13194967

We observe that the t test value of 13.45 was also computed in Section 2.7.

PROC MEANS PRINTED OUTPUT (copied from SAS)

When no statistics are specifically requested on the **PROC MEANS** statement, **MEANS** prints these statistics for each variable in the **VAR** statement:

1. The name of the variable
2. N, the number of nonmissing values for the variable
3. The MEAN or average of the variable
4. The STANDARD DEVIATION of the variable
5. The MINIMUM VALUE of the variable
6. The MAXIMUM VALUE of the variable

If there is space available on the output (depending on the computer terminal used and **LINESIZE** specified), **MEANS** also prints these statistics:

7. STD ERROR OF MEAN, the standard of the mean
8. The SUM of all the values for a variable
9. The VARIANCE of each variable
10. CV, or the coefficient of variation expressed as a percentage

If statistics are specifically requested on the **PROC** statement, only those statistics are printed. In addition to the statistics above, upon request MEANS can print these statistics:

11. N MISSING, the number of missing values for each variable
12. The RANGE of each variable
13. The UNCORRECTED SS, the raw sum of squares (not adjusted for the mean)
14. The CORRECTED SS, the sum of squares adjusted for the mean
15. SKEWNESS
16. KURTOSIS
17. T, Student's t value for testing the hypothesis that the population means is 0
18. PR> |T|, the probability of a greater absolute value for Student's t under the hypothesis that the mean is zero

Problems

2.1 For the following data on tensile strength in psi determine the mean and variance of this sample:

Tensile Strength (psi)	Frequency
18,461	2
18,466	12
18,471	15
18,476	10
18,481	8
18,486	3
Total	50

$s^2 = 41.724$

14
29
39
47
58

18 472.9
8 6

2.2 Using the results of Problem 2.1, test the hypothesis that the mean tensile strength of the population sampled is 18,470 psi (assume that $\sigma^2 = 40$, $\alpha = 0.05$).

one sided

2.3 Plot the operating characteristic curve for Problem 2.2.

2.4 Determine how large a sample would be needed to detect a 10-psi increase in the mean tensile strength of Problem 2.2 if $\alpha = 0.05$, $\beta = 0.02$, and $\sigma^2 = 40$.

2.5 Repeat Problem 2.4 for the detection of a 10-psi shift in the mean in either direction.

2.6 For a sample mean of 124 g based on 16 observations from a population whose variance is 25 g^2 set up 90 percent confidence limits on the mean of the population.

2.7 For a sample variance of 62 based on 12 observations test the hypothesis that the population variance is 40. Use a one-sided test and a 1 percent significance level.

2.8 For Problem 2.7 set up 90 percent confidence limits (two-sided) on σ^2.

2.9 Two samples are taken, one from each of two machines. For this process

$$n_1 = 8 \qquad \bar{Y}_1 = 42.2 \text{ g} \qquad s_1^2 = 10 \text{ g}^2$$
$$n_2 = 15 \qquad \bar{Y}_2 = 44.5 \text{ g} \qquad s_2^2 = 18 \text{ g}^2$$

Test these results for a significant difference in variances.

2.10 Test the results in Problem 2.9 for a significance difference in means.

2.11 A group of 10 students was pretested before instruction and posttested after six weeks of instruction with the following achievement scores:

Student	Before	After	Differences = d
1	14	17	3
2	12	16	4
3	20	21	1
4	8	10	2
5	11	10	−1
6	15	14	−1
7	17	20	3
8	18	22	4
9	9	14	5
10	7	12	5

Is there evidence of an improvement in achievement over this six-week period?

2.12 To test the hypothesis that the defective fraction p of a process is 0.20, a sample of 100 pieces was drawn at random.
   a. Use an $\alpha = 0.05$ and set up a critical region for the number of observed defectives to test this hypothesis against a two-sided alternative.
   b. If the process now shifts to a 0.10 fraction defective, find the probability of an error of the second kind for this alternative.
   c. Without repeating the work, would you expect the β error to be the same if the process shifted to 0.30? Explain.

2.13 Given the following sample data on a random variable, Y: 12, 8, 14, 20, 26, 26, 20, 21, 18, 24, 30, 21, 18, 16, 10, 20, assume $\sigma = 7$, and $\alpha = 0.05$, and test each of the following hypotheses:
   a. $H_0: \mu = 12$

   $H_1: \mu > 12$
   b. $H_0: \mu = 16$

   $H_1: \mu \neq 16$

c. $H_0: \mu = 18$

$H_1: \mu > 18$

2.14 For test 1 in Problem 2.13 evaluate the power of the test with respect to several alternative hypotheses from $\mu = 13$ to $\mu = 24$. Plot the power curve.

2.15 Repeat the results of Problem 2.14 with $\alpha = 0.01$. Plot on the same graph and compare the two power curves.

2.16 Given the following data on the current flow in amperes through a cereal-forming machine: 8.2, 8.3, 8.2, 8.6, 8.0, 8.4, 8.5, 7.3, test the hypothesis of the standard deviation $\sigma = 0.35$ ampere versus the alternative that $\sigma > 0.35$ ampere.

2.17 For the data in Problem 2.16 test the hypothesis that the true mean current flow is $\mu = 7.0$.

2.18 The percent moisture content in a puffed cereal where samples are taken from two different "guns" showed

Gun I: 3.6, 3.8, 3.6, 3.3, 3.7, 3.4

Gun II: 3.7, 3.9, 4.2, 4.9, 3.6, 3.5, 4.0

Test the hypothesis of equal variances and equal means. Use any assumptions you believe appropriate.

2.19 Pretest data for experimental and control groups on course content in a special vocational-industrial course indicated:

Experimental: $\bar{Y}_1 = 9.333$, $s_1 = 4.945$, $n_1 = 12$

Control: $\bar{Y}_2 = 8.375$, $s_2 = 1.187$, $n_2 = 8$

Make any statistical tests that the data might suggest. Comment on the results.

2.20 The WISC performance scores of 10 boys and 8 girls were recorded as follows:

Boys: 101, 86, 72, 129, 99, 118, 104, 125, 90, 107

Girls: 97, 107, 94, 101, 90, 108, 108, 86

Test whether or not there is a significant difference in the means of boys and girls. Take $\alpha = 0.10$.

2.21 An occupational therapist conducted a study to evaluate the relative merits of two prosthetic devices designed to facilitate manual dexterity. The therapist assigned 21 patients with identical handicaps to wear one or the other of the two devices while performing a certain task. Eleven patients wore device A and 10 wore device B. The researcher recorded the time each patient required to perform a certain task, with the following results:

$$\bar{y}_A = 65 \text{ seconds} \qquad s_A^2 = 81$$

$$\bar{y}_B = 75 \text{ seconds} \qquad s_B^2 = 64$$

Do these data provide sufficient evidence to indicate that device A is more effective than device B?

2.22 An anthropologist believes that the proportion of individuals in two populations with double occipital hair whorls is the same. To see whether there is any reason to doubt this hypothesis, the anthropologist takes independent random samples from each of the two populations and determines the number in each sample with this characteristic. Results showed:

Population	n	Number with Characteristic
1	100	23
2	120	32

Is there reason to doubt this hypothesis?

2.23 A psychologist randomly selected 10 wives and their husbands from among the residents of an urban area, and asked them to complete a questionnaire designed to measure their level of satisfaction with the community in which they lived. Do the following results indicate that the husbands are better satisfied with the community than are their wives?

	Scores									
Wife	33	57	32	54	52	34	60	40	59	39
Husband	44	60	55	68	40	48	57	49	47	52

2.24 If two samples randomly selected from two independent normal populations give

$$n_1 = 9 \qquad \bar{Y}_1 = 16.0 \qquad s_1^2 = 5.0$$
$$n_2 = 4 \qquad \bar{Y}_2 = 12.0 \qquad s_2^2 = 2.0$$

is there enough evidence to claim that the mean of population 1 is greater than the mean of population 2?

2.25 The standard medical treatment for a certain ailment has proved to be about 85 percent effective. One hundred patients were given an alternative treatment with the result that 78 showed marked improvement. Is the new treatment actually inferior or could the difference noted be reasonably attributed to chance?

Single-Factor Experiments with No Restrictions on Randomization

3.1 Introduction

In this and several subsequent chapters single-factor experiments will be considered. In this chapter no restrictions will be placed on the randomization so that the design will be completely randomized. Many of the techniques of analysis for a completely randomized single-factor experiment can be applied with little alteration to more complex experiments.

For example, the single factor could be steel manufacturers, where the main interest of an analyst centers on the effect of several different manufacturers on the hardness of steel purchased from them. It could be temperature in an instance where the experimenter is concerned about the effect of temperature on penicillin yield. Whenever only one factor is varied, whether the levels be quantitative or qualitative, fixed or random, the experiment is referred to as a *single-factor experiment,* and the symbol τ_j will be used to indicate the effect of the jth level of the factor. τ_j suggests that the general factor may be thought of as a "treatment" effect.

If the order of experimentation applied to the several levels of the factor is completely random, so that any material to which the treatments might be applied is considered approximately homogeneous, the design is called a *completely randomized design.* The number of observations for each level of the treatment or factor will be determined from cost considerations and the power of the test. The model then becomes

$$Y_{ij} = \mu + \tau_j + \varepsilon_{ij}$$

where Y_{ij} represents the ith observation ($i = 1, 2, \ldots, n_j$) on the jth treatment ($j = 1, 2, \ldots, k$ levels). For example, Y_{23} represents the second observation using level 3 of the factor. μ is a common effect for the whole experiment, τ_j represents the effect of the jth treatment, and ε_{ij} represents the random error present in the ith observation on the jth treatment. When the effect is fixed, the model is called a *fixed model;* when random, a *random model.*

The error term ε_{ij} is usually considered a normally and independently distributed (NID) random effect whose mean value is zero and whose variance is the same for all treatments or levels. This is expressed as: ε_{ij}'s are NID $(0, \sigma_e^2)$ where σ_e^2 is the common variance within all treatments. μ is always a fixed parameter, and $\tau_1, \tau_2, \ldots, \tau_j, \ldots, \tau_k$ are considered to be fixed parameters if the levels of treatments are fixed. It is also assumed that

$$\sum_{j=1}^{k} \tau_j = 0$$

If the k levels of treatments are chosen at random, the τ_j's are assumed NID $(0, \sigma_\tau^2)$. Whether the levels are fixed or random depends upon how these levels are chosen in a given experiment.

The analysis of a single-factor completely randomized experiment usually consists of a one-way analysis-of-variance test where, if the effects are fixed, the hypothesis $H_0: \tau_j = 0$ for all j is tested. If this hypothesis is true, then no treatment effects exist and each observation Y_{ij} is made up of its population mean μ and a random error ε_{ij}. After an analysis of variance (ANOVA), many other tests may be made, and some of these will be shown in the example that follows.

■ **EXAMPLE 3.1** In the manufacture of clothing a wear-testing machine is used to measure the resistance to abrasion of different fabrics. The dependent variable Y gives the loss of weight of the material in grams after a specified number of cycles and the problem is to determine whether or not there is a difference in the average weight loss among four competing fabrics. Here there is but one factor of interest: the fabric at levels A, B, C, and D. These are qualitative and fixed levels of this single factor. It was agreed to test four samples of each fabric and completely randomize the order of testing of the 16 samples. The mathematical model for this example then is

$$Y_{ij} = \mu + \tau_j + \varepsilon_{ij}$$

with $i = 1, 2, \ldots, 4$ and $j = 1, 2, \ldots, 4$ since there are four samples for each of four treatments (fabrics). The data are shown in Table 3.1.

Table 3.1 Fabric Wear Resistance Data

	Fabric		
A	*B*	*C*	*D*
1.93	2.55	2.40	2.33
2.38	2.72	2.68	2.40
2.20	2.75	2.31	2.28
2.25	2.70	2.28	2.25

An analysis of variance performed on these data gave the results in Table 3.2.

Table 3.2 Fabric Data ANOVA

Source of Variation	df	SS	MS	F	Pr > F
Between fabrics: τ_j	3	0.5201	0.1734	8.53	0.0026
Within fabrics or error: ε_{ij}	12	0.2438	0.0203		
Totals	15	0.7639			

To test H_0: $\tau_j = 0$ for all $j = 1, 2, 3$, and 4, the test statistic is

$$F_{3,12} = \frac{0.1734}{0.0203} = 8.53$$

which is significant at the 1 percent significance level since Statistical Table D shows 5.95 as the 1 percent F $(F_{0.99})$ for 3 and 12 degrees of freedom. We can then reject the hypothesis and claim that there is a considerable difference in average wear resistance among the four fabrics. ∎

3.2 Analysis of Variance Rationale

To review the basis for the F test in a one-way analysis of variance, k populations, each representing one level of treatment, can be considered with observations as shown in Table 3.3.

Table 3.3 Population Layout for One-Way ANOVA

	Treatment					
	1	2	$\cdots$	j	$\cdots$	k
	Y_{11}	Y_{12}		Y_{1j}		Y_{1k}
	Y_{21}	Y_{22}		Y_{2j}		Y_{2k}
	Y_{31}	Y_{32}		—		—
	—	—		—		—
	Y_{i1}	Y_{i2}		Y_{ij}		Y_{ik}
Population means	μ_1	μ_2	$\cdots$	μ_j	$\cdots$	μ_k

Since each observation could be returned to the population and measured, there could be an infinite number of observations taken on each population, so the average or $E(Y_{i1}) = \mu_1$, and so on. μ will represent the average Y_{ij} over all populations, or $E(Y_{ij}) = \mu$. In the model τ_j, the treatment effect can also be indicated by $\mu_j - \mu$, and then the model is either

$$Y_{ij} = \mu + \tau_j + \varepsilon_{ij}$$

or

$$Y_{ij} \equiv \mu + (\mu_j - \mu) + (Y_{ij} - \mu_j)$$

This last expression is seen to be an identity true for all values of Y_{ij}. Expressed another way,

$$Y_{ij} - \mu \equiv (\mu_j - \mu) + (Y_{ij} - \mu_j) \tag{3.1}$$

Since these means are unknown, random samples are drawn from each population and estimates can be made of the treatment means and the grand mean. If n_j observations are taken for each treatment where the numbers need not be equal, a sample layout would be as shown in Table 3.4.

Table 3.4 Sample Layout for One-Way ANOVA

	Treatment						
	1	2	$\cdots$	j	$\cdots$	k	
	Y_{11}	Y_{12}	$\cdots$	Y_{1j}	$\cdots$	Y_{1k}	
	Y_{21}	Y_{22}	$\cdots$	Y_{2j}	$\cdots$	Y_{2k}	
	$\vdots$	$\vdots$		$\vdots$		$\vdots$	
	Y_{i1}	Y_{i2}	$\cdots$	Y_{ij}		Y_{ik}	
	$\vdots$	$\vdots$		$\vdots$		$\vdots$	
	Y_{n_11}	$\vdots$		Y_{n_jj}		$\vdots$	
		Y_{n_22}				Y_{n_kk}	
Totals	$T_{.1}$	$T_{.2}$	$\cdots$	$T_{.j}$	$\cdots$	$T_{.k}$	$T_{..}$
Number	n_1	n_2	$\cdots$	n_j	$\cdots$	n_k	N
Means	$\bar{Y}_{.1}$	$\bar{Y}_{.2}$	$\cdots$	$\bar{Y}_{.j}$	$\cdots$	$\bar{Y}_{.k}$	$\bar{Y}_{..}$

Here the use of the "dot notation" indicates a summing over all observations in the sample. $T_{.j}$ represents the total of the observations taken under treatment j, n_j represents the number of observations taken for treatment j, and $\bar{Y}_{.j}$ is the observed mean for treatment j. $T_{..}$ represents the grand total of all observations taken where

$$T_{..} = \sum_{j=1}^{k} \sum_{i=1}^{n_j} Y_{ij} = \sum_{j=1}^{k} T_{.j}$$

and

$$N = \sum_{j=1}^{k} n_j$$

and $\bar{Y}_{..}$ is the mean of all N observations. Note too that

$$\bar{Y}_{..} = \sum_{j=1}^{k} n_j \bar{Y}_{.j} / N$$

If these sample statistics are substituted for their corresponding population parameters in Equation (3.1), we get a sample equation (also an identity) of the form

$$Y_{ij} - \bar{Y}_{..} \equiv (\bar{Y}_{.j} - \bar{Y}_{..}) + (Y_{ij} - \bar{Y}_{.j}) \tag{3.2}$$

This equation states that the deviation of any observation from the grand mean can be broken into two parts: the deviation of the observation from its own treatment mean plus the deviation of the treatment mean from the grand mean.

If both sides of Equation (3.2) are squared and then added over both i and j, we have

$$\sum_{j=1}^{k} \sum_{i=1}^{n_j} (Y_{ij} - \bar{Y}_{..})^2 = \sum_{j=1}^{k} \sum_{i=1}^{n_j} (\bar{Y}_{.j} - \bar{Y}_{..})^2 + \sum_{j=1}^{k} \sum_{i=1}^{n_j} (Y_{ij} - \bar{Y}_{.j})^2$$

$$+ 2 \sum_{j=1}^{k} \sum_{i=1}^{n_j} (\bar{Y}_{.j} - \bar{Y}_{..})(Y_{ij} - \bar{Y}_{.j}) \tag{3.3}$$

Examining the last expression on the right, we find that

$$\sum_{j=1}^{k} \sum_{i=1}^{n_j} (\bar{Y}_{.j} - \bar{Y}_{..})(Y_{ij} - \bar{Y}_{.j}) = \sum_{j=1}^{k} (\bar{Y}_{.j} - \bar{Y}_{..}) \left[\sum_{i=1}^{n_j} (Y_{ij} - \bar{Y}_{.j}) \right]$$

The term in brackets is seen to equal zero, as the sum of the deviations about the mean within a given treatment equals zero. Hence

$$\sum_{j=1}^{k} \sum_{i=1}^{n_j} (Y_{ij} - \bar{Y}_{..})^2 = \sum_{j=1}^{k} \sum_{i=1}^{n_j} (\bar{Y}_{.j} - \bar{Y}_{..})^2 + \sum_{j=1}^{k} \sum_{i=1}^{n_j} (Y_{ij} - \bar{Y}_{.j})^2 \tag{3.4}$$

This may be referred to as "the fundamental equation of analysis of variance," and it expresses the idea that the total sum of squares of deviations from the grand mean is equal to the sum of squares of deviations between treatment means and the grand mean plus the sum of squares of deviations within treatments. In Chapter 2 an unbiased estimate of population variance was determined by dividing the sum of squares $\sum_{i=1}^{n} (Y_i - \bar{Y})^2$ by the corresponding number of degrees of freedom, $n - 1$. If the hypothesis being tested in analysis of variance is true, namely, that $\tau_j = 0$ for all j, or that there is no treatment effect, then $\mu_1 = \mu_2 = \cdots \mu_j \cdots = \mu_k$ and all there is in the model is the population mean μ and random error ε_{ij}. Then any one of the three terms in Equation (3.4) may be used to give an unbiased estimate of this common population variance. For example, dividing the left-hand term by its degrees of freedom, $N - 1$ will yield an unbiased estimate of population variance σ^2. Within the jth treatment, $\sum_i (Y_{ij} - \overline{Y}_{.j})^2$ divided by $n_j - 1$ df would yield an unbiased estimate of the variance within the jth treatment. If the variances within the k treatments are really all alike, their estimates may be pooled to give $\sum_{j=1}^{k} \sum_{i=1}^{n_j} (Y_{ij} - \overline{Y}_j)^2$ with degrees of freedom

$$\sum_j^k (n_j - 1) = N - k$$

which will give another estimate of the population variance. Still another estimate can be made by first estimating the variance $\sigma_{\overline{Y}}^2$ between means drawn from a common population with $\sigma_{Y}^2 = n_j \sigma_{\overline{Y}_{.j}}^2$. An unbiased estimate for $\sigma_{\overline{Y}}^2$ is given by $\sum_{j=1}^{k} (\overline{Y}_{.j} - \overline{Y}_{..})^2 / (k - 1)$ so that an unbiased estimate of

$$\sigma_Y^2 = \sum_{j=1}^{k} n_j (\overline{Y}_{.j} - \overline{Y}_{..})^2 / (k - 1)$$

is found by summing the first term on the right-hand side of Equation (3.4) over i and dividing by $k - 1$ df. Thus there are three unbiased estimates of σ^2 possible from the data in a one-way ANOVA if the hypothesis is true. Now all three are not independent since the sum of squares is additive in Equation (3.4). However, it can be shown that if each of the terms (sum of squares) on the right of Equation (3.4) is divided by its proper degrees of freedom, it will yield two independent chi-square distributed unbiased estimates of σ^2 when H_0 is true. If two such independent unbiased estimates of the same variance are compared, their ratio can be shown to be distributed as F with $k - 1$, $N - k$ df. If, then, H_0 is true, the test of the hypothesis can be made using a critical region of the F distribution with the observed F at $k - 1$ and $N - k$ df given by

$$F_{k-1, N-k} = \frac{\sum_{j=1}^{k} n_j (\overline{Y}_{.j} - \overline{Y}_{..})^2 / (k - 1)}{\sum_{j=1}^{k} \sum_{i=1}^{n_j} (Y_{ij} - \overline{Y}_{.j})^2 / (N - k)} \tag{3.5}$$

The critical region is usually taken as the upper tail of the F distribution, rejecting H_0 if $F \geq F_{1-\alpha}$ where α is the area above $F_{1-\alpha}$. In this F ratio the sum of squares between treatments is always put into the numerator, and then a significant F will indicate that the differences between means has something in it besides the estimate of variance. It probably indicates that there is a real difference in treatment means ($\mu_1, \mu_2, \ldots$) and that H_0 should be rejected. These unbiased estimates of population variance, sums of squares divided by df, are also referred to as *mean squares*.

The actual computing of sums of squares indicated in Equation (3.5) is much easier if they are first expanded and rewritten in terms of treatment totals. These computing formulas are given in Table 3.5. Applying the formulas in

Table 3.5 to the problem data in Table 3.1, Table 3.6 shows treatment totals $T_{.j}$'s and the sums of squares of the Y_{ij}'s for each treatment along with the results added across all treatments.

Table 3.5 One-Way ANOVA

Source	df	SS	MS
Between treatments τ_j	$k - 1$	$\sum_{j=1}^{k} n_j(\bar{Y}_{.j} - \bar{Y}_{..})^2$ $= \sum_{j=1}^{k} \dfrac{T_{.j}^2}{n_j} - \dfrac{T_{..}^2}{N}$	$SS_{treatment} / (k - 1)$
Within treatments or error ε_{ij}	$N - k$	$\sum_{j=1}^{k} \sum_{i=1}^{n_j} (Y_{ij} - \bar{Y}_{.j})^2$ $= \sum_{j=1}^{k} \sum_{i=1}^{n_j} Y_{ij}^2 - \sum_{j=1}^{k} \dfrac{T_{.j}^2}{n_j}$	$SS_{error} / (N - k)$
Totals	$N - 1$	$\sum_{j=1}^{k} \sum_{i=1}^{n_j} (Y_{ij} - \bar{Y}_{..})^2$ $= \sum_{j=1}^{k} \sum_{i=1}^{n_j} Y_{ij}^2 - \dfrac{T_{..}^2}{N}$	

Table 3.6 Fabric Wear Resistance Data

	Fabric				
	A	B	C	D	
	1.93	2.55	2.40	2.33	
	2.38	2.72	2.68	2.40	
	2.20	2.75	2.31	2.28	
	2.25	2.70	2.28	2.25	
$T_{.j}$:	8.76	10.72	9.67	9.26	$T_{..} = 38.41$
n_j:	4	4	4	4	$N = 16$
$\sum_{i=1}^{n_j} Y_{ij}^2$:	19.2918	28.7534	23.4769	21.4498	$\sum_{j}^{k} \sum_{i}^{n_j} Y_{ij}^2 = 92.9719$

The sums of squares can be computed quite easily from this table. The total sum of squares states, "Square each observation, add over all observations, and subtract the correction term." The last is the grand total squared and divided by the total number of observations

$$SS_{total} = \sum_{j=1}^{k} \sum_{i=1}^{n_j} Y_{ij}^2 - \frac{T_{..}^2}{N} = 92.9719 - \frac{(38.41)^2}{16} = 0.7639$$

The sum of squares between treatments is found by totaling n_j observations for each treatment, squaring this total, dividing by the number of observations, adding for all treatments, and then subtracting the correction term:

$$SS_{treatment} = \sum_{j=1}^{k} \frac{T_{.j}^2}{n_j} - \frac{T_{..}^2}{N}$$

$$= \frac{(8.76)^2}{4} + \frac{(10.72)^2}{4} + \frac{(9.67)^2}{4} + \frac{(9.26)^2}{4} - \frac{(38.41)^2}{16} = 0.5201$$

The sum of squares for error is then determined by subtraction:

$$SS_{error} = SS_{total} - SS_{treatment} = 0.7639 - 0.5201 = 0.2438$$

These results are displayed as in Table 3.2, and the F test is run on the $H_0: \tau_j = 0$ as shown before.

The computation of these sums of squares can be simplified in two ways. One could code the data by subtracting a convenient constant such as 2.00 and then multiplying all Y's by 100 to eliminate the decimals. The subtraction of a constant will not affect the sums of squares but multiplying by 100 will multiply all SS terms by $(100)^2$ or 10,000. However, since the test statistic F is the ratio of two mean squares, the F will be unaffected by this coding scheme.

With the advent of the computer, this analysis of variance can be run on an SAS program, which will be shown in Section 3.9.

3.3 After ANOVA—What?

After concluding, as in the problem above, that there is a difference in treatment means, we must ask more questions such as: Which treatment is the best? Does the mean wear resistance of fabric A differ from that of C? Does the mean of A and B together differ from that of C and D? To answer questions following an analysis of variance, one must consider the assumptions made when the experiment was planned originally. The decision tree of Figure 3.1 may be helpful in outlining some tests after analysis of variance.

In the next sections we examine some of the tests indicated in Figure 3.1.

3.4 Tests on Means

In considering qualitative levels of treatments as in Example 3.1, any comparison of means will depend upon whether the decision on what means to compare was made before the experiment was run or after the results were examined.

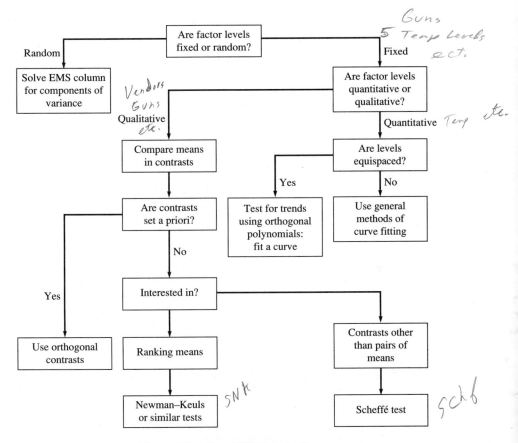

Figure 3.1 After ANOVA, what?

Tests on Means Set Prior to Experimentation—Orthogonal Contrasts

If the above decision is made prior to the running of an experiment, such comparisons can usually be set without disturbing the risk α of the original ANOVA. This means that the contrasts must be chosen with care, and the number of such contrasts should not exceed the number of degrees of freedom between the treatment means. The method usually used here is called the *method of orthogonal contrasts*.

In comparing treatments 1 and 2 it should seem quite logical to examine the difference in their means or their totals since all treatment totals are based on the same sample size n. One such comparison might be $T_{.1} - T_{.2}$. If, on the other hand, one wishes to compare the first and second treatment with the third, the third treatment total should be weighted by a factor of 2 since it is based on only four observations and the treatment totals of 1 and 2 together represent eight observations. Such a comparison would then be $T_{.1} + T_{.2} - 2T_{.3}$. Note that for each of the two comparisons or contrasts, the sum of the coefficients of the treatment totals always adds to zero:

$$T_{.1} - T_{.2} \text{ has coefficients } +1 - 1 = 0$$
$$T_{.1} + T_{.2} - 2T_{.3} \text{ has coefficients } +1 + 1 - 2 = 0$$

Hence a *contrast* C_m is defined for any linear combination of treatment totals as follows:

$$C_m = c_{1m}T_{.1} + c_{2m}T_{.2} + \cdots + c_{jm}T_{.j} + \cdots + c_{km}T_{.k}$$

where

$$c_{1m} + c_{2m} + \cdots + c_{jm} + \cdots + c_{km} = 0$$

or, more compactly,

$$C_m = \sum_{j=1}^{k} c_{jm}T_{.j}$$

where

$$\sum_{j=1}^{k} c_{jm} = 0$$

In dealing with several contrasts it is highly desirable to have the contrasts independent of each other. When the contrasts are independent—one having no projection on any of the others—they are said to be orthogonal contrasts, and independent tests of hypotheses can be made by comparing the mean square of each such contrast with the mean square of the error term in the experiment. Each contrast carries one degree of freedom.

It may be recalled that when two straight lines, say

$$c_{11}X + c_{21}Y = b_1$$
$$c_{12}X + c_{22}Y = b_2$$

are perpendicular, or orthogonal, to each other, the slope of one line is the negative reciprocal of the slope of the other line. Changing the two lines above into slope-intercept form gives

$$Y = (-c_{11}/c_{21})X + b_1/c_{21}$$
$$Y = (-c_{12}/c_{22})X + b_2/c_{22}$$

To be orthogonal,

$$-c_{11}/c_{21} = -(-c_{22}/c_{12})$$

or

$$c_{11}c_{12} = -c_{21}c_{22}$$

or

$$c_{11}c_{12} + c_{21}c_{22} = 0$$

The sum of the products of the corresponding coefficients (on X and on Y) must therefore add to zero. This idea may be extended to a general definition. Two contrasts C_m and C_q are said to be *orthogonal contrasts*, provided

$$\sum_{j=1}^{k} c_{jm}c_{jq} = 0$$

for equal n's.

The sums of squares for a contrast are given by

$$SS_{C_m} = \frac{C_m^2}{n \sum_{j=1}^{k} c_{jm}^2}$$

To apply this procedure to the problem above, three orthogonal contrasts may be set up since there are 3 df between treatments. One such set of three might be

$$C_1 = T_{.1} \qquad\qquad - T_{.4}$$
$$C_2 = \qquad T_{.2} - T_{.3}$$
$$C_3 = T_{.1} - T_{.2} - T_{.3} + T_{.4}$$

C_1 is a contrast to compare the mean of the first treatment with the fourth, C_2 compares the second treatment with the third, and C_3 compares the average of treatments 1 and 4 with the average of 2 and 3. The coefficients of the $T_{.j}$'s for the three contrasts are given in Table 3.7.

Table 3.7 Orthogonal Coefficients

	$T_{.1}$	$T_{.2}$	$T_{.3}$	$T_{.4}$
C_1	+1	0	0	−1
C_2	0	+1	−1	0
C_3	+1	−1	−1	+1

It can be seen from Table 3.7 that the sum of coefficients c_{jm} adds up to zero for each contrast, and that the sum of products of coefficients of each pair of contrasts is also zero.

For the example given in Section 3.2, using the data in Table 3.6, the contrasts are

$$C_1 = +1(8.76) + 0(10.72) + 0(9.67) - 1(9.26) = -0.50$$
$$C_2 = 0(8.76) + 1(10.72) - 1(9.67) + 0(9.26) = 1.05$$
$$C_3 = +1(8.76) - 1(10.72) - 1(9.67) + 1(9.26) = -2.37$$

The corresponding sums of squares are

$$SS_{C_1} = \frac{(-0.50)^2}{4(2)} = 0.0312$$

$$SS_{C_2} = \frac{(1.05)^2}{4(2)} = 0.1378$$

$$SS_{C_3} = \frac{(-2.37)^2}{4(4)} = 0.3511$$

$$\text{Total} = 0.5201$$

Note that these three orthogonal sums of squares add up to the treatment sums of squares. Each contrast has 1 df, so one may test hypotheses implied by the contrasts as follows:

$$H_{0_1}: \tau_1 = \tau_4 \quad \text{or} \quad \mu_1 = \mu_4 \qquad F_{1,12} = \frac{0.0312 \, / \, 1}{0.0203} = 1.54$$

$$H_{0_2}: \tau_2 = \tau_3 \quad \text{or} \quad \mu_2 = \mu_3 \qquad F_{1,12} = \frac{0.1378 \, / \, 1}{0.0203} = 6.76$$

$$H_{0_3}: \tau_1 + \tau_4 = \tau_2 + \tau_3 \qquad F_{1,12} = \frac{0.3511 \, / \, 1}{0.0203} = 17.30$$

or

$$\mu_{1+4} = \mu_{2+3}$$

Since the F for 1 and 12 df is 4.75 at the 5 percent significance level and 9.33 at the 1 percent significance level, one can reject H_{0_2} at the 5 percent level and conclude that the mean wear resistance of fabric 2 (or B) differs from that of fabric 3 (or C). At the 1 percent level one rejects H_{0_3} and concludes that the mean wear resistances of fabrics 1 and 4 (A and D) differ from that of 2 and 3 (B and C).

If a computer program is used to run the analysis of variance, it can be programmed to compute any desired contrasts.

As this method of orthogonal contrasts is used quite often in experimental design work, the definitions and formulas are given below for the case of unequal numbers of observations per treatment.

C_m is a *contrast* if

$$\sum_{j=1}^{k} n_j c_{jm} = 0$$

and C_m and C_q are *orthogonal contrasts* if

$$\sum_{j=1}^{k} n_j c_{jm} c_{jq} = 0$$

The *sum of squares for such contrasts* is given by

$$SS_{C_m} = \frac{C_m^2}{\sum_{j=1}^{k} n_j c_{jm}^2}$$

Tests on Means After Experimentation—Multiple Comparisons

If the decision on what comparisons to make is withheld until after the data are examined, comparisons may still be made, but the α level is altered because such decisions are not taken at random but are based on observed results. Several methods have been introduced to handle such situations [18], but only the Student–Newman–Keuls (SNK) range test [16] and Scheffé's test [25] are described here for balanced data.

Range Test   After the data have been compiled the following steps are taken.

1. Arrange the k means in order from low to high.
2. Enter the ANOVA table and take the error mean square with its degrees of freedom.
3. Obtain the standard error of the mean for each treatment

$$s_{\bar{Y}_{.j}} = \sqrt{\frac{\text{error mean square}}{\text{number of observations in } \bar{Y}_{.j}}}$$

   where the error mean square is the one used as the denominator in the F test on means $\bar{Y}_{.j}$.
4. Enter a Studentized range table (Statistical Table E) of significant ranges at the α level desired, using $n_2 = $ degrees of freedom for error mean square and $p = 2, 3, \ldots, k$, and list these $k - 1$ ranges.
5. Multiply these ranges by $s_{\bar{Y}_{.j}}$ to form a group of $k - 1$ least significant ranges.
6. Test the observed ranges between means, beginning with largest versus smallest, which is compared with the least significant range for $p = k$; then test largest versus second smallest with the least significant range for $p = k - 1$; and so on. Continue this for second largest versus smallest, and so forth, until all $k(k - 1) / 2$ possible pairs have been tested. The sole exception to this rule is that no difference between two means can be declared significant if the two means concerned are both contained in a subset with a nonsignificant range.

To see how this works, consider the data of Example 3.1 as given in Table 3.6. Following the steps given above:

1. $k = 4$ means are   2.19   2.32   2.42   2.68
   for treatments: A   D   C   B

2. From Table 3.2 the error mean square is 0.0203 with 12 df.

3. Standard error of a mean is

$$s_{\bar{Y}_{.j}} = \sqrt{\frac{0.0203}{4}} = 0.0712$$

4. From Statistical Table E, at the 5 percent level the significant ranges are, for $n_2 = 12$,

p:	2	3	4
Ranges:	3.08	3.77	4.20

5. Multiplying by the standard error of 0.0712, the least significant ranges (LSR) are

p:	2	3	4
LSR:	0.22	0.27	0.30

6. Largest versus smallest: B versus A, $0.49 > 0.30$.
Largest versus next smallest: B versus D, $0.36 > 0.27$.
Largest versus next largest: B versus C, $0.26 > 0.22$.
Second largest versus smallest: C versus A, $0.23 < 0.27$.
Second largest versus next largest: C versus D, $0.10 < 0.22$.
Third largest versus smallest: D versus A, $0.13 < 0.22$.

From the results in step 6 one sees that B differs significantly from A, D, and C, but A, D, and C do not differ significantly from each other. These results are shown in Figure 3.2, where any means underscored by the same line are not significantly different.

In practice, several configurations of Figure 3.2 are encountered. For five means, several possible outcomes may be seen as in Figure 3.3.

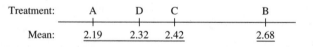

Figure 3.2 Newman–Keuls test means.

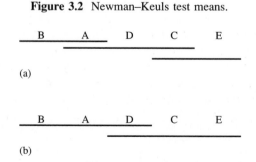

Figure 3.3 Some possible means.

If one seeks a minimum response as the "best" treatment in Figure 3.3 (a), B looks best, but it is no better than A. However, we would recommend B since its mean is significantly less than D, C, or E (no common line), whereas A is not significantly less than D or C.

In Figure 3.3 (b), B or A means are less than C or E means, whereas D is not, so we would choose either B or A as best.

Some experience with various combinations of means will give reasonable inferences when using the Newman–Keuls procedure.

Scheffé's Test   Since the Newman–Keuls test is restricted to comparing pairs of means and it is often desirable to examine other contrasts that represent combinations of treatments, many experimenters prefer a test devised by Scheffé [25]. Scheffé's method uses the concept of contrasts presented earlier but the contrasts need not be orthogonal. In fact, any and all conceivable contrasts may be tested for significance. Since comparing means in pairs is a special case of contrasts, Scheffé's scheme is more general than the Newman–Keuls. However, since the Scheffé method must be valid for such a large set of possible contrasts, it requires larger observed differences to be significant than some of the other schemes. To see how Scheffé's method applies, the following steps are taken after the data have been compiled.

1. Set up all contrasts of interest to the experimenter and compute their numeric values.

2. Determine the significant F statistic for the ANOVA just performed based on α and degrees of freedom $k - 1, N - k$.

3. Compute $A = \sqrt{(k - 1)F}$, using the F from step 2.

4. Compute the standard error of each contrast to be tested. This standard error is given by

$$s_{C_m} = \sqrt{(\text{error mean square}) \sum_j n_j c_{jm}^2}$$

5. If a contrast C_m is numerically larger than A times s_{C_m}, it is declared significant. Or, if $|C_m| > A s_{C_m}$, reject the hypothesis that the true contrast among means is zero.

Applying this technique to the coded data of Tables 3.1, 3.2, and 3.6, proceed as follows.

1. Consider two contrasts:

$$C_1 = T_{.1} - T_{.2} \qquad\qquad = 8.76 - 10.72 \qquad\qquad = -1.96$$
$$C_2 = 3T_{.1} - T_{.2} - T_{.3} - T_{.4} = 26.28 - 10.72 - 9.67 - 9.26 = -3.37$$

(Note that these are not orthogonal. The first compares treatment A with B and the second compares the mean of treatment A with the mean of the other three treatments.)

2. Since $\alpha = 0.05$, $k - 1 = 3$, $N - k = 12$, $F = 3.49$.

3. $A = \sqrt{3(3.49)} = 3.24$.

4. $s_{C_1} = \sqrt{0.0203[4(1)^2 + 4(-1)^2]} = 0.40$ and $As_{C_1} = 1.30$.

   $s_{C_2} = \sqrt{0.0203[4(3)^2 + 4(-1)^2 + 4(-1)^2 + 4(-1)^2]} = 0.99$ and $As_{C_2} = 3.21$.

5. Since $|C_1| = 1.96$ is > 1.30, this contrast is significant.
   Since $|C_2| = 3.37$ is > 3.21, this contrast is also significant.

3.5 Confidence Limits on Means

After an analysis of variance it is often desirable to set confidence limits on a treatment mean. The $100(1 - \alpha)$ percent confidence limits on $\mu_{.j}$ are given by

$$\bar{Y}_{.j} \pm t_{1-\alpha/2} \sqrt{\frac{\text{mean square used to test treatment mean square}}{n_j}} \tag{3.6}$$

where the mean square used to test treatment mean square is the error mean square in a one-way ANOVA but may be a different mean square in more complex analyses. The degrees of freedom used with Student's t statistic are the degrees of freedom that correspond to the mean square used for testing the treatment mean.

In Example 3.1, 95 percent confidence limits on the mean wear resistance for fabric B, μ_2, would be

$$\bar{Y}_{.2} \pm t_{.975} \sqrt{\frac{0.0203}{4}}$$

with 12 df on t. By substituting and taking the t value from Statistical Table B, the limits are $2.68 \pm 2.18(0.0712)$ or 2.68 ± 0.16 or from 2.52 to 2.84.

3.6 Components of Variance

In Example 3.1 the levels of the factor were considered as fixed, since only four fabrics were available and a decision was desired on the effect of these four fabrics only. If, however, the levels of the factor are random (such as operators, days, or samples where the levels in the experiment might have been chosen at random from a large number of possible levels), the model is called a *random model,* and inferences are to be extended to all levels of the population (of which the observed four levels are random samples). In a random model, the experimenter is not usually interested in testing hypotheses, setting confidence limits, or making contrasts in means, but in estimating components of variance. How much of the variance in the experiment might be considered as due to true differences in treatment means, and how much might be due to random error about these means?

■ **EXAMPLE 3.2** To see how to analyze and interpret a component of variance or random model, consider the following problem.

A company supplies a customer with several hundred batches of a raw material every year. The customer is interested in a high yield from the raw material in terms of percent usable chemical. He usually makes three sample determinations of yield from each batch in order to control the quality of the incoming material. He expects and gets some variation between determinations on a given batch, but he suspects that there may be significant batch-to-batch variation as well.

To check this, he selects five batches at random from several batches available and runs three yield determinations per batch. His 15 yield determinations are completely randomized. The mathematical model is again

$$Y_{ij} = \mu + \tau_j + \varepsilon_{i(j)}$$

except that in this experiment the k levels of the treatment (batches) are chosen at random, rather than being fixed levels. The data are shown in Table 3.8.

Table 3.8 Chemical Yield by Batch Data

		Batch		
1	2	3	4	5
74	68	75	72	79
76	71	77	74	81
75	72	77	73	79

An ANOVA performed on these data gave the results in Table 3.9.

Table 3.9 Chemical Yield ANOVA

Source	df	SS	MS	EMS
Between batches τ_j	4	147.74	36.94	$\sigma_\varepsilon^2 + n\sigma_\tau^2$
Error $\varepsilon_{i(j)}$	10	17.99	1.80	σ_ε^2
Totals	14	165.73		

The F test gives $F_{4,\,10} = 36.94 / 1.80 = 20.5$, which is highly significant. Since these batches are but a random sample of batches, we may be interested in how much of the variance in the experiment might be attributed to batch differences and how much to random error. To help answer these questions, another column has been added to the ANOVA table (Table 3.9). This is the expected mean square (EMS) column. It can be derived by inserting the mathematical model into the operational formulas for mean squares in Table 3.5 and then computing the expected values of these mean squares. The results will depend upon whether a fixed or random model is used as will be discussed in greater length in the next part of this section.

The interpretation of Table 3.9 then is that the error variance σ_ε^2 is best estimated as 1.80, its corresponding mean square. Also, the mean square 36.94 is an estimate of $\sigma_\varepsilon^2 + 3\sigma_\tau^2$. If the numerical mean squares are set equal to the variance components that they are estimating,

$$s_\varepsilon^2 = 1.80 \qquad s_\varepsilon^2 + 3s_\tau^2 = 36.94$$

where s^2 is used since these are estimates of the corresponding σ^2's.

Solving these expressions,

$$s_\tau^2 = \frac{36.94 - 1.80}{3} = 11.71$$

The total variance can be estimated as

$$s_{\text{total}}^2 = s_\tau^2 + s_\varepsilon^2 = 11.71 + 1.80 = 13.51$$

and 11.71/13.51 or 86.7 percent of the total variance is attributable to batch differences and only 1.80/13.51 or 13.3 percent is attributable to errors within batches.

It is interesting to note that the above estimate of total variance of 13.51 would give a standard deviation of 3.68. One might therefore expect all the data in such a small experiment ($N = 15$) to fall within four standard deviations or within $4(3.68) = 14.72$. The actual range is 13 (high, 81—low, 68). This breakdown of total variance into its two components then seems quite reasonable.

Some readers may wonder why the total sum of squares is never divided by its degrees of freedom to estimate the total variance. If the methods of the next section are used to determine the EMS for the total, it is found to be a biased estimate of the total variance. In fact, a derivation will show that the EMS for the total in a one-way ANOVA is equal to

$$\text{EMS}_{\text{total}} = \frac{n(k-1)}{N-1}\sigma_\tau^2 + \sigma_\varepsilon^2$$

Since the coefficient of σ_τ^2 is always less than one, this EMS will be less than the total variance given above as $\sigma_\tau^2 + \sigma_\varepsilon^2$. ■

Expected Mean Square Derivation

For a single-factor experiment

$$Y_{ij} = \mu + \tau_j + \varepsilon_{ij} \tag{3.7}$$

From Table 3.5 the sum of squares for treatments with equal n's is

$$SS_{treatment} = \sum_{j=1}^{k} n(\bar{Y}_{.j} - \bar{Y}_{..})^2$$

From model Equation (3.7)

$$\bar{Y}_{.j} = \sum_{i=1}^{n} Y_{ij} / n = \sum_{i=1}^{n} (\mu + \tau_j + \varepsilon_{ij}) / n$$

$$= \frac{n\mu}{n} + \frac{n\tau_j}{n} + \sum_{i=1}^{n} \varepsilon_{ij} / n$$

$$= \mu + \tau_j + \sum_{i=1}^{n} \varepsilon_{ij} / n \tag{3.8}$$

Also,

$$\bar{Y}_{..} = \sum_{j=1}^{k} \sum_{i=1}^{n} Y_{ij} / nk = \sum_{j}^{k} \sum_{i}^{n} (\mu + \tau_j + \varepsilon_{ij}) / nk$$

$$= \frac{nk\mu}{nk} + n\sum_{j}^{k} \tau_j / nk + \sum_{j}^{k} \sum_{i}^{n} \varepsilon_{ij} / nk$$

$$= \mu + \sum_{j}^{k} \tau_j / k + \sum_{j}^{k} \sum_{i}^{n} \varepsilon_{ij} / nk \tag{3.9}$$

Subtracting Equation (3.9) from Equation (3.8) gives

$$\bar{Y}_{.j} - \bar{Y}_{..} = \tau_j - \sum_{j=1}^{k} \tau_j / k + \sum_{i=1}^{n} \varepsilon_{ij} / n - \sum_{i}^{n} \sum_{j}^{k} \varepsilon_{ij} / nk$$

Squaring gives

$$(\bar{Y}_{.j} - \bar{Y}_{..})^2 = \left(\tau_j - \sum_{j=1}^{k} \tau_j / k\right)^2 + \frac{1}{n^2}\left(\sum_{i}^{n} \varepsilon_{ij} - \sum_{i}^{n} \sum_{j}^{k} \varepsilon_{ij} / k\right)^2$$

$$(+ \text{ cross products})$$

Multiplying by n and summing over j gives

$$SS_{treatment} = \sum_{j=1}^{k} n(\bar{Y}_{.j} - \bar{Y}_{..})^2$$

$$= n \sum_{j}^{k} \left(\tau_j - \sum_{j}^{k} \tau_j / k\right)^2$$

$$+ \frac{n}{n^2} \sum_{j}^{} \left(\sum_{i}^{n} \varepsilon_{ij} - \sum_{i}^{n} \sum_{j}^{k} \varepsilon_{ij} / k\right)^2 + n \sum_{j}^{} (\text{cross product})$$

The expected value operator may now be applied to this $SS_{\text{treatment}}$:

$$E(SS_{\text{treatment}}) = nE\left[\sum_j^k \left(\tau_j - \sum_j^k \tau_j/k\right)^2\right] + \frac{1}{n}E\left[\sum_j^k \left(\sum_i^n \varepsilon_{ij} - \sum_i^n\sum_j^k \varepsilon_{ij}/k\right)^2\right]$$

as it can be shown that the expected value of the cross-product term equals zero. If the treatment levels are fixed,

$$\sum_{j=1}^k \tau_j = \sum_{j=1}^k (\mu_j - \mu) = 0$$

and the $E(SS_{\text{treatment}})$ becomes

$$E(SS_{\text{treatment}}) = n\sum_{j=1}^k \tau_j^2 + \frac{1}{n}(nk - n)\sigma_\varepsilon^2$$

since errors are random and $\sum_{j=1}^k \tau_j^2$ is a constant. The $E(MS_{\text{treatment}}) = E[SS_{\text{treatment}}/(k-1)]$, so

$$E(MS_{\text{treatment}}) = \left[n\sum_{j=1}^k \tau_j^2/(k-1)\right] + \frac{n(k-1)}{n(k-1)}\sigma_\varepsilon^2$$

If, on the other hand, treatment levels are random and their variance is σ_τ^2

$$E(MS_{\text{treatment}}) = \frac{n(k-1)\sigma_\tau^2}{k-1} + \sigma_\varepsilon^2$$

the EMS term corresponding to the treatments is then either

$$\sigma_\varepsilon^2 + n\left[\sum_j \tau_j^2/(k-1)\right]$$

or

$$\sigma_\varepsilon^2 + n\sigma_\tau^2$$

depending upon whether treatments are fixed or random. When fixed, we shall designate $\sum_j \tau_j^2/(k-1)$ as ϕ_τ, so the fixed treatment EMS is $\sigma_\varepsilon^2 + n\phi_\tau$.

For the error mean square,

$$SS_{\text{error}} = \sum_{j=1}^k \sum_{i=1}^n (Y_{ij} - \bar{Y}_{.j})^2$$

Subtracting Equation (3.8) from Equation (3.7) gives

$$Y_{ij} - \bar{Y}_{.j} = \varepsilon_{ij} - \sum_{i=1}^n \varepsilon_{ij}/n$$

Squaring and adding gives

$$\sum_{j}^{k} \sum_{i}^{n} (Y_{ij} - \bar{Y}_{.j})^2 = \sum_{j}^{k} \sum_{i}^{n} \left(\varepsilon_{ij} - \sum_{i} \varepsilon_{ij} / n \right)^2$$

Taking the expected value, we have

$$E(\text{SS}_{\text{error}}) = E \left[\sum_{j}^{k} \sum_{i}^{n} \left(\varepsilon_{ij} - \sum_{i} \varepsilon_{ij} / n \right)^2 \right]$$

$$= \sum_{j}^{k} E \left[\sum_{i} \left(\varepsilon_{ij} - \sum_{i} \varepsilon_{ij} / n \right)^2 \right]$$

$$= \sum_{j}^{k} (n - 1)\sigma_{\varepsilon}^2$$

$$= k(n - 1)\sigma_{\varepsilon}^2$$

and

$$E(\text{MS}_{\text{error}}) = E[\text{SS}_{\text{error}} / k(n - 1)] = \sigma_{\varepsilon}^2$$

as shown in Table 3.9.

3.7 Summary and Assumptions

In this chapter consideration has been given to

Experiment	Design	Analysis
I. Single factor	Completely randomized	One-way ANOVA

In applying the ANOVA techniques, certain assumptions should be kept in mind:

1. The process is in control, that is, it is repeatable.
2. The population distribution being sampled is normal.
3. The variance of the errors within all k levels of the factor is homogeneous.

Many texts [4], [10] discuss these assumptions and what may be done if they are not met in practice.

Several studies have shown that lack of normality in the dependent variable Y does not seriously affect the analysis when the number of observations per treatment is the same for all treatments. To check for homogeneous variances within treatments, several tests have been suggested. One quick check is to examine the ranges of the observations within each treatment. For Example 3.1 these are 0.45, 0.20, 0.40, and 0.15. If the average range $\bar{R}$ is multiplied by a factor D_4, which is found in most quality control texts, and all ranges are less than $D_4\bar{R}$, it is quite safe to assume homogeneous variances. Here $\bar{R} = 0.30$ and $D_4\bar{R} = 0.68$ based on samples of four. Since all four ranges are well below 0.68, homogeneity of variance may be reasonably assured. Values of D_4 for a few selected sample sizes are given in Table 3.10.

Table 3.10 D_4 Values for Sample Size n

If	$n = 2$	3	4	5	6	7	8	9	10
	$D_4 = 3.267$	2.575	2.282	2.115	2.004	1.924	1.864	1.816	1.777

With some measured variables it may be that the means of the treatments and the variances within the treatments are related in some way. For example, a Poisson variable has its mean and variance equal. It can be shown mathematically that a necessary and sufficient condition for normality of a variable is that means and variances drawn from that population are independent of each other. If it is known or observed that sample means and sample variances are not independent but are related to each other, it may be possible to "break" this correlation by a suitable transformation of the original variable Y. Table 3.11 gives several suggested transformations to try if certain specified relationships hold.

Table 3.11 Transformations in ANOVA

If	Is Proportional to	Transform Y_{ij} to
s_j^2	$\bar{Y}_{.j}$	$\sqrt{Y_{ij}}$ (Poisson case)
s_j^2	$\bar{Y}_{.j}(1 - \bar{Y}_{.j})$	arc sin $\sqrt{Y_{ij}}$ (binomial case)
s_j	$\bar{Y}_{.j}$	log Y_{ij} or log $(Y_{ij} + 1)$
s_j	$\bar{Y}_{.j}^2$	$1 / Y_{ij}$

Computer programs can be instructed simply to change the original variable Y_{ij} to some transformed variable as shown above and carry out the ANOVA on the transformed variable.

3.8 Analysis of Residuals

The residuals in an experiment are what remains after subtracting the estimated effects in the model from the observed Y values. In a single-factor experiment, the residuals are $Y_{ij} - \hat{Y}_{ij}$, where $\hat{Y}_{ij}$ is the value of Y_{ij} predicted from the model. Here

$$E(Y_{ij}) = \mu + \tau_j \qquad \text{or} \qquad E(Y_{ij}) = \mu + (\mu_j - \mu) = \mu_j$$

or the true treatment mean. This is estimated by $\hat{Y}_{ij} = \bar{Y}_{.j}$, the sample treatment mean, and the residual of Y_{ij} is $Y_{ij} - \hat{Y}_{ij} = Y_{ij} - \bar{Y}_{.j}$, which is simply the error estimate in this single-factor case. However, as more factors are added to the model, the residuals are usually considered to be the differences between the observed Y_{ij}'s and the predicted Y_{ij}'s based on just the statistically significant terms in the model.

In the light of Section 3.7, the residuals are normally and independently distributed with homogeneous variances. As noted in Section 3.1, we assumed that errors were NID $(0, \sigma_e^2)$.

Many practitioners feel that no ANOVA is complete without an analysis of these residuals. It is therefore often desirable to examine these residuals for normality, independence, and homogeneous variance. From Table 3.1 and the four fabric means 2.19, 2.68, 2.42, and 2.32, the residuals can be found $Y_{ij} - \bar{Y}_{.j}$ for each of the j fabrics; these residuals are shown in Table 3.12.

Table 3.12 Residuals of Table 3.1

	Fabric		
A	*B*	*C*	*D*
−0.26	−0.13	−0.02	0.01
0.19	0.04	0.26	0.08
0.01	0.07	−0.11	−0.04
0.06	0.02	−0.14	−0.07

Graphic methods are often useful in examining these residuals. First we might plot them in a dot diagram as in Figure 3.4. The figure shows the distribution of these residuals from −0.26 to +0.26 and some clustering near the mean of 0.0. This may help in assuming they came from a normal population. Such a graph may sometimes indicate a strong nonnormal situation if it exists.

One might plot these 16 residuals on normal probability paper, where a straight-line plot indicates a normal population. This can be seen in Figure 3.5.

We could get some idea of the homogeneity of the variance of these residuals around their own treatment means from a plot of each treatment separately as seen in Figure 3.6. This "look test" can only be a rough one with small samples, but it does seem reasonable in the light of our test by ranges in Section 3.6.

To test whether the residuals are independent of each other, we might plot these versus the time the data were collected. To do so, we must keep track of

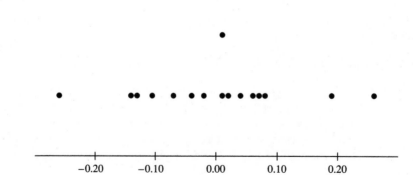

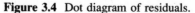

Figure 3.4 Dot diagram of residuals.

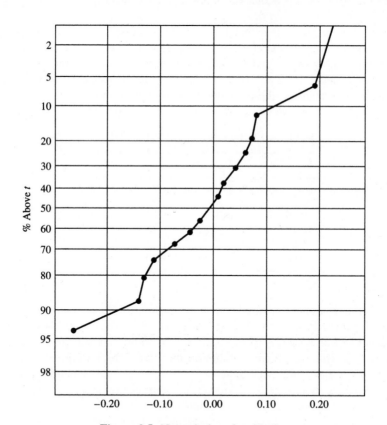

Figure 3.5 Normal plot of residuals.

the order in which the experiment was run. If the experiment has been run in a completely randomized order, we do not expect any pattern to emerge when the residuals are plotted against the time that they were recorded. If we take the order of our 16 observations to be −0.26, 0.19, 0.01, . . . , −0.07, we obtain

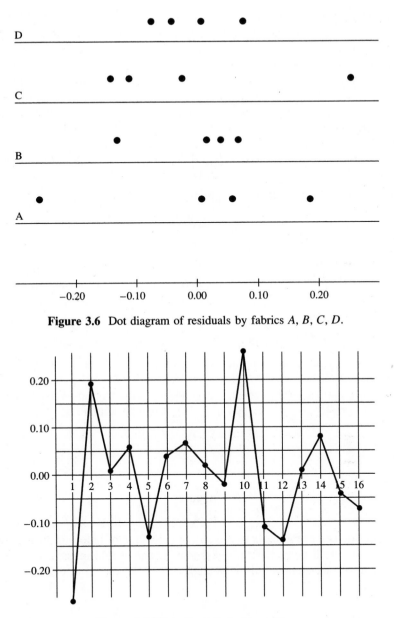

Figure 3.6 Dot diagram of residuals by fabrics A, B, C, D.

Figure 3.7 Plot of residuals versus time.

the graph of Figure 3.7. Here the pattern looks quite like a random walk. If, on the other hand, some definite pattern is seen, it is a good indication that the data were not collected in a random manner, and the order of data collection should be examined.

In addition to, or in place of, these graphic techniques some sample statistics are often associated with residual analyses. One such statistic is the autocorrelation coefficient, which indicates how the residuals are correlated with themselves. To compute this, we correlate the observed residuals with the same residuals moved one or more (say k) positions from the originals. That is, if $k = 1$, we would run a correlation between the series $-0.26, 0.19, 0.01, \ldots$ and the series $0.19, 0.01, 0.06, \ldots$. This is said to be a series with "lag" 1.

A formula for such an autocorrelation coefficient for lag k is

$$r_k = \frac{\sum (Y_i - \bar{Y})(Y_{i+k} - \bar{Y})}{\sum (Y_i - \bar{Y})^2} \tag{3.10}$$

When this formula is applied to residuals, $\bar{Y} = 0$ and for lag 1, it reads

$$r_1 = \frac{\sum Y_i Y_{i+1}}{\sum Y_i^2} \tag{3.11}$$

Applying this to our residuals we obtain

$$r_1 = \frac{(-0.26)(0.19) + (0.19)(0.01) + \cdots}{(-0.26)^2 + (0.19)^2 + \cdots}$$

$$r_1 = \frac{-0.0755}{+0.2439} = -0.31$$

which is not significantly different from zero with a sample of 16. (See [23], Table 19.)

Another statistic used to check independence is the Durbin–Watson statistic. If applied to residuals as in Equation (3.11), it can be written as

$$\text{D–W} = \frac{\sum (Y_i - Y_{i+1})^2}{\sum Y_i^2} \tag{3.12}$$

In our example,

$$\text{D–W} = \frac{0.5663}{0.2438} = 2.32$$

If this statistic is 2 or more, there is a strong indication of independence of these residuals. Should it be numerically less than 1, one should investigate whether the data were indeed randomly collected. Values between 1 and 2 lead to no conclusion.

3.9 SAS Programs for ANOVA and Tests after ANOVA

PROC ANOVA can be used to test $H_0: \mu_1 = \mu_2 \cdots = \mu_K$ using the analysis-of-variance procedure for a balanced case. Our example will be very much like previous ones.

ANOVA and SNK Test

■ **EXAMPLE 3.3** Recall the data for Example 3.1 in Table 3.1.

Table 3.13 Fabric Wear Resistance Data

	Fabric		
A	*B*	*C*	*D*
1.93	2.55	2.40	2.33
2.38	2.72	2.68	2.40
2.20	2.75	2.31	2.28
2.25	2.70	2.28	2.25

Should our test indicate that a difference exists, we will want to run another test called the Student–Newman–Keuls (SNK) to indicate those means that may be different. ■

PROC ANOVA testing of multiple means requires a **CLASS** statement to identify the classification variables and a **MODEL** statement to identify the dependent and independent variables.

Our program is as follows:

```
DATA RAW;
```
We can name our data set anything we want at this point.

```
INPUT FABRIC $ WEARLOSS @@;
CARDS;
 A 1.93 A 2.38 A 2.20 A 2.25
 B 2.55 B 2.72 B 2.75 B 2.70
 C 2.40 C 2.68 C 2.31 C 2.28
 D 2.33 D 2.40 D 2.28 D 2.25
 ;
```
See Chapter 2 for these symbols.

If we want to see what the data set looks like or to audit it for errors, we could have it printed by using the procedure **PROC PRINT.**

```
PROC PRINT;
```
This causes the printing of the most recently created data set.

```
PROC ANOVA;
CLASS FABRIC;
MODEL WEARLOSS=FABRIC;
```
The model statement identifies the dependent variable (WEARLOSS) and the independent variable (FABRIC) separated by a required = sign. Also note the similarity of the

model statement to the mathematical model of the problem. This will become more apparent as our problems become more complex. The Y_{ij} is the WEARLOSS, τ_j is the FABRIC, and μ and ε_{ij} are omitted.

Since we want an SNK analysis performed we must use an option statement as follows:

MEANS FABRIC/SNK; This statement will trigger the SNK or many other optional analyses to be performed on the independent variable stated (FABRIC).

Our final program looks like

```
DATA RAW;
INPUT FABRIC $ WEARLOSS @@;
CARDS;
 A 1.93 A 2.38 A 2.20 A 2.25
 B 2.55 B 2.72 B 2.75 B 2.70
 C 2.40 C 2.68 C 2.31 C 2.28
 D 2.33 D 2.40 D 2.28 D 2.25
;
PROC PRINT;
PROC ANOVA;
CLASS FABRIC;
MODEL WEARLOSS=FABRIC;
MEANS FABRIC/SNK;
TITLE'ONE WAY ANOVA';
RUN;
```

Our output is as follows:

SAS

OBS	FABRIC	WEARLOSS
1	A	1.93
2	A	2.38
3	A	2.20
4	A	2.25
5	B	2.55
6	B	2.72
7	B	2.75
8	B	2.70
9	C	2.40
10	C	2.68

```
11        C        2.31
12        C        2.28
13        D        2.33
14        D        2.40
15        D        2.28
16        D        2.25
              ONEWAY   ANOVA
    ANALYSIS  OF VARIANCE   PROCEDURE
       CLASS  LEVEL INFORMATION
    CLASS        LEVELS     VALUES
    FABRIC          4       A B C D
NUMBER OF OBSERVATIONS IN DATA SET = 16
```

The output continues on page 78.

PROC ANOVA PRINTED OUTPUT (copied from SAS)

ANOVA first prints a table that includes

1. The name of each variable in the **CLASS** statement
2. The number of different values or LEVELS of the CLASS variables
3. The VALUES of the CLASS variables
4. The number of observations in the data set and the number of observations excluded from the analysis because of missing values, if any

ANOVA then prints an analysis-of-variance table for each dependent variable in the **MODEL** statement. This table breaks down

5. The CORRECTED TOTAL sum of squares for the dependent variable
6. Into the portion attributed to the MODEL
7. And the portion attributed to ERROR.
8. The MEAN SQUARE term is the
9. SUM OF SQUARES divided by the
10. DEGREES OF FREEDOM (DF).
11. The MEAN SQUARE for ERROR is an estimate of σ^2, the variance of the true errors.
12. The F VALUE is the ratio produced by dividing MS(MODEL), the mean square for the model, by MS(ERROR), the mean square for error. It tests how well the model as a whole (adjusted for the mean) accounts for the dependent variable's behavior. This F test is a test that all parameters except the intercept are zero.
13. The significance probability associated with the F statistic, labeled PR > F.

ONEWAY ANOVA

ANALYSIS OF VARIANCE PROCEDURE

DEPENDENT VARIABLE: WEARLOSS

SOURCE	DF	SUM OF SQUARES	MEAN SQUARE	F VALUE	PR > F	R-SQUARE	C.V.
MODEL	3	0.52011875	0.17337292	8.53	0.0026	0.680878	5.9372
ERROR	12	0.24377500	0.02031458			ROOT MSE	WEARLOSS MEAN
CORRECTED TOTAL	15	0.76389375				0.14252924	2.40062500

SOURCE	DF	ANOVA SS	F VALUE	PR > F
FABRIC	3	0.52011875	8.53	0.0026

ONEWAY ANOVA

ANALYSIS OF VARIANCE PROCEDURE

STUDENT-NEWMAN-KEULS TEST FOR VARIABLE: WEARLOSS
NOTE: THIS TEST CONTROLS THE TYPE I EXPERIMENTWISE ERROR RATE UNDER THE COMPLETE NULL HYPOTHESIS BUT NOT UNDER PARTIAL NULL HYPOTHESES

ALPHA=0.05 DF=12 MSE=.0203146

NUMBER OF MEANS	2	3	4
CRITICAL RANGE	0.219578	0.268881	0.299213

MEANS WITH THE SAME LETTER ARE NOT SIGNIFICANTLY DIFFERENT.

SNK	GROUPING	MEAN	N	FABRIC
	A	2.6800	4	B
	B	2.4175	4	C
	B	2.3150	4	D
	B	2.1900	4	A

14. R-SQUARE, R^2, measures how much variation in the dependent variable can be accounted for by the model. R^2, which can range from 0 to 1, is the ratio of the sum of squares for the model divided by the sum of squares for the corrected total. In general, the larger the R^2 value, the better the model fits the data.

15. C.V., the coefficient of variation, is often used to describe the amount of variation in the population. The C.V. is 100 times the standard deviation of the dependent variable, STD DEV, divided by the MEAN. The coefficient of variation is often a preferred measure because it is unitless.

16. ROOT MSE estimates the standard deviation of the dependent variable and is computed as the square root of MS(ERROR), the mean square of the error term.

17. The MEAN of the dependent variable.

For each effect (or source of variation) in the model, ANOVA then prints

18. DF, degrees of freedom

19. ANOVA SS, the sum of squares

20. The F VALUE for testing the hypothesis that the group means for that effect are equal

21. PR > F, the significance probability value associated with the F VALUE

GLM (General Linear Model)

PROC GLM can be used for many different analyses, including simple and multiple regression; analyses of variance, especially for unbalanced data; analysis of covariance; response-surface models; weighted regression; polynomial regression; partial correlation; and contrasts, a customized hypotheses test. We will introduce only a few of the many options and statements available as we need them.

As with **PROC ANOVA**, **CLASS** and **MODEL** statements are required, with many options available with each. An option available with **PROC GLM** and not with **PROC ANOVA** is the generation of contrasts as described below. Also, predicted values and resulting residuals of our model can be calculated as displayed in a second example by adding P as an option in the **MODEL** statement.

Orthogonal Contrasts

Contrasts cannot be obtained with **PROC ANOVA,** but they can be with **PROC GLM.** This procedure uses the same format as ANOVA, as shown in the following program with the addition of **CONTRAST** statements that must come after the **MODEL** statement.

There is no limit to the number of contrasts requested, but it will be the analyst's responsibility to determine the orthogonality and the proper coefficients to obtain the desired results.

A label of 20 characters or less may be used to identify the contrast if placed within single quotes after **CONTRAST.** Following the label, the effect that appears in the **MODEL** statement to be contrasted follows. Allowing a space after the effect, the coefficients are entered with spaces between each one and ending with a semicolon (;).

```
DATA A;
INPUT FABRIC WEARLOSS @@;
CARDS;
 1 1.93 1 2.38 1 2.20 1 2.25
 2 2.55 2 2.92 2 2.75 2 2.70
 3 2.40 3 2.68 3 2.31 3 2.28
 4 2.33 4 2.40 4 2.28 4 2.25
 ;
PROC GLM;
CLASS FABRIC;
MODEL WEARLOSS = FABRIC;
CONTRAST 'ONE VS FOUR' FABRIC 1 0 0 -1;
CONTRAST 'TWO VS THREE' FABRIC 0 1 -1 0;
CONTRAST 'ONE AND FOUR VS TWO AND THREE' FABRIC 1 -1 -1 1;
RUN;
```

The output continues on page 81.

The analysis that follows the program gives the same F tests on the orthogonal contrasts as shown in Section 3.4.

A Second Example

This example illustrates the many uses of **PROC GLM.** The effect of three different tube types on cathode warm-up time in seconds was to be studied. The three tube types are A, B, and C, and eight observations are made on each type of tube. The data are reported in the following program:

```
DATA A;
INPUT TUBE $ WARMTIME @@;
CARDS;
 A 19 A 23 A 26 A 18 A 20 A 20 A 18 A 35
 B 20 B 20 B 32 B 27 B 40 B 24 B 22 B 18
 C 16 C 15 C 18 C 26 C 19 C 17 C 19 C 18
 ;
```

GENERAL LINEAR MODELS PROCEDURE

DEPENDENT VARIABLE: WEARLOSS

SOURCE	DF	SUM OF SQUARES	MEAN SQUARE	F VALUE	PR > F	R-SQUARE	C.V.
MODEL	3	0.52011875	0.17337292	8.53	0.0026	0.680878	5.9372
ERROR	12	0.24377500	0.02031458		ROOT MSE		WEARLOSS MEAN
CORRECTED TOTAL	15	0.76389375			0.14252924		2.40062500

SOURCE	DF	TYPE I SS	F VALUE	PR > F
FABRIC	3	0.52011875	8.53	0.0026

CONTRAST	DF	SS	F VALUE	PR > F
ONE VS FOUR	1	0.03125000	1.54	0.2386
TWO VS THREE	1	0.13781250	6.78	0.0230
ONE AND FOUR VS TWO	1	0.35105625	17.28	0.0013

```
TITLE 'COMPARISON OF THE WARM-UP TIMES OF
      3 TYPES OF TUBES';
PROC PRINT;
PROC GLM;
CLASS TUBE;
MODEL WARMTIME = TUBE/P;
RUN;
```

The printouts follow:

```
COMPARISON OF THE WARM-UP TIMES OF 3 TYPES OF TUBES
         OBS      TUBE      WARMTIME
          1        A          19
          2        A          23
          3        A          26
          4        A          18
          5        A          20
          6        A          20
          7        A          18
          8        A          35
          9        B          20
         10        B          20
         11        B          32
         12        B          27
         13        B          40
         14        B          24
         15        B          22
         16        B          18
         17        C          16
         18        C          15
         19        C          18
         20        C          26
         21        C          19
         22        C          17
         23        C          19
         24        C          18
COMPARISON OF THE WARM-UP TIMES OF 3 TYPES OF TUBES
         GENERAL LINEAR MODELS PROCEDURE

               CLASS LEVEL INFORMATION

           CLASS     LEVELS      VALUES

           TUBE        3          A B C
        NUMBER OF OBSERVATIONS IN DATA SET = 24
```

COMPARISON OF THE WARMUP TIMES OF 3 TYPES OF TUBES
GENERAL LINEAR MODELS PROCEDURE

DEPENDENT VARIABLE: WARMTIME

SOURCE	DF	SUM OF SQUARES	MEAN SQUARE	F VALUE	PR > F	R-SQUARE	C.V.
MODEL	2	190.08333333	95.0416667	2.86	0.0797	0.214098	26.1021
ERROR	21	697.75000000	33.2619048			ROOT MSE	WARMTIME MEAN
CORRECTED TOTAL	23	887.83333333				5.76421638	22.08333333

SOURCE	DF	TYPE I SS	F VALUE	PR > F	TYPE III SS	F VALUE	PR > F
TUBE	2	190.08333333	2.86	0.0797	190.08333333	2.86	0.0797

OBSERVATION	OBSERVED VALUE	PREDICTED VALUE	RESIDUAL
1	19.00000000	22.37500000	-3.37500000
2	23.00000000	22.37500000	0.62500000
3	26.00000000	22.37500000	3.62500000
4	18.00000000	22.37500000	-4.37500000
5	20.00000000	22.37500000	-2.37500000
6	20.00000000	22.37500000	-2.37500000
7	18.00000000	22.37500000	-4.37500000
8	35.00000000	22.37500000	12.62500000
9	20.00000000	25.37500000	-5.37500000
10	20.00000000	25.37500000	-5.37500000
11	32.00000000	25.37500000	6.62500000
12	27.00000000	25.37500000	1.62500000
13	40.00000000	25.37500000	14.62500000
14	24.00000000	25.37500000	-1.37500000
15	22.00000000	25.37500000	-3.37500000
16	18.00000000	25.37500000	-7.37500000
17	16.00000000	18.50000000	-2.50000000
18	15.00000000	18.50000000	-3.50000000
19	18.00000000	18.50000000	-0.50000000
20	26.00000000	18.50000000	7.50000000
21	19.00000000	18.50000000	0.50000000
22	17.00000000	18.50000000	-1.50000000
23	19.00000000	18.50000000	0.50000000
24	18.00000000	18.50000000	-0.50000000

SUM OF RESIDUALS	0.00000000
SUM OF SQUARED RESIDUALS	697.75000000
SUM OF SQUARED RESIDUALS - ERROR SS	-0.00000000
FIRST ORDER AUTOCORRELATION	-0.06872537
DURBIN-WATSON D	2.12076765

Using GLM on these data with the '/P' designation (P for predicted values) on page 82 causes the residuals to be printed with each observation as shown on page 83. We also note the two statistics discussed in Section 3.8 on residuals. Here $r_1 = -0.0687$, a very low correlation, and D–W = 2.12, a high value. These results are a pretty good indication of the independence of the residuals.

Printed Output

The GLM procedure produces the following printed output by default:

1. The overall analysis-of-variance table breaks down the CORRECTED TO-TAL sum of squares for the dependent variable

2. into the portion attributed to the MODEL

3. and the portion attributed to ERROR

4. The MEAN SQUARE term is the

5. SUM OF SQUARES divided by the

6. DEGREES OF FREEDOM (D.F.)

7. The MEAN SQUARE for ERROR, MS(ERROR), is an estimate of σ^2, the variance of the true errors.

8. The F VALUE is the ratio produced by dividing MS(MODEL) by MS(ERROR). It tests how well the model as a whole (adjusted for the mean) accounts for the dependent variable's behavior. An F test is a joint test that all parameters except the intercept are zero.

9. A small significance probability, PR > F, indicates that some linear function of the parameters is significantly different from zero.

10. R-SQUARE, R^2, measures how much variation in the dependent variable can be accounted for by the model. R^2, which can range from 0 to 1, is the ratio of the sum of squares for the model divided by the sum of squares for the corrected total. In general, the larger the value of R^2, the better the model's fit.

11. C.V., the coefficient of variation, which describes the amount of variation in the population, is 100 times the standard deviation estimate of the dependent variable, ROOT MSE, divided by the MEAN. The coefficient of variation is often a preferred measure because it is unitless.

12. ROOT MSE estimates the standard deviation of the dependent variable (or, equivalently, the error term) and equals the square root of MS(ERROR).

13. MEAN is the sample mean of the dependent variable.

These tests are used primarily in analysis-of-variance applications:

14. The TYPE I SS measures incremental sums of squares for the model as each variable is added.

15. The TYPE III SS is the sum of squares that results when that variable is added last to the model.

16. The F VALUE and PR $>$ F values for TYPE I and TYPE III tests in this section are based on the corresponding SS's.

17. This section of the output gives the ESTIMATES for the model PARAME-TERs—the intercept and the coefficients.

18. T FOR H_0: PARAMETER $= 0$ is Student's t value for testing the null hypothesis that the parameter (if it is estimable) equals zero.

19. The significance level, PR $> |T|$, is the probability of getting a larger value of t if the parameter is truly equal to zero. A very small value for this probability leads to the conclusion that the independent variable contributes significantly to the model.

20. The STD ERROR OF ESTIMATE is the standard error of the value of the parameter.

Components of Variance

PROC NESTED can be used to evaluate the components of variance of a nested or hierarchical random model. It requires a **CLASS** statement and will perform an analysis of variance for all numeric values included in the **CLASS** statement unless a **VAR** statement is used to identify the variables to be analyzed.

The data must be entered or previously sorted by the order in which they are given in the **CLASS** statement. The second effect is assumed to be nested within the first effect, the third effect is assumed to be nested within the second effect, and so on.

An example of the use of this procedure follows, using the data from Table 3.8. A description of the printed output is also included.

```
DATA A;
INPUT BATCH YIELD @@;
CARDS;
 1 74 1 76 1 75 2 68 2 71 2 72 3 73 3 77 3 77
 4 72 4 74 4 73 5 79 5 81 5 79
 ;
TITLE 'COMPONENTS OF VARIANCE';
TITLE2 'CHEMICAL YIELD BY BATCH DATA';
TITLE3 'TABLE 3.8';
PROC NESTED;
CLASSES BATCH;
VAR YIELD;
RUN;
```

Printed Output (copied from SAS)

For each effect in the model, NESTED prints

1. COEFFICIENTS OF EXPECTED MEAN SQUARES, the coefficients of the variance components making up the expected mean square.

For every dependent variable, NESTED prints an analysis-of-variance table containing the following items:

2. VARIANCE SOURCE, sources of variation
3. D.F., degrees of freedom
4. SUM OF SQUARES
5. MEAN SQUARES
6. VARIANCE COMPONENT, estimates of variance components
7. PERCENT, the percentage associated with a source of variance. The value is 100 times the ratio of that source's estimated variance component to the total variance component.

The following statistics are printed below each analysis-of-variance table:

8. MEAN, the overall mean
9. STANDARD DEVIATION
10. COEFFICIENT OF VARIATION of the response variable, based on the error mean square.

For each pair of dependent variables, NESTED prints an analysis-of-covariance table (unless ANOVA is specified). For each source of variation, this table includes

11. D.F., the degrees of freedom
12. SUM OF PRODUCTS
13. MEAN PRODUCTS
14. COVARIANCE COMPONENT, the estimate of the covariance component
15. VARIANCE COMPONENT CORRELATION, the covariance component correlation
16. MEAN SQUARE CORRELATION

```
            COMPONENTS OF VARIANCE
         CHEMICAL YIELD BY BATCH DATA
               FROM TABLE 3.8
COEFFICIENTS OF EXPECTED MEAN SQUARES
SOURCE              BATCH              ERROR
BATCH                 3                  1
ERROR                 0                  1
```

```
                 ANALYSIS OF VARIANCE YIELD
VARIANCE               SUM OF      MEAN       VARIANCE
SOURCE     D.F.       SQUARES     SQUARES     COMPONENT      PERCENT

TOTAL       14      165.73333   11.838095    13.511111        100

BATCH        4      147.73333   36.933333    11.711111      86.6776

ERROR       10           18        1.8           1.8       13.3224

MEAN                              74.8666667
STANDARD DEVIATION                 1.34164079
COEFFICIENT OF VARIATION           1.79204023
```

Problems

3.1 Assuming a completely randomized design, do a one-way analysis of variance on the following data in order to familiarize yourself with the technique.

	Factor A **Level**				
	1	2	3	4	5
Measurement	8	4	1	4	10
	6	−2	2	6	8
	7	0	0	5	7
	5	−2	−1	5	4
	8	3	−3	4	9

3.2 The cathode warm-up time in seconds was determined for three different tube types using eight observations on each type of tube. The order of experimentation was completely randomized. The results were

	Tube Type					
	A		B		C	
Warm-up time	19	20	20	40	16	19
(seconds)	23	20	20	24	15	17
	26	18	32	22	18	19
	18	35	27	18	26	18

Do an analysis of variance on these data and test the hypothesis that the three tube types require the same average warm-up time.

3.3 For Problem 3.2 set up orthogonal contrasts between the tube types and test your contrasts for significance.

3.4 Set up 95 percent confidence limits for the average warm-up time for tube type C in Problem 3.2.

3.5 Use the Newman–Keuls range method to test for differences between tube types.

3.6 The following data are on the pressure in a torsion spring for several settings of the angle between the legs of the spring in a free position.

	Angle of Legs of Spring (degrees)				
	67	71	75	79	83
Pressure (psi)	83	84	86	89	90
	85	85	87	90	92
		85	87	90	
		86	87	91	
		86	88		
		87	88		
			88		
			88		
			88		
			89		
			90		

Assuming a completely randomized design, complete a one-way analysis of variance for this experiment and state your conclusion concerning the effect of angle on the pressure in the spring.

3.7 Set up orthogonal contrasts for the angles in Problem 3.6.

3.8 Show that the expanded forms for the sums of squares in Table 3.5 are correct.

3.9 Assume that the levels of factor A in Problem 3.1 were chosen at random and determine the proportion of variance attributable to differences in level means and the proportion due to error.

3.10 Verify the results given in Table 3.9 based on the data of Table 3.8.

3.11 Set up two or more contrasts for the data of Problem 3.2 and test their significance by the Scheffé method.

3.12 Since the Scheffé method is not restricted to equal sample sizes, set up several contrasts for Problem 3.6 and test for significance by the Scheffé method.

3.13 It is suspected that the environmental temperature in which batteries are activated affects their activated life. Thirty homogeneous batteries were tested, six at each of five temperatures, and the data shown below were obtained. Analyze and interpret the data.

	Temperature (°C)				
	0	25	50	75	100
Activated life	55	60	70	72	65
(seconds)	55	61	72	72	66
	57	60	73	72	60
	54	60	68	70	64
	54	60	77	68	65
	56	60	77	69	65

3.14 A highway research engineer wishes to determine the effect of four types of sub-grade soil on the moisture content in the top soil. He takes five samples of each type of subgrade soil and the total sum of squares is computed as 280, whereas the sum of squares among the four types of subgrade soil is 120.

a. Set up an analysis-of-variance table for these results.

b. Set up a mathematical model to describe this problem, define each term in the model, and state the assumptions made on each term.

c. Set up a test of the hypothesis that the four types of subgrade soil have the same effect on moisture content in the top soil.

d. Set up a set of orthogonal contrasts for this problem.

e. Explain briefly how to set up a test on means after the analysis of variance for these data.

f. Set up an expression for 90 percent confidence limits on the mean of type 2 subgrade soil. Insert all numerical values that are known.

3.15 Data collected on the effect of four fixed types of television tube coating on the conductivity of the tubes are given as

	Coating		
I	II	III	IV
56	64	45	42
55	61	46	39
62	50	48	45
59	55	39	43
60	56	43	41

Do an analysis of variance on these data and test the hypothesis that the four coatings yield the same average conductivity.

3.16 For Problem 3.15 set up orthogonal contrasts between tube coatings and test them for significance.

3.17 Use a Newman–Keuls range test on the data of Problem 3.15 and discuss any significant results that you find.

3.18 Set up contrasts and test using Scheffé's method to answer the following questions: Do the means of coating I and II differ? the means of II and IV? the mean of I and II versus the mean of III and IV?   *SAs won't do*

3.19 Four bonding machines (*A*, *B*, *C*, *D*) are used in a certain plant to bond circuit wires onto a board. In order to determine which bonder does the best job, samples of 10 are taken at random from each machine, and the strength needed to break the bond (in grams) is determined. The data are

	Bonder		
A	*B*	*C*	*D*
204	197	264	248
181	223	226	138
201	206	228	273
203	232	249	220
214	213	246	186
262	207	255	304
246	259	186	330
230	223	237	268
256	195	236	295
288	197	240	276

Analyze these data and make your recommendations to management.

3.20 To determine the effect of several teaching methods on student achievement, 30 students were assigned to five treatment groups with six students per group. The treatments given for the semester are as follows:

Treatment	Description
1	Current textbook
2	Textbook *A* with teacher
3	Textbook *A* with machine
4	Textbook *B* with teacher
5	Textbook *B* with machine

At the end of a semester of instruction, achievement scores were recorded and some of the statistics were found to be

	Treatment				
	1	2	3	4	5
Totals	120	600	720	240	420
Source		df	SS	MS	F	$F_{0.95}$
Between treatments			340		
Error					
Total			465		

a. Complete the ANOVA table.
b. Write the mathematical model assumed here and state the hypothesis to be tested.
c. Test the hypothesis and state your conclusion.

3.21 Set up one set of orthogonal contrasts that might seem reasonable from the treatment descriptions in Problem 3.20.

3.22 Test whether or not the mean achievement under treatment 1 differs significantly from the mean achievement under treatment 5 in a Newman–Keuls sense for Problem 3.20.

3.23 Determine the standard error of the contrast $4T_{.1} - T_{.2} - T_{.3} - T_{.4} - T_{.5}$ based on Problem 3.20 data.

3.24 Three fertilizers are tried on 27 plots of land in a random fashion such that each fertilizer is applied to nine plots. The total yield for each fertilizer type is given by

	Type		
	1	2	3
$T_{.j}$	240	320	180

and the ANOVA table is

Source	df	SS	MS
Between fertilizers	2	1096	548
Error	24	1440	60
Total	26	2536	

a. Set up one set of orthogonal contrasts that might be used on these data.
b. For your first contrast, determine the sum of squares due to this contrast.
c. For your second contrast, find its standard error.
d. If one wished to compare types 1 and 2 and also the average of 1 and 3 versus 2, which method would you recommend and why?

3.25 Birth weights in pounds of five Poland China pigs were recorded from each of six randomly chosen litters. The ANOVA printout showed:

Source	df	SS	MS
Between litters	5	6.0	1.20
Within litters	24	9.6	0.40

Determine what percentage of the total variance in this experiment can be attributed to litter-to-litter differences.

3.26 To determine whether there is a difference in leakage between three vendors' (A, B, C) capacitors, six samples were randomly drawn from each vendor and the following leakage readings (in milliamperes) were found:

A	B	C
7.3	10.7	10.5
8.0	10.2	10.1
8.1	10.2	10.8
8.5	10.7	11.6
8.4	9.9	11.4
7.5	11.0	10.8

Analyze these data and make recommendations.

3.27 If vendor A in Problem 3.26 is the present supplier and vendors B and C are competing for the job, set up two reasonable orthogonal contrasts, test them, and discuss the results.

Single-Factor Experiments—   **4**
Randomized Block and
Latin Square Designs

4.1 Introduction

Consider the problem of determining whether or not different brands of tires exhibit different amounts of tread loss after 20,000 miles of driving. A fleet manager wishes to consider four brands that are available and make some decision about which brand might show the least amount of tread wear after 20,000 miles. The brands to be considered are A, B, C, and D, and although driving conditions might be simulated in the laboratory, he wants to try these four brands under actual driving conditions. The variable to be measured is the difference in maximum tread thickness on a tire between the time it is mounted on the wheel of a car and after it has completed 20,000 miles on this car. The measured variable Y_{ij} is this difference in thickness in mils (0.001 in.), and the only factor of interest is brands, say, τ_j where $j = 1$, 2, 3, and 4.

Since the tires must be tried on cars and since some measure of error is necessary, more than one tire of each brand must be used and a set of four of each brand would seem quite practical. This means 16 tires, 4 each of 4 different brands, and a reasonable experiment would involve at least 4 cars. Designating the cars as I, II, III, and IV, one might put brand A's four tires on car I, brand B's on car II, and so on, with a design as shown in Table 4.1.

Table 4.1 Design 1 for Tire Brand Test

	Car			
	I	II	III	IV
Brand	A	B	C	D
distribution	A	B	C	D
	A	B	C	D
	A	B	C	D

One look at this design shows its fallibility, since averages for brands are also averages for cars. If the cars travel over different terrains, using different drivers, any apparent brand differences are also car differences. This design is called completely confounded, since we cannot distinguish between brands and cars in the analysis.

A second attempt at design might be to try a completely randomized design, as given in Chapter 3. Assigning the 16 tires to the four cars in a completely random manner might give results as in Table 4.2. In Table 4.2 the loss in thickness is given for each of the 16 tires. The purpose of complete randomization here is to average out any car differences that might affect the results. The model would be

$$Y_{ij} = \mu + \tau_j + \varepsilon_{ij}$$

with

$$j = 1, \ 2, \ 3, \ 4 \qquad i = 1, \ 2, \ 3, \ 4$$

An ANOVA on these data gives the results in Table 4.3.

Table 4.2 Design 2 for Tire Brand Test

	Car			
	I	II	III	IV
Brand distribution	$C(12)$	$A(14)$	$C(10)$	$A(13)$
and loss in thickness	$A(17)$	$A(13)$	$D(11)$	$D(9)$
	$D(13)$	$B(14)$	$B(14)$	$B(8)$
	$D(11)$	$C(12)$	$B(13)$	$C(9)$

Table 4.3 ANOVA for Design 2

Source	df	SS	MS
Brands	3	30.69	10.2
Error	12	50.25	4.2
Totals	15	80.94	

The F test shows $F_{3, 12} = 2.43$, and the 5 percent critical region for F is $F_{3, 12} \geq 3.49$ (Statistical Table D), so there is no reason to reject the hypothesis of equal average tread loss among the four brands.

4.2 Randomized Complete Block Design

A more careful examination of design 2 in Table 4.2 will reveal some glaring disadvantages of the completely randomized design in this problem. One thing to be noted is that brand A is never used on car III or brand B on car I. Also, any variation within brand A may reflect variation between cars I, II, and IV. Thus the random error may not be merely an experimental error but may include variation between cars. Since the chief objective of experimental design is to reduce the experimental error, a better design might be one in which car variation is removed from error variation. Although the completely randomized design averaged out the car effects, it did not eliminate the variance among cars. A design that requires that each brand be used once on each car is a *randomized complete block design,* given in Table 4.4.

Table 4.4 Design 3: Randomized Block Design for Tire Brand Test

	Car			
	I	II	III	IV
Brand distribution	$B(14)$	$D(11)$	$A(13)$	$C(9)$
and loss in thickness	$C(12)$	$C(12)$	$B(13)$	$D(9)$
	$A(17)$	$B(14)$	$D(11)$	$B(8)$
	$D(13)$	$A(14)$	$C(10)$	$A(13)$

In this design the order in which the four brands are placed on a car is random and each car gets one tire of each brand. In this way better comparisons can be made between brands since they are all driven over approximately the same terrain, and so on. This provides a more homogeneous environment in which to test the four brands. In general, these groupings for homogeneity are called *blocks* and randomization is now restricted within blocks. This design also allows the car (block) variation to be independently assessed and removed from the error term. The model for this design is

$$Y_{ij} = \mu + \beta_i + \tau_j + \varepsilon_{ij} \qquad (4.1)$$

where β_i now represents the block effect (car effect) in the example above.

The analysis of this model is a two-way analysis of variance, since the block effect may now also be isolated. A slight rearrangement of the data in Table 4.4 gives Table 4.5.

The total sum of squares is computed as in Chapter 3

$$SS_{total} = \sum_{j}^{4} \sum_{i}^{4} Y_{ij}^2 - \frac{T_{..}^2}{N}$$

$$= 2409 - \frac{(193)^2}{16} = 80.94$$

Table 4.5 Randomized Block Design Data for Tire Brand Test

Car	Brand				$T_{i\cdot}$
	A	B	C	D	
I	17	14	12	13	56
II	14	14	12	11	51
III	13	13	10	11	47
IV	13	8	9	9	39
$T_{\cdot j}$	57	49	43	44	$T_{\cdot\cdot} = 193$
$\sum\limits_{i} Y_{ij}^2$	823	625	469	492	$\sum\limits_{j}\sum\limits_{i} Y_{ij}^2 = 2409$

The brand (treatment) sum of squares is computed as usual:

$$SS_{brands} = \sum_{j} \frac{T_{\cdot j}^2}{n} - \frac{T_{\cdot\cdot}^2}{N}$$

$$= \frac{(57)^2 + (49)^2 + (43)^2 + (44)^2}{4} - \frac{(193)^2}{16} = 30.69$$

Since the car (block) effect is similar to the brand effect but totaled across the rows of Table 4.5, the car sum of squares is computed exactly like the brand sum of squares, using row totals $T_{i\cdot}$ instead of column totals. Calling the number of treatments (brands) in general k, then

$$SS_{car} = \sum_{i=1}^{n} \frac{T_{i\cdot}^2}{k} - \frac{T_{\cdot\cdot}^2}{N}$$

$$= \frac{(56)^2 + (51)^2 + (47)^2) + (39)^2}{4} - \frac{(193)^2}{16} = 38.69$$

The error sum of squares is now the remainder after subtracting both brand and car sum of squares from the total sum of squares

$$SS_{error} = SS_{total} - SS_{brand} - SS_{car}$$
$$= 80.94 - 30.69 - 38.69 = 11.56$$

Table 4.6 is an ANOVA table for these data.

Table 4.6 ANOVA for Randomized Block
Design of Tire Brand Test

Source	df	SS	MS	EMS
Brands	3	30.69	10.2	$\sigma_\varepsilon^2 + 4\phi_\tau$
Cars	3	38.69	12.9	$\sigma_\varepsilon^2 + 4\sigma_\beta^2$
Error	9	11.56	1.3	σ_ε^2
Totals	15	80.94		

To test the hypothesis, H_0: $\mu_{.1} = \mu_{.2} = \mu_{.3} = \mu_{.4}$, the ratio is

$$F_{3,9} = \frac{10.2}{1.3} = 7.8$$

which is significantly larger than the corresponding critical F even at the 1 percent level (Statistical Table D). The hypothesis of equal brand means is thus rejected. It is to be noted that this hypothesis could not be rejected using a completely randomized design. The randomized block design that allows for removal of the block (car) effect definitely reduced the error variance estimate from 4.2 to 1.3.

It is also possible, if desired, to test the hypothesis that the average tread loss of all four cars is the same. H_1: $\mu_1. = \mu_2. = \mu_3. = \mu_4.$, and $F_{3,9} = 12.9/1.3 = 9.9$, which is also significant at the 1 percent level (Statistical Table D). Here this hypothesis is rejected and a car-to-car variation is detected.

Even though an effect due to cars (blocks) has been isolated, the main objective is still to test brand differences. Thus it is still a single-factor experiment, the blocks representing only a restriction on complete randomization due to the environment in which the experiment was conducted. Other examples include testing differences in materials that are fed into several different machines, testing differences in fertilizers that must be spread on several different plots of ground, and testing the effect of different teaching methods on several pupils. In these examples the blocks are machines, plots, and pupils, respectively, and the levels of the factors of interest can be randomized within each block.

When data are presented in tabular form there is usually no way to determine how the data were collected. Was the randomization complete over all N observations or was the experiment run in blocks with randomization restricted to within the blocks? To help in signifying the design of the experiment, it is suggested that in the case of a completely randomized design, no horizontal or vertical lines be drawn on the data, as in Table 4.2. When randomization has been restricted, either vertical or horizontal lines as shown in Tables 4.4 and

4.5, respectively, can be used to indicate this restriction on the randomization. As more complex designs are presented, this scheme may require some double lines for one restriction and single lines for a second restriction, and so on. It is hoped, however, that such a scheme will help the observer in noting just how the randomization has been restricted.

Since blocking an experiment is a very useful procedure to reduce the experimental error, a second example may be helpful as it illustrates the situation in which the same experimental units (animals, people, parts, etc.) are measured before the experiment and then again at the conclusion of the experimental treatment. These before–after, pretest–posttest designs can be treated very well as randomized block designs where the experimental units are the blocks. Interest is primarily in the effect of the treatment, and the block effects can be removed to reduce the experimental error. Some refer to these designs as repeated-measures designs because data are repeated on the same units a second time. The following example will illustrate the procedure.

■ **EXAMPLE 4.1** A study on a physical strength measurement in pounds on seven subjects before and after a specified training period gave the results shown in Table 4.7.

Table 4.7 Pretest and Posttest Strength Measures

Subjects	Pretest	Posttest
1	100	115
2	110	125
3	90	105
4	110	130
5	125	140
6	130	140
7	105	125

This problem could be handled by the methods of Chapter 2 using a t test on differences where d = posttest measure − pretest measure. Here $\bar{d} = 15.71$, $s_d = 3.45$, and $t = 12.05$, which is highly significant ($p = 0.00002$) with 6 df.

If it is considered as a single-factor experiment at two levels (pre- and post-) with seven blocks making up a randomized block design, the data and ANOVA appear as shown in Table 4.8.

Table 4.8 Strength Measure Data and Analysis

Subjects	Pretest	Posttest	$T_{i.}$
1	100	115	215
2	110	125	235
3	90	105	195
4	110	130	240
5	125	140	265
6	130	140	270
7	105	125	230
Totals $T_{.j}$	770	880	$T_{..} = 1650$
$\sum_i Y_{ij}^2:$	85,850	111,600	$\sum_j \sum_i Y_{ij}^2 = 197,450$

Source	df	SS	MS	F	Probability
Tests (treatments)	1	864.29	864.29	145	0.00002
Subjects (blocks)	6	2085.71	347.62		
Error	6	35.71	5.95		
Totals	13	2985.71			

Here the large F indicates a strong treatment effect. In fact, the F reported is equal to t^2 given by the method of differences as $t = 12.05$ when squared $(12.05)^2 = 145.2$. ■

4.3 ANOVA Rationale

For this randomized complete block design, the model is

$$Y_{ij} = \mu + \beta_i + \tau_j + \varepsilon_{ij} \tag{4.2}$$

or

$$Y_{ij} = \mu + (\mu_i - \mu) + (\mu_j - \mu) + (Y_{ij} - \mu_i - \mu_j + \mu) \tag{4.3}$$

where μ_i represents the true mean of block i. The last term can be obtained by subtracting the treatment and block deviations from the overall deviation as follows:

$$(Y_{ij} - \mu) - (\mu_i - \mu) - (\mu_j - \mu) \equiv Y_{ij} - \mu_i - \mu_j + \mu$$

Best estimates of the parameters in Equation (4.3) give the sample model (after moving $\bar{Y}_{..}$ to the left of the equation):

$$Y_{ij} - \bar{Y}_{..} = (\bar{Y}_{i.} - \bar{Y}_{..}) + (\bar{Y}_{.j} - \bar{Y}_{..}) + (\bar{Y}_{ij} - \bar{Y}_{i.} - \bar{Y}_{.j} + \bar{Y}_{..})$$

Squaring both sides and summing over $i = 1, 2, \ldots, n$ and $j = 1, 2, \ldots, k$, we have

$$\sum_{i=1}^{n}\sum_{j=1}^{k}(Y_{ij} - \bar{Y}_{..})^2 = \sum_{i}^{n}\sum_{j}^{k}(\bar{Y}_{i.} - \bar{Y}_{..})^2$$

$$+ \sum_{i}^{n}\sum_{j}^{k}(\bar{Y}_{.j} - \bar{Y}_{..})^2 + \sum_{i}^{n}\sum_{j}^{k}(Y_{ij} - \bar{Y}_{i.} - \bar{Y}_{.j} + \bar{Y}_{..})^2$$

$$+ \text{3 cross products} \tag{4.4}$$

A little algebraic work on the sums of the cross products will show that they all reduce to zero and the remaining equation becomes the fundamental equation of a two-way ANOVA. The equation states that

$$SS_{total} = SS_{block} + SS_{treatment} + SS_{error}$$

Thus, there is one sum of squares for each variable term in the model [Equation (4.2)]. Each sum of squares has associated with it its degrees of freedom, and dividing any sum of squares by its degrees of freedom will yield an unbiased estimate of population variance σ_ε^2 if the hypotheses under test are true.

The breakdown of degrees of freedom here is

$$\begin{array}{cccc} \text{total} & \text{blocks} & \text{treatments} & \text{error} \\ (nk - 1) = (n - 1) & + & (k - 1) & + & (n - 1)(k - 1) \end{array}$$

The error degrees are derived from the remainder:

$$(nk - 1) - (n - 1) - (k - 1) = nk - n - k + 1 = (n - 1)(k - 1)$$

It can be shown that each sum of squares on the right of Equation (4.4) when divided by its degrees of freedom provides mean squares that are independently chi-square distributed, so that the ratio of any two of them is distributed as F.

The sums of squares formulas given in Equation (4.4) are usually expanded and rewritten to give formulas that are easier to apply. These are shown in Table 4.9. They are the formulas applied to the data of Table 4.5 where $nk = N$.

Table 4.9 ANOVA for Randomized Block Design

Source	df	SS	MS
Between blocks β_i	$n-1$	$\sum_i^n \dfrac{T_{i.}^2}{k} - \dfrac{T_{..}^2}{nk}$	$SS_{block}/(n-1)$
Between treatments τ_j	$k-1$	$\sum_j^k \dfrac{T_{.j}^2}{n} - \dfrac{T_{..}^2}{nk}$	$SS_{treatment}/(k-1)$
Error ε_{ij}	$(n-1)(k-1)$	$\sum_i^n \sum_j^k Y_{ij}^2 - \sum_{i=1}^n \dfrac{T_{i.}^2}{k}$ $- \sum_{j=1}^k \dfrac{T_{.j}^2}{n} + \dfrac{T_{..}^2}{nk}$	$SS_{error}/(n-1)(k-1)$
Totals	$nk-1$	$\sum_i^n \sum_j^k Y_{ij}^2 - \dfrac{T_{..}^2}{nk}$	

4.4 Missing Values

Occasionally in a randomized block design an observation is lost. A vial may break, an animal may die, or a tire may disintegrate, so that there occurs one or more missing observations in the data. For a single-factor completely randomized design this presents no problem, since the analysis of variance can be run with unequal n_j's. But for a two-way analysis this means a loss of orthogonality, since for some blocks the $\sum_j \tau_j$ no longer equals zero, and for some treatment the $\sum_i \beta_i$ no longer equals zero. When blocks and treatments are orthogonal, the block totals are added over all treatments, and vice versa. If one or more observations are missing, the usual procedure is to replace the value with one that makes the sum of the squares of the errors a minimum.

In the tire brand test example suppose that the brand C tire on car III blew out and was ruined before completing the 20,000 miles. The resulting data after coding by subtracting 13 mils appear in Table 4.10, where y is inserted in place of this missing value.

Now,

$$SS_{error} = SS_{total} - SS_{treatment} - SS_{block}$$

$$= \sum_i \sum_j Y_{ij}^2 - \sum_j \dfrac{T_{.j}^2}{n} - \sum_i \dfrac{T_{i.}^2}{k} + \dfrac{T_{..}^2}{nk}$$

Table 4.10 Missing Value Example

Car	A	B	C	D	$T_{i.}$
			Brand		
I	4	1	-1	0	4
II	1	1	-1	-2	-1
III	0	0	y	-2	$y-2$
IV	0	-5	-4	-4	-13
$T_{.j}$	5	-3	$y-6$	-8	$y-12=T_{..}$

For this example,

$$SS_{error} = 4^2 + 1^2 + \cdots + y^2 + \cdots + (-4)^2$$
$$- \frac{(5)^2 + (-3)^2 + (y-6)^2 + (-8)^2}{4}$$
$$- \frac{(4)^2 + (-1)^2 + (y-2)^2 + (-13)^2}{4} + \frac{(y-12)^2}{16}$$

To find the y value that will minimize this expression, it is differentiated with respect to y and set equal to zero. As all constant terms have their derivatives zero,

$$\frac{d(SS_{error})}{dy} = 2y - \frac{2(y-6)}{4} - \frac{2(y-2)}{4} + \frac{2(y-12)}{16} = 0$$

Solving gives

$$16y - 4y + 24 - 4y + 8 + y - 12 = 0$$
$$9y = -20$$
$$y = -\frac{20}{9} = -2.2 \text{ or } 10.8 \text{ mils}$$

If this value is now used in the y position, the resulting ANOVA table is as shown in Table 4.11. This is an approximate ANOVA as the sums of squares are slightly biased. The resulting ANOVA is not too different from before, but the degrees of freedom for the error term are reduced by one, since there are only 15 actual observations, and y is determined from these 15 readings.

Table 4.11 ANOVA for Tire Brand Test Example Adjusted for a Missing Value (approximate)

Source	df	SS	MS
Brands τ_j	3	28.7	9.5
Cars β_i	3	38.3	12.7
Error ε_{ij}	8	11.2	1.4
Totals	14	78.2	

This procedure can be used on any reasonable number of missing values by differentiating the error sum of squares partially with respect to each such missing value and setting it equal to zero, giving as many equations as unknown values to be solved for these missing values.

In general, for one missing value y_{ij} it can be shown that

$$y_{ij} = \frac{nT'_{i.} + kT'_{.j} - T'_{..}}{(n-1)(k-1)} \tag{4.5}$$

where the primed totals $T'_{i.}$, $T'_{.j}$, and $T'_{..}$ are the totals indicated without the missing value y. In our example,

$$y_{ij} = y_{33} = \frac{4(-2) + 4(-6) - (-12)}{(3)(3)}$$

$$y_{33} = \frac{-20}{9} = -2.2 \quad \text{(as before)}$$

4.5 Latin Squares

The reader may have wondered about a possible position effect in the problem on testing tire brands. Experience shows that rear tires get different wear than front tires and even different sides of the same car may show different amounts of tread wear. In the randomized block design, the four brands were randomized onto the four wheels of each car with no regard for position. The effect of position on wear could be balanced out by rotating the tires every 5000 miles, giving each brand 5000 miles on each wheel. However, if this is not feasible, the positions can impose another restriction on the randomization in such a way that each brand is not only used once on each car but also only once in each of the four possible positions: left front, left rear, right front, and right rear.

A design in which each treatment appears once and only once in each row (position) and once and only once in each column (cars) is called a *Latin square design.* Interest is still centered on one factor, treatments, but two restrictions are placed on the randomization. An example of one such 4×4 Latin square is shown in Table 4.12.

Table 4.12 4×4 Latin Square Design

			Car	
Position	I	II	III	IV
1	C	D	A	B
2	B	C	D	A
3	A	B	C	D
4	D	A	B	C

Such a design is only possible when the number of levels of both restrictions equals the number of treatment levels. In other words, it must be a square. It is not true that all randomization is lost in this design, as the particular Latin square to be used on a given problem may be chosen at random from several possible Latin squares of the required size. Tables of such squares are found in Fisher and Yates [12].

The analysis of the data in a Latin square design is a simple extension of previous analyses in which the data are now added in a third direction— positions. If the data of Table 4.5 were imposed on the Latin square of Table 4.12, the results could be those shown in Table 4.13.

Table 4.13 Latin Square Design Data on Tire Wear

Position				Car					$T_{..k}$
	I		**II**		**III**		**IV**		
1	C	12	D	11	A	13	B	8	44
2	B	14	C	12	D	11	A	13	50
3	A	17	B	14	C	10	D	9	50
4	D	13	A	14	B	13	C	9	49
$T_{i..}$		56		51		47		39	193

Treatment totals for A, B, C, and D brands are 57, 49, 43, and 44 as before, where the model is now

$$Y_{ijk} = \mu + \beta_i + \tau_j + \gamma_k + \varepsilon_{ijk}$$

and γ_k represents the positions' effect. Since the only new totals are for positions, a position sum of squares can be computed as

$$SS_{position} = \frac{(44)^2 + (50)^2 + (50)^2 + (49)^2}{4} - \frac{(193)^2}{16} = 6.19$$

and

$$SS_{error} = SS_{total} - SS_{brand} - SS_{car} - SS_{position}$$
$$= 80.94 - 30.69 - 38.69 - 6.19 = 5.37$$

Table 4.14 is the ANOVA table for these data.

Table 4.14 Latin Square ANOVA

Source	df	SS	MS	EMS
Brands τ_j	3	30.69	10.2	$\sigma_\varepsilon^2 + 4\phi_\tau$
Cars β_i	3	38.69	12.9	$\sigma_\varepsilon^2 + 4\sigma_\beta^2$
Positions γ_k	3	6.19	2.1	$\sigma_\varepsilon^2 + 4\phi_\gamma$
Error ε_{ijk}	6	5.37	0.9	σ_ε^2
Totals	15	80.94		

$$y = u + B_i + t_{ij} + \gamma_k + E_{ijk}$$

Once again another restriction placed on the randomization has further reduced the experimental error by identifying another possible source of variation, although the position effect is not significant at the 5 percent level. This further reduction of error variance is attained at the expense of degrees of freedom, since now the estimate of σ_ε^2 is based on only 6 df instead of 9 df as in the randomized block design. This means less precision in estimating this error variance. But the added restrictions should be made if the environmental conditions suggest them. After discovering that position had no significant effect, some investigators might "pool" the position sum of squares with the error sum of squares and obtain a more precise estimate of σ_ε^2, namely, 1.3, as given in Table 4.6. However, there is a danger in "pooling," as it means "accepting" a hypothesis of no position effect, and the investigator has no idea about the possible error involved in "accepting" a hypothesis. Naturally, if the degrees of freedom on the error term are reduced much below that of Table 4.14, there will have to be some pooling to get a reasonable yardstick for assessing other effects.

4.6 Interpretations

After any ANOVA one usually wishes to investigate the single factor further for a more detailed interpretation of the effect of the treatments. Here, after the

Latin square ANOVA of Table 4.14, we are probably concerned about which brand to buy. Using a Newman–Keuls approach, the results are as follows:

1. Means: 10.75   11.00   12.25   14.25
   For brands: C D B A

2. $s_e^2 = 0.9$ with 6 df.

3. $s_{\bar{Y}.j.} = \sqrt{0.9/4} = 0.47$

4. For $p =$ 2   3   4
   Tabled ranges: 3.46   4.34   4.90

5. LSRs: 1.63   2.04   2.30

6. Testing averages:

$$A - C = 3.50 > 2.30^*$$
$$A - D = 3.25 > 2.04^*$$
$$A - B = 2.00 > 1.63^*$$
$$B - C = 1.50 < 2.04$$

Thus only A differs from the other brands in tread wear. A has the largest tread wear and so is the poorest brand to buy. One could recommend brand B, C, or D, whichever is the cheapest or whichever is "best" according to some criterion other than tread wear.

4.7 Graeco–Latin Squares

In some experiments still another restriction may be imposed on the randomization. The design may be a Graeco–Latin square such as Table 4.15 exhibits.

Table 4.15 Graeco–Latin Square Design

Position	Car			
	I	II	III	IV
1	$A\alpha$	$B\beta$	$C\gamma$	$D\delta$
2	$B\gamma$	$A\delta$	$D\alpha$	$C\beta$
3	$C\delta$	$D\gamma$	$A\beta$	$B\alpha$
4	$D\beta$	$C\alpha$	$B\delta$	$A\gamma$

In this design the third restriction is at levels α, β, γ, δ, and not only do these each appear once and only once in each row and each column, but they appear once and only once with each level of treatment A, B, C, or D. The model for this would be

$$Y_{ijkm} = \mu + \beta_i + \tau_j + \gamma_k + \omega_m + \varepsilon_{ijkm}$$

where ω_m is the effect of the latest restriction with levels α, β, γ, and δ. An outline of the analysis appears in Table 4.16.

Table 4.16 Graeco–Latin ANOVA Outline

Source	df
β_i	3
τ_j	3
γ_k	3
ω_m	3
ε_{ijkm}	3
Total	15

Such a design may not be very practical, as only 3 df are left for the error variance, as a small number of degrees of freedom will require a very large F for significance.

4.8 Extensions

In some situations it is not possible to place all treatments in every block of an experiment. This leads to a design known as an *incomplete block design*. Sometimes a whole row or column of a Latin square is missing and the resulting incomplete Latin square is called a *Youden square*. These two special designs are discussed in Chapter 16.

4.9 Summary

Experiment	Design	Analysis
I. Single factor	1. Completely randomized $Y_{ij} = \mu + \tau_j + \varepsilon_{ij}$	1. One-way ANOVA
	2. Complete randomized block $Y_{ij} = \mu + \tau_j + \beta_i + \varepsilon_{ij}$	2. Two-way ANOVA
	3. Complete Latin square $Y_{ijk} = \mu + \beta_i + \tau_j + \gamma_k + \varepsilon_{ijk}$	3. Three-way ANOVA
	4. Graeco–Latin square $Y_{ijkm} = \mu + \beta_i + \tau_j + \gamma_k + \omega_m + \varepsilon_{ijkm}$	4. Four-way ANOVA

It might be emphasized once again that, so far, interest has been centered on a single factor (treatments) for the discussion in Chapters 3 and 4. The special designs simply represent restrictions on the randomization. In the next chapter, two or more factors are considered.

4.10 SAS Programs for Randomized Blocks and Latin Squares

PROC ANOVA can be used for randomized complete block and Latin square designs if there are no missing values in the data. Since we are now considering more than one-factor experiments, the changes to our program will only occur in the number of variables referred to.

This example will progress through the three analysis stages as presented in the text. The data, however, will be as entered in Table 4.13. The table is recreated with four different cars (I, II, III, IV), four positions of tires on each car (1, 2, 3, 4), and four different brands (A, B, C, D). The numbers within the array are the results of the tests measured in difference in tire thicknesses in mils (0.001 in.).

Latin Square Design Data on Tire Wear

Position	I	II	III	IV
1	C(12)	D(11)	A(13)	B(8)
2	B(14)	C(12)	D(11)	A(13)
3	A(17)	B(14)	C(10)	D(9)
4	D(13)	A(14)	B(13)	C(9)

Our program for a one-way ANOVA is

```
DATA NEW;
INPUT CAR POSITION BRAND $ RDG;        for reading
CARDS;
  1   1   C    12
  1   2   B    14
  1   3   A    17
  1   4   D    13
  2   1   D    11
  2   2   C    12
  2   3   B    14
  2   4   A    14
  3   1   A    13
  3   2   D    11
  3   3   C    10
```

```
3   4   B   13
4   1   B   8
4   2   A   13
4   3   D   9
4   4   C   9
;
TITLE'CAR TIRE WEAR STUDY';
PROC PRINT;
PROC ANOVA;
CLASS BRAND;
MODEL RDG=BRAND;
MEANS BRAND/SNK;
RUN;
```

Our output is as follows:

```
        CAR   TIRE   WEAR STUDY
OBS     CAR    POSITION    BRAND    RDG
  1      1        1          C      12
  2      1        2          B      14
  3      1        3          A      17
  4      1        4          D      13
  5      2        1          D      11
  6      2        2          C      12
  7      2        3          B      14
  8      2        4          A      14
  9      3        1          A      13
 10      3        2          D      11
 11      3        3          C      10
 12      3        4          B      13
 13      4        1          B       8
 14      4        2          A      13
 15      4        3          D       9
 16      4        4          C       9
```

```
        CAR TIRE WEAR STUDY
      ANALYSIS  OF VARIANCE PROCEDURE
          CLASS  LEVEL INFORMATION
        CLASS      LEVELS    VALUES
        BRAND         4      A B C D
NUMBER OF OBSERVATIONS IN DATA SET = 16
```

CAR TIRE WEAR STUDY

ANALYSIS OF VARIANCE PROCEDURE

DEPENDENT VARIABLE: RDG

SOURCE	DF	SUM OF SQUARES	MEAN SQUARE	F VALUE	PR > F	R-SQUARE	C.V.
MODEL	3	30.68750000	10.22916667	2.44	0.1145	0.379151	16.9645
ERROR	12	50.25000000	4.18750000			ROOT MSE	RDG MEAN
CORRECTED TOTAL	15	80.93750000				2.04633819	12.06250000

SOURCE	DF	ANOVA SS	F VALUE	PR > F
BRAND	3	30.68750000	2.44	0.1145

CAR TIRE WEAR STUDY

ANALYSIS OF VARIANCE PROCEDURE

STUDENT-NEWMAN-KEULS TEST FOR VARIABLE: RDG
NOTE: THIS TEST CONTROLS THE TYPE I EXPERIMENTWISE ERROR RATE
UNDER THE COMPLETE NULL HYPOTHESIS BUT NOT UNDER PARTIAL
NULL HYPOTHESES

ALPHA=0.05   DF=12   MSE=4.1875

NUMBER OF MEANS	2	3	4
CRITICAL RANGE	3.15256	3.86041	4.29589

MEANS WITH THE SAME LETTER ARE NOT SIGNIFICANTLY DIFFERENT.

SNK	GROUPING	MEAN	N	BRAND
	A	14.250	4	A
	A			
	A	12.250	4	B
	A			
	A	11.000	4	D
	A			
	A	10.750	4	C

The results are the same as displayed in the text, and at the 5 percent level there is no reason to reject the hypothesis of equal tread loss among the brands.

A randomized complete block design can be accomplished quickly by changing only three of our commands to reflect an additional variable. The **CLASS, MODEL,** and **MEANS** statements become

```
CLASS BRAND CAR;
MODEL RDG=BRAND CAR;
MEANS BRAND CAR/SNK;
```

As you can see, the more complicated problem of two or more factors is handled quite easily by recognizing these variables in the **CLASS, MODEL,** and **MEANS** statements.

The analysis is now a two-way ANOVA with blocking on cars. Although we are primarily interested in brand differences, the added variable removes the car-to-car variability from the error term at the expense of some loss in degrees of freedom. We can see from the following output that this was beneficial because we now have a significant difference at the 1 percent level. For added information we include an SNK analysis of car differences by adding this variable to the **MEANS** statement. Our output is on page 112.

Although not required for making the decision on tire brands, we extend our analysis to a third variable to illustrate a Latin square design. Again, only three of our commands are changed (**CLASS, MODEL,** and **MEANS**).

```
CLASS BRAND CAR POSITION;
MODEL RDG=BRAND CAR POSITION;
MEANS BRAND CAR POSITION/SNK;
```

Except for μ and ε_{ij}, note the similarity of the model statement to the mathematical model of the problem.

Our output now is on pages 113–114.

These tests on car and position means are probably unnecessary because our chief concern is brand effect. They are simply included here if anyone is interested in such means.

Since positions are nonsignificant, one could "pool" position effect with error and revert to the randomized block as an adequate analysis.

CAR TIRE WEAR STUDY

ANALYSIS OF VARIANCE PROCEDURE

DEPENDENT VARIABLE: RDG

SOURCE	DF	SUM OF SQUARES	MEAN SQUARE	F VALUE	PR > F	R-SQUARE	C.V.
MODEL	6	69.37500000	11.56250000	9.00	0.0022	0.857143	9.3965
ERROR	9	11.56250000	1.28472222			ROOT MSE	RDG MEAN
CORRECTED TOTAL	15	80.93750000				1.13345588	12.06250000

SOURCE	DF	ANOVA SS	F VALUE	PR > F
BRAND	3	30.68750000	7.96	0.0067
CAR	3	38.68750000	10.04	0.0031

CAR TIRE WEAR STUDY

ANALYSIS OF VARIANCE PROCEDURE

STUDENT-NEWMAN-KEULS TEST FOR VARIABLE: RDG

NOTE: THIS TEST CONTROLS THE TYPE I EXPERIMENTWISE ERROR RATE UNDER THE COMPLETE NULL HYPOTHESIS BUT NOT UNDER PARTIAL NULL HYPOTHESES

ALPHA=0.05 DF=9 MSE=1.28472

NUMBER OF MEANS	2	3	4
CRITICAL RANGE	1.81314	2.23767	2.50205

MEANS WITH THE SAME LETTER ARE NOT SIGNIFICANTLY DIFFERENT.

SNK	GROUPING		MEAN	N	BRAND
	A		14.2500	4	A
	B		12.2500	4	B
	B				
	B		11.0000	4	D
	B				
	B		10.7500	4	C

CAR TIRE WEAR STUDY

ANALYSIS OF VARIANCE PROCEDURE

STUDENT-NEWMAN-KEULS TEST FOR VARIABLE: RDG

NOTE: THIS TEST CONTROLS THE TYPE I EXPERIMENTWISE ERROR RATE UNDER THE COMPLETE NULL HYPOTHESIS BUT NOT UNDER PARTIAL NULL HYPOTHESES

ALPHA=0.05 DF=9 MSE=1.28472

NUMBER OF MEANS	2	3	4
CRITICAL RANGE	1.81314	2.23767	2.50205

MEANS WITH THE SAME LETTER ARE NOT SIGNIFICANTLY DIFFERENT.

SNK	GROUPING		MEAN	N	CAR
	A		14.0000	4	1
	A				
	A	B	12.7500	4	2
		B			
		B	11.7500	4	3
		C	9.7500	4	4

CAR TIRE WEAR STUDY
ANALYSIS OF VARIANCE PROCEDURE

DEPENDENT VARIABLE: RDG

SOURCE	DF	SUM OF SQUARES	MEAN SQUARE	F VALUE	PR > F	R-SQUARE	C.V.
MODEL	9	75.56250000	8.39583333	9.37	0.0066	0.933591	7.8465
ERROR	6	5.37500000	0.89583333		ROOT MSE		RDG MEAN
CORRECTED TOTAL	15	80.93750000			0.94648472		12.06250000

SOURCE	DF	ANOVA SS	F VALUE	PR > F
BRAND	3	30.68750000	11.42	0.0068
CAR	3	38.68750000	14.40	0.0038
POSITION	3	6.18750000	2.30	0.1769

CAR TIRE WEAR STUDY
ANALYSIS OF VARIANCE PROCEDURE

STUDENT-NEWMAN-KEULS TEST FOR VARIABLE: RDG

NOTE: THIS TEST CONTROLS THE TYPE I EXPERIMENTWISE ERROR RATE UNDER THE COMPLETE NULL HYPOTHESIS BUT NOT UNDER PARTIAL NULL HYPOTHESES

ALPHA=0.05   DF=6   MSE=0.895833

NUMBER OF MEANS	2	3	4
CRITICAL RANGE	1.63761	2.05357	2.31683

MEANS WITH THE SAME LETTER ARE NOT SIGNIFICANTLY DIFFERENT.

SNK	GROUPING	MEAN	N	BRAND
	A	14.2500	4	A
	B	12.2500	4	B
	B			
	B	11.0000	4	D
	B			
	B	10.7500	4	C

113

CAR TIRE WEAR STUDY

ANALYSIS OF VARIANCE PROCEDURE

STUDENT—NEWMAN—KEULS TEST FOR VARIABLE: RDG
NOTE: THIS TEST CONTROLS THE TYPE I EXPERIMENTWISE ERROR RATE
UNDER THE COMPLETE NULL HYPOTHESIS BUT NOT UNDER PARTIAL
NULL HYPOTHESES

ALPHA=0.05   DF=6   MSE=0.895833

NUMBER OF MEANS	2	3	4
CRITICAL RANGE	1.63761	2.05357	2.31683

MEANS WITH THE SAME LETTER ARE NOT SIGNIFICANTLY DIFFERENT.

SNK	GROUPING		MEAN	N	CAR
		A	14.0000	4	1
		A			
	B	A	12.7500	4	2
	B				
	B		11.7500	4	3
		C	9.7500	4	4

CAR TIRE WEAR STUDY

ANALYSIS OF VARIANCE PROCEDURE

STUDENT—NEWMAN—KEULS TEST FOR VARIABLE: RDG
NOTE: THIS TEST CONTROLS THE TYPE I EXPERIMENTWISE ERROR RATE
UNDER THE COMPLETE NULL HYPOTHESIS BUT NOT UNDER PARTIAL
NULL HYPOTHESES

ALPHA=0.05   DF=6   MSE=0.895833

NUMBER OF MEANS	2	3	4
CRITICAL RANGE	1.63761	2.05357	2.31683

MEANS WITH THE SAME LETTER ARE NOT SIGNIFICANTLY DIFFERENT.

SNK	GROUPING	MEAN	N	POSITION
	A	12.5000	4	2
	A			
	A	12.5000	4	3
	A			
	A	12.2500	4	4
	A			
	A	11.0000	4	1

Problems

4.1 The effects of four types of graphite coaters on light box readings are to be studied. As these readings might differ from day to day, observations are to be taken on each of the four types every day for three days. The order of testing of the four types on any given day can be randomized. The results are

Day	Graphite Coater Type			
	M	*A*	*K*	*L*
1	4.0	4.8	5.0	4.6
2	4.8	5.0	5.2	4.6
3	4.0	4.8	5.6	5.0

Analyze these data as a randomized block design and state your conclusions.

4.2 Set up one set of orthogonal contrasts among coater types and analyze for Problem 4.1.

4.3 Use the Newman–Keuls range test to compare four coater-type means for Problem 4.1.

4.4 If the reading on type *K* in Problem 4.1 for the second day was missing, what missing value should be inserted and what is the analysis now?

4.5 Set up at least four contrasts in Problem 4.1 and use the Scheffé method to test the significance of these.

4.6 In a research study at Purdue University on metal-removal rate, five electrode shapes *A, B, C, D*, and *E* were studied. The removal was accomplished by an electric discharge between the electrode and the material being cut. For this experiment five holes were cut in five workpieces, and the order of electrodes was arranged so that only one electrode shape was used in the same position on each of the five workpieces. Thus, the design was a Latin square design, with workpieces (strips) and positions on the strip as restrictions on the randomization. Several variables were studied, one of which was the Rockwell hardness of metal where each hole was to be cut. The results were as follows:

Strip	Position				
	1	2	3	4	5
I	A(64)	B(61)	C(62)	D(62)	E(62)
II	B(62)	C(62)	D(63)	E(62)	A(63)
III	C(61)	D(62)	E(63)	A(63)	B(63)
IV	D(63)	E(64)	A(63)	B(63)	C(63)
V	E(62)	A(61)	B(63)	C(63)	D(62)

Analyze these data and test for an electrode effect, position effect, and strip effect on Rockwell hardness.

4.7 The times in hours necessary to cut the holes in Problem 4.6 were recorded as follows:

	Position				
Strip	**1**	**2**	**3**	**4**	**5**
I	A(3.5)	B(2.1)	C(2.5)	D(3.5)	E(2.4)
II	E(2.6)	A(3.3)	B(2.1)	C(2.5)	D(2.7)
III	D(2.9)	E(2.6)	A(3.5)	B(2.7)	C(2.9)
IV	C(2.5)	D(2.9)	E(3.0)	A(3.3)	B(2.3)
V	B(2.1)	C(2.3)	D(3.7)	E(3.2)	A(3.5)

Analyze these data for the effect of electrodes, strips, and positions on time.

4.8 Analyze the electrode effect further in Problem 4.7 and make some statement as to which electrodes are best if the shortest cutting time is the most desirable factor.

4.9 For an $m \times m$ Latin square, prove that a missing value may be estimated by

$$Y_{ijk} = \frac{m(T'_{i..} + T'_{.j.} + T'_{..k}) - 2T'_{...}}{(m-1)(m-2)}$$

where the primes indicate totals for row, treatment, and column containing the missing value and $T'_{...}$ is the grand total of all actual observations.

4.10 A composite measure of screen quality was made on screens using four lacquer concentrations, four standing times, four acryloid concentrations (A, B, C, D), and four acetone concentrations (α, β, Γ, Δ). A Graeco–Latin square design was used with data recorded as follows:

	Lacquer Concentration			
Standing Time	$\frac{1}{2}$	**1**	$1\frac{1}{2}$	**2**
30	$C\beta(16)$	$B\Gamma(12)$	$D\Delta(17)$	$A\alpha(11)$
20	$B\alpha(15)$	$C\Delta(14)$	$A\Gamma(15)$	$D\beta(14)$
10	$A\Delta(12)$	$D\alpha(6)$	$B\beta(14)$	$C\Gamma(13)$
5	$D\Gamma(9)$	$A\beta(9)$	$C\alpha(8)$	$B\Delta(9)$

Do a complete analysis of these data.

4.11 Explain why a three-level Graeco–Latin square is not a feasible design.

4.12 In a chemical plant, five experimental treatments are to be used on a basic raw material in an effort to increase the chemical yield. Since batches of raw material may differ, five batches are chosen at random. Also, the order of the experiments may affect yield as well as may the operator who performs the experiment. Considering order and operators as further restrictions on randomization, set up a suitable design for this experiment that will be least expensive to run. Outline its analysis.

4.13 Three groups of students are to be tested for the percentage of high-level questions asked by each group. As questions can be on various types of material, six lessons are taught to each group and a record is made of the percentage of high-level questions asked by each group on all six lessons. Show a data layout for this situation and outline its ANOVA table.

4.14 Data from the results of Problem 4.13 were as follows:

	Group		
Lesson	*A*	*B*	*C*
1	13	18	7
2	16	25	17
3	28	24	14
4	26	13	15
5	27	16	12
6	23	19	9

Do an analysis of these data and state your conclusions.

4.15 For Problem 4.14 analyze further in an attempt to determine which group asks the highest percentage of high-level questions.

4.16 Thirty students were pretested for science achievement and were subsequently posttested using the same test after 15 weeks of special instruction. A *t* test on the difference scores of those students was found to be 2.3. Consider this problem as a randomized block design and show a data layout and outline its ANOVA table with the proper degrees of freedom. Also indicate what *F* value should be exceeded if one were to conclude that the training made a difference at the 5 percent significance level.

4.17 A student decides to investigate the accuracy of five scales in various local commercial places. She decides to account for the variations in her weight at different times of day and also on different days. She will only collect data on Saturdays, which she has free. The design will be a Latin square so she chooses the five times of day as 9:00 a.m., 11:00 a.m., 1:00 p.m., 3:00 p.m., and 5:00 p.m. One possible Latin square design is as follows:

			Time		
Day	9	11	1	3	5
1	A	B	C	D	E
2	B	C	D	E	A
3	C	D	E	A	B
4	D	E	A	B	C
5	E	A	B	C	D

a. Give the model and write out the ANOVA table.

b. Suppose she does a Scheffé test on the times of day and finds that the 3 and 5 are not significantly different. If she is going to do the experiment again on some other scales, should she do a simpler randomized block design with five days as blocks and make observations at 3 p.m. and 5 p.m.? Give reasons for your decision.

4.18 Workers at an agricultural experiment station conducted an experiment to compare the effect of four different fertilizers on the yield of a cane crop. They divided four blocks of soil into four plots of equal size and shape and assigned the fertilizers to the plots at random such that each fertilizer was applied once in each block. Data were collected on yield in hundred weights per plot. The analysis gave a total sum of squares of 540 with fertilizer sum of squares at 210 and block sum of squares at 258. Set up an ANOVA table for these results.

4.19 From the data of Problem 4.18 state the hypothesis to be tested in terms of the mathematical model, run the test, and state your conclusions.

4.20 In Problem 4.18 if the observed mean yield of cane for fertilizer B was 41.25 hundred weight, what is its standard error?

4.21 In Problem 4.18 the original data show that all readings are divisible by 5. If one codes the data by subtracting 40 and dividing by 5, set up an expression for 95 percent confidence limits on the mean of fertilizer B in terms of the coded data.

4.22 Write a contrast to be used in comparing the mean of fertilizers A and B with the mean of fertilizer D and also show its standard error for Problem 4.18.

4.23 Set up three contrasts in which one contrast is orthogonal to the other two but these two are *not* orthogonal to each other.

4.24 a. What is the main advantage of a randomized block design over a completely randomized design?

b. Under what conditions would the completely randomized design have been better?

4.25 Material is analyzed for weight in grams from three vendors—A, B, C—by three different inspectors—I, II, III—and using three different scales—1, 2, 3. The experiment is set up on a Latin square plan with results as follows:

Inspector	Scale		
	1	2	3
I	A = 16	B = 10	C = 11
II	B = 15	C = 9	A = 14
III	C = 13	A = 11	B = 13

Analyze the experiment to test for a difference in weight between vendors, between inspectors, and between scales.

4.26 A school district was concerned about its history curriculum. They found two new curricula that proponents said would improve students' achievement in history as compared to the present curriculum that they were using. In order to check these out, three high schools (I, II, III) were chosen at random in the district and curricula A, B, and C (their present one) were used in each school. It was also decided to try these three curricula at three different grade levels (9, 10, 11). An experiment was designed in which each curriculum was tried once and only once in each school and once and only once at each grade level. The criterion used to measure achievement was the change in test scores from the beginning to the end of a semester. Devise a data layout for this experiment and outline its ANOVA.

4.27 The changes in test scores in Problem 4.26 are as follows:

Schools	Curricula	Grade	Score
I	A	9	28
I	C	10	31
I	B	11	30
II	B	9	31
II	A	10	30
II	C	11	32
III	C	9	30
III	B	10	30
III	A	11	29

Do a complete analysis of these data and state your recommendations.

4.28 In order to check on a possible laboratory bias in the reported ash content of coal, 10 samples of coal were split in half and then sent at random to each of two laboratories (L_1 and L_2). The laboratories reported the following ash content data:

Sample	Lab 1	Lab 2
1	5.47	5.13
2	5.31	5.46
3	5.46	5.54
4	5.55	5.54
5	5.93	6.00
6	5.97	5.99
7	6.32	6.43
8	6.02	6.13
9	5.87	5.87
10	5.58	5.60

Try two different methods to analyze these data and show that the results agree. What is your conclusion?

4.29 To study aluminum thickness, samples were taken from five machines. This was then repeated three times. Results were

Replication	Machine				
	1	2	3	4	5
I	175	95	180	170	155
II	190	185	180	200	190
III	185	165	175	195	200

Analyze these data and state your conclusions.

Factorial Experiments   **5**

5.1 Introduction

In the preceding three chapters all of the experiments involved only one factor and its effect on a measured variable. Several different designs were considered, but all of these represented restrictions on the randomization in which interest was still centered on the effect of a single factor.

Suppose there are now two factors of interest to the experimenter, for example, the effect of both temperature and altitude on the current flow in an integrated circuit. One traditional method is to hold altitude constant and vary the temperature and then hold temperature constant and change the altitude, or, in general, hold all factors constant except one and take current flow readings for several levels of this one factor, then choose another factor to vary, holding all others constant, and so forth. To examine this type of experimentation, consider a very simple example in which temperature is to be set at 25°C and 55°C only and altitudes of 0 K (K = 10,000 feet) and 3 K (30,000 feet) are to be used. If one factor is to be varied at a time, the altitude may be set for sea level or 0 K and the temperature varied from 25°C to 55°C. Suppose the current flow changed from 210 mA to 240 mA with this temperature increase. Now there is no way to assess whether or not this 30-mA increase is real or due to chance. Unless there is available some previous estimate of the error variability, the experiment must be repeated in order to obtain an estimate of the error or chance variability within the experiment. If the experiment is now repeated and the readings are 205 mA and 230 mA as the temperature is varied from 25°C to 55°C, it seems obvious, without any formal statistical analysis, that there is a real increase in current flow, since for each repetition the increase is large compared with the variation in current flow within a given temperature. Graphically these results appear as in Figure 5.1.

Four experiments have now been run to determine the effect of temperature at 0 K altitude only. To check the effect of altitude, the temperature can be held at 25°C and the altitude varied to 3 K by adjustment of pressure in the laboratory.

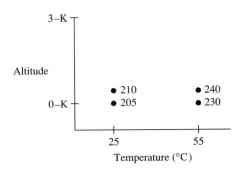

Figure 5.1   Temperature effect on current flow.

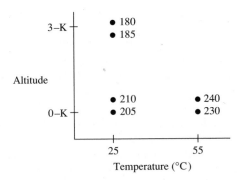

Figure 5.2   Temperature and altitude effect on current flow.

Using the results already obtained at 0 K (assuming they are representative), two observations of current flow are now taken at 3 K with the results shown in Figure 5.2.

From these experiments the temperature increase is seen to increase the current flow an average of

$$\frac{30 \text{ mA} + 25 \text{ mA}}{2} = 27.5 \text{ mA}$$

and the increase in altitude decreases the current flow on the average of

$$\frac{30 \text{ mA} + 20 \text{ mA}}{2} = 25 \text{ mA}$$

This information is gained after six experiments have been performed and no information is available on what would happen at a temperature of 55°C and an altitude of 3 K.

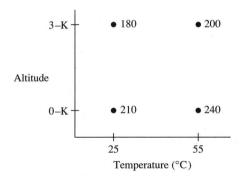

Figure 5.3 Factorial arrangement of temperature and altitude effect on current flow.

An alternative experimental arrangement would be a factorial arrangement where each temperature level is combined with each altitude and only four experiments are run. Results of four such experiments might appear as in Figure 5.3.

With this experiment, one estimate of temperature effect on current flow is 240 − 210 mA = 30 mA at 0 K, and another estimate is 200 − 180 mA = 20 mA at 3 K. Hence two estimates can be made of temperature effect [average (30 + 20) / 2 = 25 mA] using all four observations without necessarily repeating any observation at the same point. Using the same four observations, two estimates of altitude effect can be determined: 180 − 210 = −30 mA at 25°C, and 200 − 240 = −40 mA at 55°C, an average decrease of 35 mA for a 3-K increase in altitude. Here, with just four observations instead of six, valid comparisons have been made on both temperature and altitude, and in addition some information has been obtained as to what happens at 55°C and altitude 3 K.

From this simple example, some of the advantages of a factorial experiment can be seen:

1. More efficiency is possible than with one-factor-at-a-time experiments (here four sixths or two thirds the amount of experimentation).

2. All data are used in computing both effects. (Note that all four observations are used in determining the average effect of temperature and the average effect of altitude.)

3. Some information is gleaned on possible interaction between the two factors. (In the example, the increase in current flow of 20 mA at 3 K was about the same order of magnitude as the 30 mA increase at 0 K. If these increases had differed considerably, interaction might be said to be present.)

These advantages are even more pronounced as the number of levels of the two factors is increased. A *factorial experiment* is one in which all levels of a

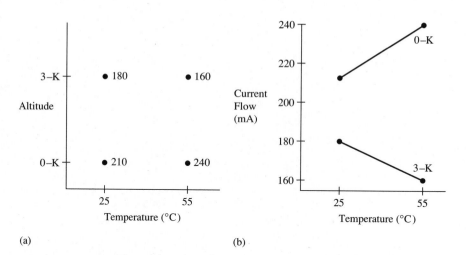

Figure 5.4 Interaction in a factorial experiment.

given factor are combined with all levels of every other factor in the experiment. Thus, if four temperatures are considered at three altitudes, a 4 × 3 factorial experiment would be run requiring 12 different experimental conditions. In the example above, if the current flow were 160 mA for 55°C and 3 K, the results could be shown as in Figure 5.4.

From Figure 5.4(a) note that as the temperature is increased from 25 to 55°C at 0 K, the current flow increases by 30 mA, but at 3 K for the same temperature increase the current flow *decreases* by 20 mA. When a change in one factor produces a different change in the response variable at one level of another factor than at other levels of this factor, there is an *interaction* between the two factors. This is also observable in Figure 5.4(b) as the two altitude lines are not parallel. If the data of Figure 5.3 are plotted, we get the relations pictured in Figure 5.5.

It is seen that the lines are much more nearly parallel and no interaction is said to be present. An increase in temperature at 0 K produces about the same increase in current flow (30 mA) as at 3 K (20 mA). Now a word of warning is necessary. Lines can be made to look nearly parallel or quite diverse depending on the scale chosen; therefore, it is necessary to run statistical tests in order to determine whether or not the interaction is statistically significant. It is only possible to test the significance of such interaction if more than one observation is taken for each experimental condition. The graphic procedure shown above is given merely to provide some insight into what interaction is and how it might be displayed for explaining the factorial experiment.

In addition to the advantages of a factorial experiment, it may be that when the response variable Y is a yield and one is concerned with which combination of two independent variables X_1 and X_2 will produce a maximum yield, the

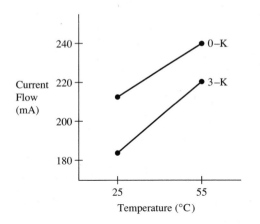

Figure 5.5 No-interaction temperature diagram in a temperature–altitude study.

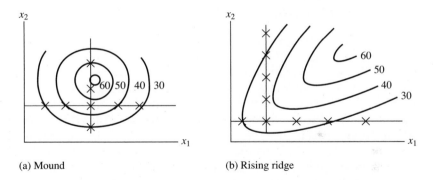

(a) Mound   (b) Rising ridge

Figure 5.6 Yield contours using one-factor-at-a-time method.

factorial will give a combination near the maximum whereas the one-factor-at-a-time procedure will not do so. Figure 5.6 shows two cases in which the contours indicate the yields.

In case (a), X_2 is set and X_1 is varied until the maximum yield is found, then X_1 is held there while X_2 is varied and a maximum is found near the top of the mound where the yield is around 60 units. But in case (b), the same procedure misses the maximum by a long way, giving about a 37-unit yield.

If, now, a factorial arrangement is used, one first notes the extremes of each independent variable and may also choose one or more points between these extremes. If this is done for the diagrams in Figure 5.6, Figure 5.7 shows that in both cases the maximum is reached. In some cases the maximum may not have been reached but the factorial will indicate the direction to be followed in

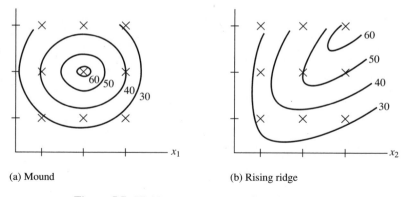

(a) Mound   (b) Rising ridge

Figure 5.7   Yield contours using a factorial method.

the next experiment to get closer to such a maximum. This concept is presented in Chapter 16.

5.2 Factorial Experiments

To see how a factorial experiment is set up and analyzed, consider Example 5.1.

■ **EXAMPLE 5.1**   To determine the effect of exhaust index (in seconds) and pump heater voltage (in volts) on the pressure inside a vacuum tube (in microns of mercury), three exhaust indexes and two voltages are chosen at fixed levels. It was decided to run two experiments at each of these six treatment conditions (three exhaust indexes × two voltages). The order for running the 12 experiments was completely randomized. This can be accomplished easily by labeling the six treatment conditions numbers 1 through 6 and then tossing a die to decide the order in which the experiments are to be run.

Table 5.1 shows the order of one such complete randomization for the 12 experiments. In such a procedure some treatment combinations may be run twice before others are run once, but this is the nature of complete randomization. If one were to randomize six runs for each combination and then randomize six more, the randomization will have been restricted and no longer a completely randomized design. In such a procedure it is also true that some numbers may come up again on the die even though that combination already has been run twice. In these cases the third time is ignored and the flipping continues until all treatment combinations are run twice. This is, of course, a slight restriction to keep the number of observations per treatment constant. The resulting experiment may be labeled a 3 × 2 factorial experiment with two observations per treatment with all $N = 12$ observations run in a completely randomized order. Of course, this random order can be determined before the experiment is actually run and it is important to supervise the experiment to make sure that it is actually run in the order prescribed.

Table 5.1 Data Layout for Vacuum Tube
Pressure Experiment

Pump Heater Voltage (volts)	Exhaust Index (seconds) 60		90		150	
127	1⌋	4 9	2⌋	1 11	3⌋	6 12
220	4⌋	2 8	5⌋	3 5	6⌋	7 10

The mathematical model for this experiment can be written as

$$Y_{ijk} = \mu + \tau_{ij} + \varepsilon_{k(ij)} \tag{5.1}$$

where

$i = 1, 2, 3$ for the three exhaust indexes

$j = 1, 2$ for the two voltages

$k = 1, 2$ for the two observations in each i, j treatment combination

$\tau_{ij} =$ the six treatment effects

$\varepsilon_{k(ij)} =$ the error within each of the six treatments

The notation $k(ij)$ indicates that the errors are unique to each i, j combination or are nested within each i, j. Following the analysis procedures of Chapter 3, Table 5.2 shows the resulting data in a between-treatments, within-treatments format after multiplying all results by 1000 to eliminate decimal points.

Table 5.2 Vacuum Tube Pressure as a One-Way Experiment

	EI_{60}, V_{127}	EI_{90}, V_{127}	EI_{150}, V_{127}	EI_{60}, V_{220}	EI_{90}, V_{220}	EI_{150}, V_{220}
	48 58	28 33	7 15	62 54	14 10	6 9
$T_{ij.}$	106	61	22	116	24	15
$\sum_k Y_{ijk}^2$	5668	1873	274	6760	296	117

$$T_{..} = 344$$

$$\sum_i \sum_j \sum_k Y_{ijk}^2 = 14{,}988$$

If the data of Table 5.2 are subjected to a one-way ANOVA, the results are as given in Table 5.3.

Table 5.3 One-Way ANOVA for Vacuum Tube Pressure Experiment

Source	df	SS
Between treatments	5	4987.67
Within treatments	6	139.00
Totals	11	5126.67

In Table 5.3 the treatment effects are really combinations of exhaust indexes and voltages, and if rearranged to show these main effects, the data might appear as in Table 5.4.

Table 5.4 Two-Way Layout for Vacuum Tube Pressure Experiment

Pump Heater Voltage	Exhaust Index			$T_{.j.}$
	60	90	150	
127	48	28	7	189
	58	33	15	
220	62	14	9	155
	54	10	6	
$T_{i..}$	222	85	37	$T_{...} = 344$

If the sums of squares for the two main effects are computed from the marginal totals of Table 5.4, we find

$$SS_{EI} = \frac{(222)^2 + (85)^2 + (37)^2}{4} - \frac{(344)^2}{12} = 4608.17$$

$$SS_V = \frac{(189)^2 + (155)^2}{6} - \frac{(344)^2}{12} = 96.33$$

These two add to 4704.50, which is less than the sum of squares between treatments of 4987.67. It is this difference between the treatment SS and the SS of the two main effects that shows the interaction between the two main effects. Here

$$SS_{EI \times V \text{ interaction}} = SS_{\text{treatments}} - SS_{EI} - SS_V$$
$$= 4987.67 - 4608.17 - 96.33 = 283.17$$

Since the treatments carried 5 df, the interaction has $5 - 2 - 1 = 2$ df. These results may now be displayed in the ANOVA of Table 5.5. In this final table, each main effect (EI and V) and their interaction can be tested for significance by comparing each mean square with the error mean square. Exhaust index is seen to be highly significant. Voltage is not significant at the 5 percent significance level but the interaction is significant at the 5 percent level.

Table 5.5 Two-Way ANOVA for Vacuum Tube Pressure Experiment

Source	df	SS	MS	F	Probability
Between EI's	2	4608.17	2304.08	99.5	0.001
Between V's	1	96.33	96.33	4.2	0.088
EI by V interaction	2	283.17	141.58	6.1	0.036
Error (within treatments)	6	139.00	23.17		
Totals	11	5126.67			

The mathematical model of Equation (5.1) can now be expanded to read

$$Y_{ijk} = \mu + \quad \tau_{ij} \quad + \varepsilon_{k(ij)}$$

$$Y_{ijk} = \mu + \overbrace{E_i + V_j + EV_{ij}} + \varepsilon_{k(ij)} \tag{5.2}$$

where E_i represents the exhaust index effect in this problem, V_j the voltage effect, and EV_{ij} the interaction. Note that whereas in Chapter 4 randomization restrictions came from the error term, these effects come from the treatment term. In Equation (5.2) one tests the following hypotheses:

$$H_{0_1}: E_i = 0 \qquad \text{for all } i$$
$$H_{0_2}: V_j = 0 \qquad \text{for all } j$$
$$H_{0_3}: EV_{ij} = 0 \qquad \text{for all } i \text{ and } j$$

as both main effects are fixed effects. ∎

5.3 Interpretations

Since the interaction is significant in this example, one should be very cautious in interpreting the main effects. A significant interaction here means that the effect of exhaust index on vacuum tube pressure at one voltage is different from its effect at the other voltage. This can be seen graphically by plotting the six treatment means as shown in Figure 5.8. Note that the lines in the figure are not parallel. If one wishes to optimize the response variable a reasonable procedure for interpreting such an interaction is to run a Newman–Keuls test on the six means. Had the main effects not interacted one could treat each set of main effect means separately. Here testing main effects is not recommended because the results depend on how these main effects combine. Using the six means, Newman–Keuls gives

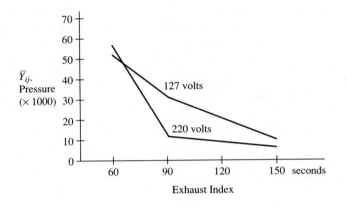

Figure 5.8 Vacuum tube current flow interaction plot.

1. Means are   7.5   11   12   30.5   53   58
   for treatments: E_3V_2 E_3V_1 E_2V_2 E_2V_1 E_1V_1 E_1V_2

2. $s_\varepsilon^2 = 23.17$ with 6 df from Table 5.5.

3. $s_{\bar{Y}_{ij.}} = \sqrt{23.17/2} = 3.40$.

4. For p: 2   3   4   5   6
   5 percent ranges: 3.46   4.34   4.90   5.31   5.63

5. LSR: 11.76, 14.76, 16.66, 18.05, 19.14

6. Checking means:

$$58 - 7.5 = 50.5 \;>\; 19.14*$$
$$58 - 11 = 47 \quad > 18.05^*$$
$$58 - 12 = 46 \quad > 16.66^*$$
$$58 - 30.5 = 27.5 > 14.76^*$$
$$58 - 53 = \;\;5 \quad < 11.76$$
$$53 - 7.5 = 45.5 > 18.05^*$$
$$53 - 11 = 42 \quad > 16.66^*$$
$$53 - 12 = 41 \quad > 14.76^*$$
$$53 - 30.5 = 22.5 > 11.76$$
$$30.5 - 7.5 = 23 \quad > 16.66^*$$
$$30.5 - 11 = 19.5 > 14.76^*$$
$$30.5 - 12 = 18.5 > 11.76^*$$
$$12 - 7.5 = \;\;4.5 < 14.76$$

Hence there are three groups of means:

$$\underline{7.5, \ 11, \ 12} \qquad \underline{30.5} \qquad \underline{53, \ 58}$$

If one is looking for the lowest pressure, any one of the three combinations in the first group will minimize the pressure in the vacuum tube. Thus one can recommend a 150-second exhaust index at either voltage or a 90-second exhaust index at 220 volts. From a practical point of view this really gives two choices: Pump at 127 volts for 150 seconds or at 220 volts for 90 seconds, whichever is cheaper.

It might be noted that in the model for this problem, Equation (5.2), it is assumed that the errors $\varepsilon_{k(ij)}$ are NID $(0,\sigma_e^2)$. This means that the variances within each of the six treatment combinations or "cells" are assumed to have come from normal populations with equal variances. In the data of Table 5.4 the six ranges are 10, 5, 8, 8, 4, and 3, which average 6.33. From Table 3.15, $D_4 \bar{R} = 20.7$ and all six ranges are well within this maximum range.

It may be of interest to note that whenever there are but two observations per treatment, the sum of squares for the error may be computed directly. When $n = 2$, the sum of squares of the ranges divided by 2 equals the SS_{error}. Here $\sum R^2 = 278$ and $\sum R^2 / 2 = 139$, which equals the error sum of squares of Table 5.5.

5.4 ANOVA Rationale

For a two-factor factorial experiment with n observations per cell, run as a completely randomized design, a general model would be

$$Y_{ijk} = \mu + A_i + B_j + AB_{ij} + \varepsilon_{k(ij)} \tag{5.3}$$

where A and B represent the two factors, $i = 1, 2, \ldots, a$ levels of factor A, $j = 1, 2, \ldots, b$ levels of factor B, and $k = 1, 2, \ldots, n$ observations per cell. In terms of population means this becomes

$$Y_{ijk} - \mu \equiv (\mu_i - \mu) + (\mu_j - \mu) + (\mu_{ij} - \mu_i - \mu_j + \mu) + (Y_{ijk} - \mu_{ij}) \tag{5.4}$$

where μ_{ij} represents the true mean of the i, j cell or treatment combination. Justification for the interaction term in the model comes from subtracting A and B main effects from the cell effect as follows:

$$(\mu_{ij} - \mu) - (\mu_i - \mu) - (\mu_j - \mu) = \mu_{ij} - \mu_i - \mu_j + \mu$$

If each mean is now replaced by its sample estimate, the resulting sample model is

$$Y_{ijk} - \bar{Y}_{\ldots} = (\bar{Y}_{i..} - \bar{Y}_{\ldots}) + (\bar{Y}_{.j.} - \bar{Y}_{\ldots}) + (\bar{Y}_{ij.} - \bar{Y}_{i..} - \bar{Y}_{.j.} + \bar{Y}_{\ldots}) + (Y_{ijk} - \bar{Y}_{ij.})$$

If this expression is now squared and summed over $i, j,$ and k, all cross products vanish, and the results give

$$\sum_i^a \sum_j^b \sum_k^n (Y_{ijk} - \bar{Y}_{...})^2 = \sum_i^a \sum_j^b \sum_k^n (\bar{Y}_{i..} - \bar{Y}_{...})^2 + \sum_i^a \sum_j^b \sum_k^n (\bar{Y}_{.j.} - \bar{Y}_{...})^2$$

$$+ \sum_i^a \sum_j^b \sum_k^n (\bar{Y}_{ij.} - \bar{Y}_{i..} - \bar{Y}_{.j.} + \bar{Y}_{...})^2$$

$$+ \sum_i^a \sum_j^b \sum_k^n (Y_{ijk} - \bar{Y}_{ij.})^2$$

which again expresses the idea that the total sum of squares can be broken down into the sum of squares between means of factor A, plus the sum of squares between means of factor B, plus the sum of squares of $A \times B$ interaction, plus the error sum of squares (or within cell sum of squares). Each sum of squares is seen to be independent of the others. Hence, if any such sum of squares is divided by its associated degrees of freedom, the results are independently chi-square distributed, and F tests may be run.

The degree-of-freedom breakdown would be

$$(abn - 1) \equiv (a - 1) + (b - 1) + (a - 1)(b - 1) + ab(n - 1)$$

the interaction being cell df $= (ab - 1)$ minus the main effect df, $(a - 1)$ and $(b - 1)$ or $(ab - 1) - (a - 1) - (b - 1) = ab - a - b + 1 = (a - 1) \times (b - 1)$, and within each cell the degrees of freedom are $n - 1$ and there are ab such cells giving $ab(n - 1)$ df for error. An ANOVA table can now be set up expanding and simplifying the sum of squares expressions using totals (Table 5.6).

The formulas for sum of squares in Table 5.6 provide good computational formulas for a two-way ANOVA with replication. The error sum of squares might be rewritten as

$$SS_{error} = \sum_i^a \sum_j^b \left[\sum_k^n Y_{ijk}^2 - \frac{T_{ij.}^2}{n} \right]$$

which points up the fact that the sum of squares within each of the $a \times b$ cells is being pooled or added for all such cells. This depends on the assumption that the variance within all cells came from populations with equal variance. The interaction sum of squares can also be rewritten as

$$\left(\sum_i^a \sum_j^b \frac{T_{ij.}^2}{n} - \frac{T_{...}^2}{nab} \right) - \left(\sum_i^a \frac{T_{i..}^2}{nb} - \frac{T_{...}^2}{nab} \right) - \left(\sum_j^b \frac{T_{.j.}^2}{na} - \frac{T_{...}^2}{nab} \right)$$

which shows again that interaction is calculated by subtracting the main effect sum of squares from the cell sum of squares.

Table 5.6 General ANOVA for Two-Factor Factorial with n Replications per Cell

Source	df	SS	MS
Factor A_i	$a - 1$	$\sum_{i}^{a} \dfrac{T_{i..}^2}{nb} - \dfrac{T_{...}^2}{nab}$	Each SS divided by its df
Factor B_j	$b - 1$	$\sum_{j}^{b} \dfrac{T_{.j.}^2}{na} - \dfrac{T_{...}^2}{nab}$	
$A \times B$ interaction	$(a - 1)(b - 1)$	$\sum_{i}^{a}\sum_{j}^{b} \dfrac{T_{ij.}^2}{n} - \sum_{i}^{a} \dfrac{T_{i..}^2}{nb}$ $- \sum_{j}^{b} \dfrac{T_{.j.}^2}{na} + \dfrac{T_{...}^2}{nab}$	
Error $\varepsilon_{k(ij)}$	$ab(n - 1)$	$\sum_{i}^{a}\sum_{j}^{b}\sum_{k}^{n} Y_{ijk}^2 - \sum_{i}^{a}\sum_{j}^{b} \dfrac{T_{ij.}^2}{n}$	
Totals	$abn - 1$	$\sum_{i}^{a}\sum_{j}^{b}\sum_{k}^{n} Y_{ijk}^2 - \dfrac{T_{...}^2}{nab}$	

■ **EXAMPLE 5.2**   To extend the factorial idea a bit further, consider a problem with three factors. Such a problem was presented in Chapter 1 on the effect of tool type, angle of bevel, and type of cut on power consumption for ceramic tool cutting. Reference to this problem will point out the phases of experiment, design, and analysis as followed in Example 5.1.

It is a $2 \times 2 \times 2$ factorial experiment with four observations per cell run in a completely randomized manner. The mathematical model is

$$Y_{ijkm} = \mu + T_i + B_j + TB_{ij} + C_k + TC_{ik} + BC_{jk} + TBC_{ijk} + \varepsilon_{m(ijk)}$$

where TBC_{ijk} represents a three-way interaction.

The data for this example are given in Table 1.2 and the ANOVA table in Table 1.3. That this analysis is a simple extension of the methods used on Example 5.1 will be shown with the coded data from Table 1.2 (see Table 5.7).

Table 5.7 shows the total for each small cell and the sum of the squares of the cell observations. These results will be useful in doing the ANOVA. By this time we should be able to set up the steps in the analysis without recourse to formulas in dot notation.

First the total sum of squares: Add the squares of all readings (the circled numbers) and subtract a correction term. The correction term is the grand total -13 squared, divided by the number of observations (32):

$$SS_{total} = 441 - \frac{(-13)^2}{32} = 435.72$$

Table 5.7 Coded Ceramic Tool Data of Table 1.2, Code: 2(X–28.0)

	Tool Type				
	1		**2**		
	Bevel Angle		Bevel Angle		
Type of Cut	15°	30°	15°	30°	
Continuous	2	1	0	3	
	−3	1	1	8	
	5　(42)	4　(99)	0　(37)	2　(77)	25
	−2	9	−6	0	
	2	15	−5	13	
Interrupted	0	−2	−7	−1	
	−6	2	−6	0	
	−3　(54)	−1　(10)	0　(101)	−2　(21)	−38
	−3	−1	−4	−4	
	−12	−2	−17	−7	
Totals	−10	13	−22	+6	−13

Tool type sum of squares: Add for each tool type. The totals are +3 and −16; square these and divide by the number of observations per type (16), add these results for both types, and subtract the correction term. Thus,

$$SS_{\text{tool type}} = \frac{3^2 + (-16)^2}{16} - \frac{(-13)^2}{32} = 11.28$$

Bevel angle sum of squares: Same procedure on the totals for each bevel angle, −32 and 19:

$$SS_{\text{bevel angle}} = \frac{(-32)^2 + (19)^2}{16} - \frac{(-13)^2}{32} = 81.28$$

Type of cut sum of squares: Same procedure, with cut totals of 25 and −38:

$$SS_{\text{type of cut}} = \frac{(25)^2 + (-38)^2}{16} - \frac{(-13)^2}{32} = 124.03$$

For the $T \times B$ interaction, ignore type of cut and use cell totals for the $T \times B$ cells. These are −10, 13, −22, +6:

$$SS_{T \times B \text{ interaction}} = \frac{(-10)^2 + (13)^2 + (-22)^2 + (6)^2}{8} - \frac{(-13)^2}{32} - 11.28 - 81.28$$

$$= 0.78$$

For $T \times C$ interaction, ignore bevel angle and the cell totals become 17, -14, 8, -24:

$$SS_{T \times C \text{ interaction}} = \frac{(17)^2 + (-14)^2 + (8)^2 + (-24)^2}{8}$$

$$- \frac{(-13)^2}{32} - 11.28 - 124.03$$

$$= 0.03$$

For $B \times C$ interaction, ignore tool type and the cell totals become -3, 28, -29, -9:

$$SS_{B \times C \text{ interaction}} = \frac{(-3)^2 + (28)^2 + (-29)^2 + (-9)^2}{8}$$

$$- \frac{(-13)^2}{32} - 81.28 - 124.03$$

$$= 3.78$$

For the three-way interaction $T \times B \times C$, consider the totals of the smallest cells, 2, 15, -5, 13, -12, -2, -17, and -7. From this cell sum of squares subtract *not only* the main effect sum of squares *but also* the three two-way interaction sums of squares. Thus,

$$SS_{T \times B \times C \text{ interaction}}$$

$$= \frac{(2)^2 + (15)^2 + (-5)^2 + (13)^2 + (-12)^2 + (-2)^2 + (-17)^2 + (-7)^2}{4}$$

$$- \frac{(-13)^2}{32} - 11.28 - 81.28 - 124.03 - 0.78 - 0.03 - 3.78$$

$$= 0.79$$

By subtraction

$$SS_{\text{error}} = 213.75$$

These results are displayed in Table 5.8. If they are compared with those in Table 1.3, they appear to differ considerably. Actually they give the same F test results, but in Table 5.8 the data were coded involving multiplication by 2; the data of Table 1.3 are uncoded. Multiplication by 2 will multiply the variance or mean square by 4; thus if all mean squares in Table 5.8 are divided by 4, the results are the same as in Table 1.3, e.g., on tool types $11.28 / 4 = 2.82$, and error $213.75 / 4 = 53.44$. It is worth noting that any decoding is unnecessary for determining the F ratios. However, if one wishes confidence limits on the original data or components of variance on the original data, it may be necessary to decode the results.

Table 5.8 ANOVA for Ceramic Tool Problem

Source	df	SS	MS
Tool type T_i	1	11.28	11.28
Bevel angle B_j	1	81.28	81.28
$T \times B$ interaction TB_{ij}	1	0.78	0.78
Type of cut C_k	1	124.03	124.03
$T \times C$ interaction TC_{ik}	1	0.03	0.03
$B \times C$ interaction BC_{jk}	1	3.78	3.78
$T \times B \times C$ interaction TBC_{ijk}	1	0.79	0.79
Error $\varepsilon_{m(ijk)}$	24	213.75	8.91
Totals	31	435.72	

The interpretation of the results of this example is given in Chapter 1. The purpose of presenting it again in this chapter is to show that factorial experiments with three or more factors can easily be analyzed by simple extension of the methods of this chapter.

This problem, of course, could be fed into the computer to determine Table 5.8. ■

5.5 One Observation per Treatment

Since the examples in this chapter have contained several replications within a cell, it would be well to examine a situation involving only one observation per cell. In this case $k = 1$, and the model is written as

$$Y_{ij} = \mu + A_i + B_j + AB_{ij} + \varepsilon_{ij}$$

A glance at the last two terms indicates that we cannot distinguish between the interaction and the error—they are hopelessly confounded. Then the only reasonable situation for running one observation per cell is one in which past experience generally assures us that there is no interaction. In such a case, the model is written as

$$Y_{ij} = \mu + A_i + B_j + \varepsilon_{ij}$$

It may also be noted that this model looks very much like the model for a randomized block design for a single-factor experiment (Chapter 4). In Chapter 4 that model was written as

$$Y_{ij} = \mu + \tau_j + \beta_i + \varepsilon_{ij}$$

Even though the models do look alike and an analysis would be run in the same way, this latter is a single-factor experiment—treatments are the factor—and β_i represents a restriction on the randomization. In the factorial model, there are two factors of interest, A_i and B_j, and the design is completely randomized. It is, however, assumed in the randomized block situation that there is no interaction between treatments and blocks. This is often a more reasonable assumption for blocks and treatments since blocks are often chosen at random. For a two-factor experiment an interaction between A and B may very well be present, and some external information must be available in order to assume that no such interaction exists. An experimenter who is not sure about interaction must take more than one observation per cell and test the hypotheses of no interaction.

As the number of factors increases, however, the presence of higher order interactions is much more unlikely, so it is fairly safe to assume no four-way, five-way, . . . , interactions. Even if these were present, they would be difficult to explain in practical terms.

Following is a suggested set of steps that might be taken in analyzing data from an experiment in which there are several factors and, due to the size of the experiment, only one observation can be afforded per treatment, namely $n = 1$. These steps would usually be taken with the aid of a computer:

1. Analyze the complete factorial including all main effects and all interactions in your model.

2. Examine the mean squares for all interactions and use all three-way and higher interactions as an error term on a second run of the analysis. Some three-way interactions may be left in the model if their mean squares look large compared with those to be eliminated. There is no F test here, for with $n = 1$ there is no error term. A word of caution: If some interactions are to be retained in the model, be sure to include in the model all main effects and lower-order interactions that appear in the interaction to be retained.

3. After a second run, if the "pooled" error term from the first run has adequate degrees of freedom (say df ≥ 6), interpret the F tests as shown.

4. If the degrees of freedom in the error term are not adequate, repeat the analysis, placing all effects (even main effects, sometimes) and all interactions that are not significant at a high α level—say 0.25—in a new error term. Since this "pooling" assumes acceptance of null hypotheses, a large α (≥ 0.25) may help in reducing the β error when hypotheses are accepted.

5. Repeat this procedure until you feel you have an adequate model to explain the phenomena in the experiment.

5.6 Summary

Experiment	Design	Analysis
I. Single factor		
	1. Completely randomized $Y_{ij} = \mu + \tau_j + \varepsilon_{ij}$	1. One-way ANOVA
	2. Complete randomized block $Y_{ij} = \mu + \tau_j + \beta_i + \varepsilon_{ij}$	2. Two-way ANOVA
	3. Complete Latin square $Y_{ijk} = \mu + \beta_i + \tau_j + \gamma_k + \varepsilon_{ijk}$	3. Three-way ANOVA
	4. Graeco–Latin square $Y_{ijkm} = \mu + \beta_i + \tau_j + \gamma_k + \omega_m + \varepsilon_{ijkm}$	4. Four-way ANOVA
II. Two or more factors Factorial (crossed)	1. General case for completely randomized $Y_{ijk} = \mu + A_i + B_j + AB_{ij} + \varepsilon_{k(ij)} \cdots$ for more factors	1. ANOVA with interactions

5.7 SAS Programs for Factorial Experiments

PROC ANOVA can be used to analyze factorial experiments. Consider the data of Table 5.4:

Pump Heater Voltage (V)	Exhaust Index (E)			$T._j$.
	60	90	150	
127	48	28	7	189
	58	33	15	
220	62	14	9	155
	54	10	6	
T_i..	222	85	37	$T_{...} = 344$

The only new concept is the interaction between the factors. To write a model statement to include an interaction between two factors, say A and B, one of two methods can be used: (1) With an asterisk (*) include $A*B$, giving RDG=A B A*B or (2) use an upright "pipe" between the factors giving RDG=A|B, which is equivalent to writing a model with all main effects and all possible interaction effects.

If the analysis then shows no significant interaction, one may treat each main effect separately and, if desired, run an SNK or similar test on the means of each factor if one is interested in the optimum conditions for each factor.

If, on the other hand, the interaction is significant, it makes little sense to analyze the main effects. Instead, one should determine the means of each combination of the main factor levels and then proceed to an SNK test on these "cell" means. We usually will also wish to plot these cell means to give a visual picture of the outcome.

For all of this we create a new variable EV using the **COMPRESS** commands with a "/" to assist in identifying all of the various combinations of the variables E and V. This statement must immediately follow the **INPUT** command. This new variable EV must be added to the **CLASS** statement but it is not part of the **MODEL** statement.

For plotting we use **PROC MEANS** to calculate the cell means, outputting the results to a new data set, which then is used by the **PROC PLOT** to plot the means versus levels of the independent variables. We will sort our variables first by using the following:

```
PROC SORT;
BY E V;
PROC MEANS MEAN;
VAR RDG;
BY E V;
OUTPUT OUT=CELL MEAN=CELLMEAN;
```

The program statements above sort the data (required) prior to calculation of the cell means (RDG). Finally the statement OUTPUT OUT=CELL is telling the computer to create a new data set we call CELL and, in addition to all of the data from the original data set, a new variable called CELLMEAN, which is the result of the **PROC MEANS** procedure. We do this by finishing the statement with MEAN=CELLMEAN; .

We are now ready to plot the results. We do this by requesting the procedure **PLOT** and telling it what data to use as follows:

```
PROC PLOT DATA=CELL;
```

We now tell it specifically what we want plotted on the Y and X axes and what symbols to print.

```
PLOT CELLMEAN*E=V;
            OR
PLOT CELLMEAN*V=E;
```

The first variable indicated will be the Y axis, and the second variable will be the X axis. Since there are three levels of INDEX and two levels of VOLTAGE in our data, six points will be generated and identified by the first character of the variable after the $=$. We must connect the points manually.

Our final program is

```
DATA RAW;
INPUT E V RDG @@;
EV = COMPRESS(E)|| '/' || COMPRESS(V);
CARDS;
    60    127    48
    60    127    58
    60    220    62
    60    220    54
    90    127    28
    90    127    33
    90    220    14
    90    220    10
   150    127     7
   150    127    15
   150    220     9
   150    220     6
    ;
TITLE'VACUUM TUBE EXAMPLE';
PROC PRINT;
PROC ANOVA;
CLASS E V EV;
MODEL RDG=E|V;
MEANS E V EV/SNK;
PROC SORT;
BY E V;
PROC MEANS MEAN;
VAR RDG;
BY E V;
OUTPUT OUT=CELL MEAN=CELLMEAN;
PROC PLOT DATA=CELL;
PLOT CELLMEAN*E=V;
PLOT CELLMEAN*V=E;
```

Our output is as follows:

SAS

OBS	E	V	RDG	EV
1	60	127	48	60/127
2	60	127	58	60/127
3	60	220	62	60/220
4	60	220	54	60/220
5	90	127	28	90/127
6	90	127	33	90/127
7	90	220	14	90/220
8	90	220	10	90/220
9	150	127	7	150/127
10	150	127	15	150/127
11	150	220	9	150/220
12	150	220	6	150/220

SAS

ANALYSIS OF VARIANCE PROCEDURE

CLASS LEVEL INFORMATION

CLASS	LEVELS	VALUES
E	3	60 90 150
V	2	127 220
EV	6	150/127 150/220 60/127 60/220 90/127 90/220

NUMBER OF OBSERVATIONS IN DATA SET = 12

SAS

ANALYSIS OF VARIANCE PROCEDURE

DEPENDENT VARIABLE: RDG

SOURCE	DF	SUM OF SQUARES	MEAN SQUARE	F VALUE	PR > F	R-SQUARE	C.V.
MODEL	5	4987.66666667	997.53333333	43.06	0.0001	0.972887	16.7902
ERROR	6	139.00000000	23.16666667			ROOT MSE	RDG MEAN
CORRECTED TOTAL	11	5126.6666667				4.8131 7636	28.6666667

SOURCE	DF	ANOVA SS	F VALUE	PR > F
E	2	4608.16666667	99.46	0.0001
V	1	96.33333333	4.16	0.0875
E*V	2	283.16666667	6.11	0.0357

SAS

ANALYSIS OF VARIANCE PROCEDURE

STUDENT-NEWMAN-KEULS TEST FOR VARIABLE: RDG

NOTE: THIS TEST CONTROLS THE TYPE I EXPERIMENTWISE ERROR RATE UNDER THE COMPLETE NULL HYPOTHESIS BUT NOT UNDER PARTIAL NULL HYPOTHESES

ALPHA=0.05   DF=6   MSE=23.1667

NUMBER OF MEANS	2	3
CRITICAL RANGE	8.32777	10.4431

MEANS WITH THE SAME LETTER ARE NOT SIGNIFICANTLY DIFFERENT.

SNK	GROUPING	MEAN	N	E
	A	55.500	4	60
	B	21.250	4	90
	C	9.250	4	150

SAS

ANALYSIS OF VARIANCE PROCEDURE

STUDENT—NEWMAN—KEULS TEST FOR VARIABLE: RDG
NOTE: THIS TEST CONTROLS THE TYPE I EXPERIMENTWISE ERROR RATE
      UNDER THE COMPLETE NULL HYPOTHESIS BUT NOT UNDER PARTIAL
      NULL HYPOTHESES

ALPHA=0.05   DF=6   MSE=23.1667

NUMBER OF MEANS	2
CRITICAL RANGE	6.7996

MEANS WITH THE SAME LETTER ARE NOT SIGNIFICANTLY DIFFERENT.

SNK	GROUPING	MEAN	N	V
	A	31.500	6	127
	A			
	A	25.833	6	220

SAS

ANALYSIS OF VARIANCE PROCEDURE

STUDENT—NEWMAN—KEULS TEST FOR VARIABLE: RDG
NOTE: THIS TEST CONTROLS THE TYPE I EXPERIMENTWISE ERROR RATE
      UNDER THE COMPLETE NULL HYPOTHESIS BUT NOT UNDER PARTIAL
      NULL HYPOTHESES

ALPHA=0.05   DF=6   MSE=23.1667

NUMBER OF MEANS	2	3	4	5	6
CRITICAL RANGE	11.7772	14.7687	16.662	18.0545	19.155

MEANS WITH THE SAME LETTER ARE NOT SIGNIFICANTLY DIFFERENT.

SNK	GROUPING	MEAN	N	EV
	A	58.000	2	60/220
	A			
	A	53.000	2	60/127
	B	30.500	2	90/127
	C	12.000	2	90/220
	C			
	C	11.000	2	150/127
	C			
	C	7.500	2	150/220

```
              SAS

VARIABLE            MEAN

----- E=60   V=127 ------

RDG              53.00000000

----- E=60   V=220 ------

RDG              58.00000000

----- E=90   V=127 ------

RDG              30.50000000

----- E=90   V=220 ------

RDG              12.00000000

----- E=150 V=127 ------

RDG              11.00000000

----- E=150 V=220 ------
RDG              7.500000000
```

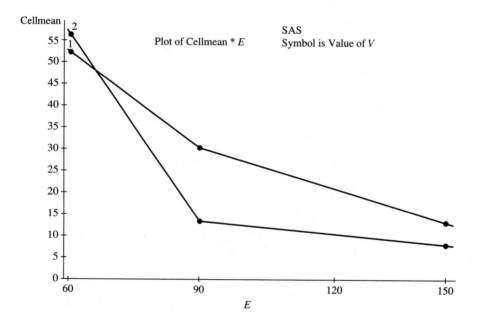

Plot of Cellmean * E

SAS
Symbol is Value of V

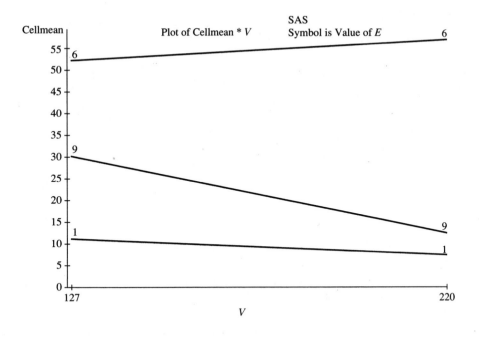

From the printout we can see the same results as shown in Section 5.3, including the SNK test and the graph of Figure 5.8.

Example 5.2 can be programmed as follows from Table 5.7:

```
DATA A;
INPUT TOOL BEVEL CUT RDG @@;
CARDS;
        1   1   1    2
        1   1   1   -3
            etc.
        2   2   2   -4
        ;
TITLE '3-WAY INTERACTION MODEL';
PROC PRINT;
PROC GLM; (or PROC ANOVA)
CLASS TOOL BEVEL CUT;
MODEL RDG=TOOL|BEVEL|CUT;
```

The output is:

```
3-WAY INTERACTION MODEL
OBS        TOOL      BEVEL      CUT      RDG
  1          1          1        1        2
  2          1          1        1       -3
  3          1          1        1        5
  4          1          1        1       -2
  5          1          1        2        0
  6          1          1        2       -6
  7          1          1        2       -3
  8          1          1        2       -3
  9          1          2        1        1
 10          1          2        1        1
 11          1          2        1        4
 12          1          2        1        9
 13          1          2        2       -2
 14          1          2        2        2
 15          1          2        2       -1
 16          1          2        2       -1
 17          2          1        1        0
 18          2          1        1        1
 19          2          1        1        0
 20          2          1        1       -6
 21          2          1        2       -7
 22          2          1        2       -6
 23          2          1        2        0
 24          2          1        2       -4
 25          2          2        1        3
 26          2          2        1        8
 27          2          2        1        2
 28          2          2        1        0
 29          2          2        2       -1
 30          2          2        2        0
 31          2          2        2       -2
 32          2          2        2       -4
```

3-WAY INTERACTION MODEL

GENERAL LINEAR MODELS PROCEDURE

DEPENDENT VARIABLE: RDG

SOURCE	DF	SUM OF SQUARES	MEAN SQUARE	F VALUE	PR > F	R-SQUARE	C.V.
MODEL	7	221.96875000	31.70982143	3.56	0.0091	0.509431	734.6053
ERROR	24	213.75000000	8.90625000		ROOT MSE		RDG MEAN
CORRECTED TOTAL	31	435.71875000			2.98433410		-0.40625000

SOURCE	DF	TYPE I SS	F VALUE	PR > F	DF	TYPE III SS	F VALUE	PR > F
TOOL	1	11.28125000	1.27	0.2715	1	11.28125000	1.27	0.2715
BEVEL	1	81.28125000	9.13	0.0059	1	81.28125000	9.13	0.0059
TOOL*BEVEL	1	0.78125000	0.09	0.7696	1	0.78125000	0.09	0.7696
CUT	1	124.03125000	13.93	0.0010	1	124.03125000	13.93	0.0010
TOOL*CUT	1	0.03125000	0.00	0.9533	1	0.03125000	0.00	0.9533
BEVEL*CUT	1	3.78125000	0.42	0.5209	1	3.78125000	0.42	0.5209
TOOL*BEVEL*CUT	1	0.78125000	0.09	0.7696	1	0.78125000	0.09	0.7696

These results agree with those of Table 5.8. No provision was made here for analyzing interaction because no significant interaction was found. Had interaction shown up, the program should be repeated and the necessary steps in Example 5.1 inserted.

Also note that if no error term is available ($n = 1$), some interactions should be used as error, in which case the model desired must be spelled out. One cannot use RDG=A|B|C, but if $A*B*C$ is to be used as error, one must write RDG=A B A*B C A*C B*C.

Problems

5.1 To determine the effect of two glass types and three phosphor types on the light output of a television tube, light output is measured by the current required in series with the tube to produce 30 foot-lamberts of light output. Thus the higher the current is in microamperes, the poorer the tube is in light output. Three observations were taken under each of the six treatment conditions and the experiment was completely randomized. The following data were recorded.

	Phosphor Type		
Glass Type	*A*	*B*	*C*
1	280	300	270
	290	310	285
	285	295	290
2	230	260	220
	235	240	225
	240	235	230

Do an ANOVA on these data and test the effect of glass type, phosphor types, and interaction on the current flow.

5.2 Plot the results of Problem 5.1 to show that your conclusions are reasonable.

5.3 For any significant effects in Problem 5.1 test further between the levels of the significant factors.

5.4 Based on the results in Problems 5.1–5.3, what glass type and phosphor type (or types) would you recommend if a low current is most desirable?

5.5 Adhesive force on gummed material was determined under three fixed humidity and three fixed temperature conditions. Four readings were made under each set of conditions. The experiment was completely randomized and the results set out in an ANOVA table as follows:

Source	df	SS	MS
Humidity		9.07	
Temperature		8.66	
$H \times T$ interaction		6.07	
Error			
Total		52.30	

Complete this table.

5.6 For the data in Problem 5.5 test all indicated hypotheses and state your conclusions.

5.7 Set up a mathematical model for the experiment in Problem 5.5 and indicate the hypotheses to be tested in terms of your model.

5.8 The object of an experiment is to determine thrust forces in drilling at different speeds and feeds, and in different materials. Five speeds and three feeds are used, and two materials with two samples are tested under each set of conditions. The order of the experiment is completely randomized and the levels of all factors are fixed. The following data are recorded on thrust forces after subtracting 200 from all readings.

Material	Feed	Speed				
		100	220	475	715	870
B_{10}	0.004	122	108	108	66	80
		110	85	60	50	60
	0.008	332	276	248	248	276
		330	310	295	275	310
	0.014	640	612	543	612	696
		500	500	450	610	610
V_{10}	0.004	192	136	122	108	136
		170	130	85	75	75
	0.008	386	333	318	472	499
		365	330	330	350	390
	0.014	810	779	810	893	1820
		725	670	750	890	890

Do a complete analysis of this experiment and state your conclusions.

5.9 Plot any results in Problem 5.8 that are significant.

5.10 Set up tests on means where suitable and draw conclusions from Problem 5.8.

5.11 In an experiment for testing rubber materials interest centered on the effect of the mix (A, B, or C), the laboratory involved (1, 2, 3, or 4), and the temperature (145,

155, 165°C) on the time in minutes to a 2-in.-lb rise above the minimum time. Assuming a completely randomized design, do an ANOVA on the following data.

		Temperature (°C)								
		145			155			165		
		Mix			Mix			Mix		
Laboratory	A	B	C	A	B	C	A	B	C
1	11.2	11.2	11.5	6.7	6.8	7.0	4.8	4.8	5.0
	11.1	11.5	11.4	6.8	6.7	7.0	4.8	4.9	4.9
2	11.8	12.3	12.3	7.3	7.5	7.5	5.3	5.4	5.3
	11.8	12.3	11.9	7.2	7.7	7.3	5.3	5.2	5.3
3	11.5	12.3	12.7	6.6	7.1	7.8	5.0	5.3	5.2
	11.6	12.0	12.5	6.9	7.2	7.3	5.0	5.0	5.0
4	11.5	11.8	12.7	7.2	6.7	7.1	4.5	4.7	4.5
	11.3	11.7	12.7	6.9	7.0	7.0	4.6	4.5	4.5

5.12 What is there in the results of Problem 5.11 that would lead you to question some assumption about the experiment?

5.13 Tomato plants were grown in a greenhouse under treatments consisting of combinations of soil type (factor A) and fertilizer type (factor B). A completely randomized two-factor design was used with two replications per cell. The following data on the yield Y (in kilograms) of tomatoes were obtained for the 30 plants under study.

Soil	Fertilizer Type (B)			
Type (A)	1	2	3	T_i.
I	5, 7	5, 5	3, 5	30
II	5, 9	1, 3	2, 2	22
III	6, 8	4, 8	2, 4	32
IV	7, 11	7, 9	3, 7	44
V	6, 9	4, 6	3, 5	33
$T._j$	73	52	36	161

Note: $\sum_k \sum_j \sum_i Y_{ijk}^2 = 1043$.

Complete the following ANOVA table.

Source	df	Sum of Squares	Mean Square
Soil type (A)			
Fertilizer type (B)			
Interaction (AB)			
Error			

5.14 Carry out tests on main effects and interactions *in an appropriate order,* using levels of significance for each test that seem appropriate to you. For each test state H_0 and H_1, show the rejection region, and give your conclusion. After doing all tests, summarize your results in words.

5.15 On the basis of your results in Problem 5.14, state how you would carry out Newman–Keuls comparisons to find out the best combination (or combinations) of soil type and fertilizer type to achieve maximum mean yield of tomatoes. Then carry out these comparisons (at an α of 0.05) and state your conclusions.

5.16 Each of the following tables represents the cell means $(\bar{Y}_{ij}.)$ for a two-factor completely randomized balanced experiment. For each table tell which (if any) of the following SS's would be zero for that table: SS_A, SS_B, SS_{AB}. (There may be none; there may be more than one.)

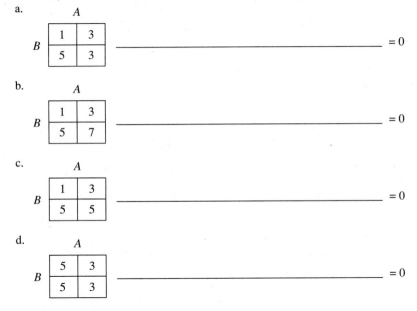

a. A

B	1	3
	5	3

_____ $= 0$

b. A

B	1	3
	5	7

_____ $= 0$

c. A

B	1	3
	5	5

_____ $= 0$

d. A

B	5	3
	5	3

_____ $= 0$

5.17 In each two-factor design table below the numbers in the cells of the table are the population means of the observations for those cells. For each table, indicate whether or not there is interaction between the factors and justify your assertion.

a.

		Factor B			
		1	2	3	4
Factor A	1	7	6	5	2
	2	9	8	7	4

Is there interaction? *Yes/No* (circle one). Explain.

b.

Factor B

		1	2	3
Factor A	1	7	6	5
	2	5	6	7

Is there interaction? *Yes/No* (circle one). Explain.

c.

Factor B

		1	2
	1	8	10
Factor A	2	7	9
	3	4	7

Is there interaction? *Yes/No* (circle one). Explain.

5.18 An industrial engineer presented two types of stimuli (two-dimensional and three-dimensional films) of two different jobs (1 and 2) to each of five analysts. Each analyst was presented each job-stimulus film twice, and the order of the whole experiment was considered completely randomized. The engineer was interested in the consistency of analyst ratings of four sequences within each job stimulus presentation. The variable recorded is the log variance of the four sequences since log variance is more likely to be normally distributed than the variance. Data showed:

	Stimulus			
	Two-Dimensional		Three-Dimensional	
Analyst	*Job 1*	*Job 2*	*Job 1*	*Job 2*
1	1.42	1.40	1.00	0.92
	1.25	1.44	0.90	0.93
2	1.59	1.83	1.46	1.43
	2.09	2.02	1.68	1.02
3	1.26	1.80	0.85	1.33
	1.48	1.75	1.43	1.48
4	1.76	0.98	1.21	0.73
	1.47	1.23	1.25	1.22
5	1.43	1.17	1.60	1.31
	1.72	1.18	2.03	1.12

Do a complete analysis and summarize your results.

5.19 Make further tests on Problem 5.18 as suggested by the ANOVA results.

5.20 The following results were reported on a study of "factors that affect the salary of high-school teachers of commercial subjects":

Source	df	SS
Sex	1	239,763
Size of school	2	423,056
Years in position	5	564,689
Sex by size interaction	2	18,459
Sex × years interaction	5	85,901
Size × years interaction	10	240,115
Sex × size × years interaction	10	151,394
Within classes	153	1,501,642

Indicate the mathematical model for the above study. Show a possible data layout. Complete the ANOVA table and comment on any significant results and interpret.

5.21 Four factors are studied for their effect on the luster of plastic film. These factors are (1) film thickness (1 or 2 mils); (2) drying conditions (regular or special); (3) length of wash (20, 30, 40, or 60 minutes); and (4) temperature of wash (92°C or 100°C). Two observations of film luster are taken under each set of conditions. Assuming complete randomization, analyze the following data.

	Regular Dry				Special Dry			
Minutes	**92°C**		**100°C**		**92°C**		**100°C**	
1-mil Thickness								
20	3.4	3.4	19.6	14.5	2.1	3.8	17.2	13.4
30	4.1	4.1	17.5	17.0	4.0	4.6	13.5	14.3
40	4.9	4.2	17.6	15.2	5.1	3.3	16.0	17.8
60	5.0	4.9	20.9	17.1	8.1	4.3	17.5	13.9
2-mil Thickness								
20	5.5	3.7	26.6	29.5	4.5	4.5	25.6	22.5
30	5.7	6.1	31.6	30.2	5.9	5.9	29.2	29.8
40	5.6	5.6	30.5	30.2	5.5	5.8	32.6	27.4
60	7.2	6.0	31.4	29.6	8.0	9.9	33.5	29.5

5.22 Plot and discuss any significant interactions found in Problem 5.21. What conditions would you recommend for maximum luster and why?

5.23 To study the effect of four types of wax applied to floors for three different lengths of polishing times it was decided to try each combination twice and to completely randomize the order in which floors received which treatments. The criterion was a gloss index.
  a. Outline an ANOVA table for this problem with proper degrees of freedom.
  b. Explain, in words, what a significant interaction would mean in this situation.

5.24 A manufacturer of combustion engines is concerned about the percentage of smoke emitted by the engine in order to meet EPA standards. An experiment was conducted involving engines with three timing levels (T), three throat diameters (D), two volume ratios in the combustion chamber (V), and two injection systems (I). Thirty-six engines were designed to represent all possible combinations of these four factors and the engines were then operated in a completely random order and percent smoke recorded to the nearest 0.1 percent. Results showed:

							T					
				1			2			3		
				D			D			D		
				1	2	3	1	2	3	1	2	3
	1	I	1	0.4	3.8	7.0	1.8	4.8	8.1	1.7	4.9	8.2
			2	0.5	1.6	4.0	0.7	3.0	5.6	1.2	3.2	10.0
V												
	2	I	1	0.7	1.8	6.0	1.4	1.0	1.5	7.0	3.1	3.4
			2	0.1	0.1	0.2	0.6	0.4	1.4	4.0	1.8	2.0

Do a complete analysis of these data and make recommendations as to which combination or combinations of factor levels should give a minimum percent of smoke emitted.

5.25 Students' scores on a psychomotor test were recorded for various sized targets at which the student must aim, different machines used, and the level of background illumination. Results are shown in the following ANOVA table.

Source	df	SS	MS
Target size (A)	3	235.200	78.400
Machine (B)	2	86.467	43.233
Level of illumination (C)	1	76.800	76.800
$A \times B$	6	104.200	17.367
$A \times C$	3	93.867	31.289
$B \times C$	2	12.600	6.300
$A \times B \times C$	6	174.333	29.056
Within sets	96	1198.000	12.478
Totals	119	1981.467	

a. Using a 5 percent significance level indicate which effects are statistically significant.
b. Assuming that you had the original data at your disposal, explain briefly the next steps you would take in analyzing the data from this experiment.
c. Devise a data layout sheet that could be used to collect each item of data for this experiment.

d. Assuming the mean score for target size A_3, machine B_2, and level of illumination C_1 is 9.6, set 90 percent confidence limits on the true mean for this treatment combination.

5.26 When Problem 3.19 on bonding machines was run, the bonding took place in three different positions (A, B, C) on the piece being bonded. Data including this position factor follow:

Position	Bonder			
	#41	#42	#43	#44
A	204	197	264	248
	181	223	226	138
	201	206	228	273
	203	232	249	220
	214	213	246	186
B	262	207	255	304
	246	259	186	330
	230	223	237	268
	256	195	236	295
	288	197	240	276
C	220	214	215	208
	232	248	176	248
	235	191	171	247
	231	197	208	220
	220	222	180	241

Analyze these data and discuss the results.

5.27 To determine the effect of cool-zone oxygen level and preheat oxygen level on the coefficient of variation in resistance on a 20-k Ω resistor, five cool-zone oxygen levels and two preheat oxygen levels were used. Three observations were taken per experiment. The results follow:

		Cool-Zone Oxygen Level (in ppm)				
		5	10	15	20	25
Preheat Oxygen (in ppm)	5	6.45	2.51	4.71	11.47	10.69
		2.71	4.42	4.91	9.31	9.25
		3.08	4.20	8.19	12.04	10.00
	25	2.86	5.43	5.35	8.92	13.13
		5.51	8.37	4.20	7.57	12.01
		5.66	3.54	6.49	7.14	11.71

Analyze these data and report your recommendations, assuming that a low coefficient of variation is desirable.

Fixed, Random, and Mixed Models   6

6.1 Introduction

In Chapter 1 it was pointed out that, in the planning stages of an experiment, the experimenter must decide whether the levels of factors considered are to be set at fixed values or are to be chosen at random from many possible levels. In the intervening chapters it has always been assumed that the factor levels were fixed. In practice it may be necessary to choose the levels of the factors at random, depending on the objectives of the experiment. Are the results to be judged for these levels alone, or are they to be extended to more levels of which those in the experiment are but a random sample? In the case of some factors such as temperature, time, or pressure, it is usually desirable to pick fixed levels, often near the extremes and at some intermediate points, because a random choice might not cover the range in which the experimenter is interested. In such cases of fixed quantitative levels, we often feel safe in interpolating between the fixed levels chosen. Other factors such as operators, days, or batches may often be only a small sample of all possible operators, days, or batches. In such cases the particular operator, day, or batch may not be very important but only whether or not operators, days, or batches in general increase the variability of the experiment.

It is not reasonable to decide after the data have been collected whether the levels are to be considered fixed or random. This decision must be made prior to the running of the experiment, and if random levels are to be used, they must be chosen from all possible levels by a random process. In the case of random levels, it will be assumed that the levels are chosen from an infinite population of possible levels. Bennett and Franklin [4] discuss a case in which the levels chosen are from a finite set of possible levels.

When all levels are fixed, the mathematical model of the experiment is called a *fixed model*. When all levels are chosen at random, the model is called a *random model*. When several factors are involved, some at fixed levels and others at random levels, the model is called a *mixed model*.

6.2 Single-Factor Models

In the case of a single-factor experiment the factor may be referred to as a *treatment effect,* as in Chapter 3; and if the design is completely randomized, the model is

$$Y_{ij} = \mu + \tau_j + \varepsilon_{ij} \tag{6.1}$$

Whether the treatment levels are fixed or random, it is assumed in this model that μ is a fixed constant and the errors are normally and independently distributed with a zero mean and the same variance, that is, ε_{ij} are NID $(0, \sigma_\varepsilon^2)$. The decision as to whether the levels of the treatment are fixed or random will affect the assumptions about the treatment term τ_j. The different assumptions and other differences will be compared in parallel columns.

Fixed Model

1. Assumptions: τ_j's are fixed constants.

$$\sum_{j=1}^{k} \tau_j = \sum_{j=1}^{k} (\mu_j - \mu) = 0$$

(These add to zero as they are the only treatment means being considered.)

Random Model

1. Assumptions: τ_j's are random variables and are

$$\text{NID}\,(0, \sigma_\tau^2)$$

(Here σ_τ^2 represents the variance among all possible τ_j's or among the true treatment means $\mu_{.j}$. The τ_j average to zero when averaged over all possible levels, but for the k levels of the experiment they usually will not average 0.)

Figure 6.1(a) shows three fixed means whose average is μ as these are the only means of concern and $\sum_j \tau_j = \sum_j (\mu_j - \mu) = 0$. Figure 6.1(b) shows three random means whose average is obviously not μ as these are but three means chosen at random from many possible means. These means and their corresponding τ_j's are assumed to form a normal distribution with a standard deviation of σ_τ.

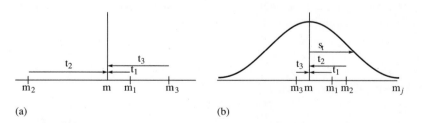

(a)   (b)

Figure 6.1   Assumed means in (a) fixed and (b) random models.

2. Analysis: Procedures as given in Chapter 3 for computing SS.

3. EMS:

Source	df	EMS
τ_j	$k - 1$	$\sigma_\varepsilon^2 + n\phi_\tau$
ε_{ij}	$k(n - 1)$	σ_ε^2

4. Hypothesis tested:

$$H_0: \tau_j = 0 \text{ (for all } j)$$

2. Analysis: Same as for fixed model.

3. EMS:

Source	df	EMS
τ_j	$k - 1$	$\sigma_\varepsilon^2 + n\sigma_\tau^2$
ε_{ij}	$k(n - 1)$	σ_ε^2

4. Hypothesis tested:

$$H_0: \sigma_\tau^2 = 0$$

The expected mean square (EMS) column turns out to be extremely important in more complex experiments as an aid in deciding how to set up an F test for significance. The EMS for any term in the model is the long-range average of the calculated mean square when the Y_{ij} from the model is substituted in algebraic form into the mean square computation. The derivation of these EMS values is often complicated, but those for the single-factor model were derived in Chapter 3, and those for two-factor models are derived in a later section of this chapter.

For the fixed model, if the hypothesis is true that $\tau_j = 0$ for all j, that is, all the k fixed treatment means are equal, then $\sum_j \tau_j^2 = 0$ and the EMS for τ_j and ε_{ij} are both σ_ε^2. Hence the observed mean squares for treatments and error mean square are both estimates of the error variance, and they can be compared by means of an F test. If this F test shows a significantly high value, it must mean that $n \sum_j \tau_j^2 / (k - 1) = n\phi_\tau$ is not zero and the hypothesis is to be rejected.

For the random model, if the hypothesis is true that $\sigma_\tau^2 = 0$, that is, the variance among all treatment means is zero, then again each mean square is an estimate of the error variance. Again, an F test between the two mean squares is appropriate.

From the two tables in step 3 above, it is seen that for a single-factor experiment there is no difference in the test to be made after the analysis, and the only difference is in the generality of the conclusions. If H_0 is rejected, there is probably a difference between the k fixed treatment means for the fixed model; for the random model there is a difference between all treatments of which the k examined are but a random sample.

6.3 Two-Factor Models

For two factors A and B the model in the general case is

$$Y_{ijk} = \mu + A_i + B_j + AB_{ij} + \varepsilon_{k(ij)}$$

with

$$i = 1, 2, \ldots, a \qquad j = 1, 2, \ldots, b \qquad k = 1, 2, \ldots, n$$

provided the design is completely randomized. In this model, it is again assumed that μ is a fixed constant and $\varepsilon_{k(ij)}$'s are NID $(0, \sigma_\varepsilon^2)$. If both A and B are at fixed levels, the model is a fixed model. If both are at random levels, the model is a random model, and if one is at fixed levels and the other at random levels, the model is a mixed model. Comparing each of these models gives:

Fixed	*Random*	*Mixed*
1. Assumptions: A_i's are fixed constants and $$\sum_{i=1}^{a} A_i = 0$$	1. Assumptions: A_i's are NID $(0, \sigma_A^2)$	1. Assumptions: A_i's are fixed and $$\sum_{i}^{a} A_i = 0$$
B_j's are fixed constants and $$\sum_{j=1}^{b} B_j = 0$$	B_j's are NID $(0, \sigma_B^2)$	B_j's are NID $(0, \sigma_B^2)$
AB_{ij}'s are fixed constants and $$\sum_{i} AB_{ij} = 0$$ $$\sum_{j} AB_{ij} = 0$$	AB_{ij}'s are NID $(0, \sigma_{AB}^2)$	AB_{ij}'s are NID $(0, \sigma_{AB}^2)$ but $$\sum_{i}^{a} AB_{ij} = 0$$ $$\sum_{j}^{b} AB_{ij} \neq 0$$ (for A fixed, B random)
2. Analysis: Procedures of Chapter 5 for sums of squares	2. Analysis: Same	2. Analysis: Same
3. EMS:	3. EMS:	3. EMS:

Source	df	EMS (Fixed)	EMS (Random)	EMS (Mixed)
A_i	$a - 1$	$\sigma_\varepsilon^2 + nb\phi_A$	$\sigma_\varepsilon^2 + n\sigma_{AB}^2 + nb\sigma_A^2$	$\sigma_\varepsilon^2 + n\sigma_{AB}^2 + nb\phi_A$
B_j	$b - 1$	$\sigma_\varepsilon^2 + na\phi_B$	$\sigma_\varepsilon^2 + n\sigma_{AB}^2 + na\sigma_B^2$	$\sigma_\varepsilon^2 + na\sigma_B^2$
AB_{ij}	$(a - 1)(b - 1)$	$\sigma_\varepsilon^2 + n\phi_{AB}$	$\sigma_\varepsilon^2 + n\sigma_{AB}^2$	$\sigma_\varepsilon^2 + n\sigma_{AB}^2$
$\varepsilon_{k(ij)}$	$ab(n - 1)$	σ_ε^2	σ_ε^2	σ_ε^2

4. Hypotheses tested:

H_1: $A_i = 0$ for all i

H_2: $B_j = 0$ for all j

H_3: $AB_{ij} = 0$ for all i and j

4. Hypotheses tested:

H_1: $\sigma_A^2 = 0$

H_2: $\sigma_B^2 = 0$

H_3: $\sigma_{AB}^2 = 0$

4. Hypotheses tested:

H_1: $A_i = 0$ for all i

H_2: $\sigma_B^2 = 0$

H_3: $\sigma_{AB}^2 = 0$

In the assumptions for the mixed model the fact that summing the interaction term over the fixed factor ($\sum_i$) is zero but summing it over the random factor ($\sum_j$) is not zero affects the expected mean squares, as seen in item 3 on page 159.

For the fixed model the mean squares for A, B, and AB are each compared with the error mean square to test the respective hypotheses, as should be clear from an examination of the EMS column when the hypotheses are true. For the random model the third hypothesis of no interaction is tested by comparing the mean square for interaction with the mean square for error, but the first and second hypotheses are each tested by comparing the mean square for the main effect (A_i or B_j) with the mean square for the interaction as seen by their expected mean square values. For a mixed model the interaction hypothesis is tested by comparing the interaction mean square with the error mean square. The random effect B_j is also tested by comparing its mean square with the error mean square. The fixed effect (A_i), however, is tested by comparing its mean square with the interaction mean square.

From these observations on a two-factor experiment, the importance of the EMS column is evident, as this column can be used to see how the tests of hypotheses should be run. It is also important to note that these EMS expressions can be determined prior to the running of the experiment. This will indicate whether or not a valid test of a hypothesis exists. In some cases the proper test indicated by the EMS column will have insufficient degrees of freedom to be sufficiently sensitive, in which case the investigator might wish to change the experiment. This would involve such changes as a choice of more levels of some factors, or changing from random to fixed levels of some factors.

6.4 EMS Rules

The two examples above have shown the importance of the EMS column in determining what tests of significance are to be run after the analysis is completed. Because of the importance of this EMS column in these and more complex models, it is often useful to have some simple method of determining these values from the model for the given experiment. A set of rules can be stated for the balanced case that will determine the EMS column very rapidly, without recourse to their derivation. The rules will be illustrated on the two-factor mixed model of Section 6.3. To determine the EMS column for any model:

1. Write the variable terms in the model as row headings in a two-way table.

A_i
B_j
AB_{ij}
$\varepsilon_{k(ij)}$

2. Write the subscripts in the model as column headings; over each subscript write F if the factor levels are fixed, R if random. Also write the number of observations each subscript is to cover.

	a	b	n
	F	R	R
	i	j	k
A_i			
B_j			
AB_{ij}			
$\varepsilon_{k(ij)}$			

3. For each row (each term in the model) copy the number of observations under each subscript, providing the subscript does not appear in the row heading.

	a	b	n
	F	R	R
	i	j	k
A_i		b	n
B_j	a		n
AB_{ij}			n
$\varepsilon_{k(ij)}$			

subscripts in ( )

4. For any parenthetical subscripts in the model, place a 1 under those subscripts that are inside the parentheses.

	a F i	b R j	n R k
A_i		b	n
B_j	a		n
AB_{ij}			n
$\varepsilon_{k(ij)}$	1	1	

5. Fill the remaining cells with a 0 or a 1, depending upon whether the subscript represents a fixed F or a random R factor.

	a F i	b R j	n R k
A_i	0	b	n
B_j	a	1	n
AB_{ij}	0	1	n
$\varepsilon_{k(ij)}$	1	1	1

n values in each combination

6. To find the expected mean square for any term in the model:

   a. Cover the entries in the column (or columns) that contain nonparenthetical subscript letters in this term in the model (for example, for A_i, cover column i; for $\varepsilon_{k(ij)}$, cover column k).

   b. Multiply the remaining numbers in each row. Each of these products is the coefficient for its corresponding term in the model, provided the subscript on the term is also a subscript on the term whose expected mean square is being determined. The sum of these coefficients multiplied by the variance of their corresponding terms (ϕ_τ or σ_τ^2) is the EMS of the term being considered (for example, for A_i, cover column i). The products of the remaining coefficients are bn, n, n, and 1, but the first n is not used, as there is no i in its term (B_j). The resulting EMS is then $bn\phi_A + n\sigma_{AB}^2 + 1 \cdot \sigma_\varepsilon^2$. For all terms, these rules give:

	a F i	b R j	n R k	EMS
A_i	0	b	n	$\sigma_\varepsilon^2 + n\sigma_{AB}^2 + nb\phi_A$
B_j	a	1	n	$\sigma_\varepsilon^2 + na\sigma_B^2$
AB_{ij}	0	1	n	$\sigma_\varepsilon^2 + n\sigma_{AB}^2$
$\varepsilon_{k(ij)}$	1	1	1	σ_ε^2

Fixed Term use Term ϕ Random use σ^2

These results are seen to be in agreement with the EMS values for the mixed model in Section 6.3. Here ϕ_A is, of course, a fixed type of variance

$$\phi_A = \frac{\sum_i A_i^2}{a - 1}$$

Although the rules seem rather involved, they become very easy to use with a bit of practice. Two examples will illustrate the concept.

■ **EXAMPLE 6.1**   The viscosity of a slurry is to be determined by four randomly selected laboratory technicians. Material from each of five mixing machines is bottled and divided in such a way as to provide two samples for each technician to test for viscosity. These are the only mixing machines of interest and the samples can be presented to the technicians in a completely randomized order.

The model here assumes four random technicians and five fixed mixing machines, and each technician measures samples of each machine twice. The model is shown as the first column of Table 6.1, and the remainder of the table shows how the EMS column is determined.

Table 6.1 EMS for Example 6.1

Source	df	4 R i	5 F j	2 R k	EMS
T_i	3	1	5	2	$\sigma_\varepsilon^2 + 10\sigma_T^2$
M_j	4	4	0	2	$\sigma_\varepsilon^2 + 2\sigma_{TM}^2 + 8\phi_M$
TM_{ij}	12	1	0	2	$\sigma_\varepsilon^2 + 2\sigma_{TM}^2$
$\varepsilon_{k(ij)}$	20	1	1	1	σ_ε^2

The proper F tests are quite obvious from Table 6.1 and all tests have adequate degrees of freedom for a reasonable test. ■

■ **EXAMPLE 6.2**   An industrial engineering student wished to determine the effect of five different clearances on the time required to position and assemble mating parts. As all such experiments involve operators, it was natural to consider a random sample of operators to perform the experiment. He also decided the part should be assembled directly in front of the operator and at arm's length from the operator. He tried four different angles, from 0° directly in front of the operator through 30°, 60°, and 90° from this position. Thus four factors were involved, any one of which might affect the time required to position and assemble the part. The experimenter decided to replicate each set up six times and to randomize completely the order of experimentation. Here operators O_i were at random levels (six being chosen), angles A_j at four fixed levels (0°, 30°, 60°, 90°), clearances C_k at five fixed levels, and locations L_m fixed either in front of or at arm's length from the operator. This is a $6 \times 4 \times 5 \times 2$ factorial experiment with six replications, run in a completely randomized design. The expected mean square values can be determined from the rules given in Section 6.4 as shown in Table 6.2.

Table 6.2 EMS for Clearance Problem

Source	df	6 R i	4 F j	5 F k	2 F m	6 R q	EMS
O_i	5	1	4	5	2	6	$\sigma_\varepsilon^2 + 240\sigma_O^2$
A_j	3	6	0	5	2	6	$\sigma_\varepsilon^2 + 60\sigma_{OA}^2 + 360\phi_A$
OA_{ij}	15	1	0	5	2	6	$\sigma_\varepsilon^2 + 60\sigma_{OA}^2$
C_k	4	6	4	0	2	6	$\sigma_\varepsilon^2 + 48\sigma_{OC}^2 + 288\phi_C$
OC_{ik}	20	1	4	0	2	6	$\sigma_\varepsilon^2 + 48\sigma_{OC}^2$
AC_{jk}	12	6	0	0	2	6	$\sigma_\varepsilon^2 + 12\sigma_{OAC}^2 + 72\phi_{AC}$
OAC_{ijk}	60	1	0	0	2	6	$\sigma_\varepsilon^2 + 12\sigma_{OAC}^2$
L_m	1	6	4	5	0	6	$\sigma_\varepsilon^2 + 120\sigma_{OL}^2 + 720\phi_L$
OL_{im}	5	1	4	5	0	6	$\sigma_\varepsilon^2 + 120\sigma_{OL}^2$
AL_{jm}	3	6	0	5	0	6	$\sigma_\varepsilon^2 + 30\sigma_{OAL}^2 + 180\phi_{AL}$
OAL_{ijm}	15	1	0	5	0	6	$\sigma_\varepsilon^2 + 30\sigma_{OAL}^2$
CL_{km}	4	6	4	0	0	6	$\sigma_\varepsilon^2 + 24\sigma_{OCL}^2 + 144\phi_{CL}$
OCL_{ikm}	20	1	4	0	0	6	$\sigma_\varepsilon^2 + 24\sigma_{OCL}^2$
ACL_{jkm}	12	6	0	0	0	6	$\sigma_\varepsilon^2 + 6\sigma_{OACL}^2 + 36\phi_{ACL}$
$OACL_{ijkm}$	60	1	0	0	0	6	$\sigma_\varepsilon^2 + 6\sigma_{OACL}^2$
$\varepsilon_{q(ijkm)}$	1200	1	1	1	1	1	σ_ε^2

From this table it is easily seen that all interactions involving operators and the operator main effect are tested against the error mean square at the bottom of the table. All interactions and main effects involving fixed factors are tested by the mean square just below them in the table. ∎

The rules given in this section are generally enough to be applied to the most complex designs, as will be seen in later chapters.

6.5 EMS Derivations

Single-Factor Experiment

The EMS expressions for a single-factor experiment were derived in Section 3.6.

Two-Factor Experiment

Using the definitions of expected values given in Chapter 2 and the procedures of Section 3.6 the EMS expressions may be derived for the two-factor experiment.

For a two-factor experiment, the model is

$$Y_{ijk} = \mu + A_i + B_j + AB_{ij} + \varepsilon_{k(ij)} \tag{6.2}$$

with

$$i = 1, 2, \ldots, a \qquad j = 1, 2, \ldots, b \qquad k = 1, 2, \ldots, n$$

The sum of squares for factor A is

$$SS_A = \sum_{i=1}^{a} nb(\bar{Y}_{i..} - \bar{Y}_{...})^2$$

From the model in Equation (6.2),

$$\bar{Y}_{i..} = \sum_{j}^{b} \sum_{k}^{n} Y_{ijk} / bn = \sum_{j}^{b} \sum_{k}^{n} (\mu + A_i + B_j + AB_{ij} + \varepsilon_{k(ij)}) / bn$$

$$\bar{Y}_{i..} = \mu + A_i + \sum_{j}^{b} B_j / b + \sum_{j}^{b} AB_{ij} / b + \sum_{j}^{b} \sum_{k}^{n} \varepsilon_{k(ij)} / bn$$

$$\bar{Y}_{...} = \sum_{i}^{a} \sum_{j}^{b} \sum_{k}^{n} Y_{ijk} / abn$$

$$= \sum_{i}^{a} \sum_{j}^{b} \sum_{k}^{n} (\mu + A_i + B_j + AB_{ij} + \varepsilon_{k(ij)}) / abn$$

$$\bar{Y}_{...} = \mu + \sum_{i}^{a} A_i / a + \sum_{j}^{b} B_j / b + \sum_{i}^{a} \sum_{j}^{b} AB_{ij} / ab + \sum_{i}^{a} \sum_{j}^{b} \sum_{k}^{n} \varepsilon_{k(ij)} / nab$$

Subtracting gives

$$\bar{Y}_{i..} - \bar{Y}_{...} = \left(A_i - \sum_{i}^{a} A_i / a \right) + \left(\sum_{j}^{b} AB_{ij} / b - \sum_{i}^{a} \sum_{j}^{b} AB_{ij} / ab \right)$$

$$+ \left(\sum_{j}^{b} \sum_{k}^{n} \varepsilon_{k(ij)} / bn - \sum_{i}^{a} \sum_{j}^{b} \sum_{k}^{n} \varepsilon_{k(ij)} / abn \right)$$

Note that the B effect cancels out of the A sum of squares as it should, since A

and B are orthogonal effects in a factorial experiment. Squaring and adding gives

$$SS_A = nb \sum_{i=1}^{a} \left(A_i - \sum_i A_i / a \right)^2$$

$$+ nb \sum_{i=1}^{a} \left(\sum_j^b AB_{ij} / b - \sum_i^a \sum_j^b AB_{ij} / ab \right)^2$$

$$+ nb \sum_{i=1}^{a} \left(\sum_j^b \sum_k^n \varepsilon_{k(ij)} / bn - \sum_i^a \sum_j^b \sum_k^n \varepsilon_{k(ij)} / abn \right)^2$$

$$+ \text{ cross-product terms}$$

Taking the expected value for a fixed model where

$$\sum_i A_i = 0 \qquad \sum_{i \text{ or } j} AB_{ij} = 0$$

the result is

$$E(SS_A) = nb \sum_i^a A_i^2 + 0 + \frac{nb}{n^2 b^2} (abn - bn) \sigma_\varepsilon^2$$

$$E(MS_A) = E[SS_A / (a-1)] = nb \sum_{i=1}^a A_i^2 / (a-1) + \sigma_\varepsilon^2 = nb\phi_A + \sigma_\varepsilon^2$$

which agrees with Section 6.3 for the fixed model.

If now the levels of A and B are random,

$$\sum_i A_i \neq 0 \qquad \sum_{i \text{ or } j} AB_{ij} \neq 0$$

$$E(SS_A) = nb(a-1)\sigma_A^2 + \frac{nb}{b^2}(ab-b)\sigma_{AB}^2 + \frac{nb}{n^2 b^2}(abn - bn)\sigma_\varepsilon^2$$

$$E(MS_A) = nb\sigma_A^2 + n\sigma_{AB}^2 + \sigma_\varepsilon^2$$

as stated in Section 6.3 for a random model.

If the model is mixed with A fixed and B random,

$$\sum_i^a A_i = 0 \qquad \sum_i^a AB_{ij} = 0$$

but

$$\sum_j^b AB_{ij} \neq 0$$

then

$$E(SS_A) = nb \sum_i^a A_i^2 + \frac{nb}{b^2}(ab-b)\sigma_{AB}^2 + \frac{nb}{n^2 b^2}(nab - nb)\sigma_\varepsilon^2$$

$$E(MS_A) = nb \sum_i^a A_i^2 / (a-1) + n\sigma_{AB}^2 + \sigma_\varepsilon^2 = nb\phi_A + n\sigma_{AB}^2 + \sigma_\varepsilon^2$$

which agrees with the value stated in Section 6.3 for a mixed model.

Using these expected value methods, one can derive all EMS values given in Section 6.3. Note that if *A* were random in the mixed model

$$\sum_{j}^{a} AB_{ij} = 0$$

then the interaction term would not appear in the factor *A* sum of squares. This is true of *B* in the mixed model of Section 6.3.

These few derivations should be sufficient to show the general method of derivation and to demonstrate the advantages of the simple rules in Section 6.4 in determining these EMS values.

6.6 The Pseudo-*F* Test

Occasionally the EMS column for a given experiment indicates that there is no exact *F* test for one or more factors in the design model. Consider the following example.

■ **EXAMPLE 6.3**   Two days in a given month were randomly selected in which to run an experiment. Three operators were also selected at random from a large pool of available operators. The experiment consisted of measuring the dry-film thickness of varnish in mils for three different gate settings: 2, 4, and 6 mils. Two determinations were made by each operator each day and at each of the three gate settings. Results are shown in Table 6.3.

Table 6.3 Dry-Film Thickness Experiment

Gate Setting	Day 1 Operator A	Day 1 Operator B	Day 1 Operator C	Day 2 Operator A	Day 2 Operator B	Day 2 Operator C
2	0.38	0.39	0.45	0.40	0.39	0.41
	0.40	0.41	0.40	0.40	0.43	0.40
4	0.63	0.72	0.78	0.68	0.77	0.85
	0.59	0.70	0.79	0.66	0.76	0.84
6	0.76	0.95	1.03	0.86	0.86	1.01
	0.78	0.96	1.06	0.82	0.85	0.98

Assuming that days and operators are random effects, gate settings are fixed, and the design is completely randomized, the analysis yields Table 6.4.

Table 6.4 Analysis of Dry-Film Thickness Experiment

Source	df	SS	MS	EMS
Days D	1	0.0010	0.0010	$\sigma_\varepsilon^2 + 6\sigma_{DO}^2 + 18\sigma_D^2$
Operators O	2	0.1121	0.0560	$\sigma_\varepsilon^2 + 6\sigma_{DO}^2 + 12\sigma_O^2$
$D \times O$ interaction	2	0.0060	0.0030**	$\sigma_\varepsilon^2 + 6\sigma_{DO}^2$
Gate setting G	2	1.5732	0.7866**	$\sigma_\varepsilon^2 + 2\sigma_{DOG}^2 + 4\sigma_{OG}^2$
				$+6\sigma_{DG}^2 + 12\phi_G$
$D \times G$ interaction	2	0.0113	0.0056	$\sigma_\varepsilon^2 + 2\sigma_{DOG}^2 + 6\sigma_{DG}^2$
$O \times G$ interaction	4	0.0428	0.0107	$\sigma_\varepsilon^2 + 2\sigma_{DOG}^2 + 4\sigma_{OG}^2$
$D \times O \times G$ interaction	4	0.0099	0.0025**	$\sigma_\varepsilon^2 + 2\sigma_{DOG}^2$
Error	18	0.0059	0.0003	σ_ε^2
Totals	35	1.7622		

**Two asterisks indicate significance at the 1 percent level.

All F tests are clear from the EMS column except for the test on gate setting, which is probably the most important factor in the experiment. Two interactions show significance. It is obvious from the results that gate setting is the most important factor, but how can it be tested? If the $D \times G$ interaction is assumed to be zero, then the gate setting can be tested against the $O \times G$ interaction term. On the other hand, if the $O \times G$ interaction is assumed to be zero, gate setting can be tested against the $D \times G$ interaction term. Although neither of these interactions is significant at the 5 percent level, both are numerically larger than the $D \times O \times G$ interaction against which they are tested. In this case any test on G is contingent upon these tests on interaction. One method for testing hypotheses in such situations was developed by Satterthwaite and is given in Bennett and Franklin [4, pp. 367–368].

The scheme consists of constructing a mean square as a linear combination of the mean squares in the experiment, where the EMS for this mean squares includes the same terms as in the EMS of the term being tested, except for the variance of that term. For example, to test the gate-setting effect G in Table 6.4 a mean square is to be constructed whose expected value is

$$\sigma_\varepsilon^2 + 2\sigma_{DOG}^2 + 4\sigma_{OG}^2 + 6\sigma_{DG}^2$$

This can be found by the linear combination

$$MS = MS_{DG} + MS_{OG} - MS_{DOG}$$

as its expected value is

$$E(MS) = \sigma_\varepsilon^2 + 2\sigma_{DOG}^2 + 4\sigma_{OG}^2 + \sigma_\varepsilon^2 + 2\sigma_{DOG}^2 + 6\sigma_{DG}^2 - \sigma_\varepsilon^2 - 2\sigma_{DOG}^2$$

$$= \sigma_\varepsilon^2 + 2\sigma_{DOG}^2 + 4\sigma_{OG}^2 + 6\sigma_{DG}^2$$

An F test can now be constructed using the mean square for gate setting as the numerator and this mean square as the denominator. Such a test is called a pseudo-F or F' test. The real problem here is to determine the degrees of freedom for the denominator mean square. According to Bennett and Franklin, if

$$MS = a_1(MS)_1 + a_2(MS)_2 + \cdots$$

and $(MS)_1$ is based on v_1 df, $(MS)_2$ is based on v_2 df, and so on, then the degrees of freedom for MS are

$$v = \frac{(MS)^2}{a_1^2[(MS)_1^2 / v_1] + a_2^2[(MS)_2^2 / v_2] + \cdots}$$

In the case of testing for the gate-setting effect above, $a_1 = 1$, $a_2 = 1$, and $a_3 = -1$ and the degrees of freedom are $v_1 = 4$, $v_2 = 2$, and $v_3 = 4$. Here $MS = 0.0107 + 0.0056 - 0.0025 = 0.0138$ and its df is

$$v = \frac{(0.0138)^2}{(1)^2[(0.0107)^2 / 4] + (1)^2[(0.0056)^2 / 2] + (-1)^2[(0.0025)^2 / 4]}$$

$$= \frac{1.9044 \times 10^{-4}}{0.4586 \times 10^{-4}} = 4.2$$

Hence the F' test is

$$F' = \frac{MS_G}{MS} = \frac{0.7866}{0.0138} = 57.0$$

with 2 and 4.2 df, which is significant at the 1 percent level of significance based on F with 2 and 4 or 2 and 5 df. ∎

6.7 Remarks

The examples in this chapter should be sufficient to show the importance of the EMS column in deciding just what mean squares should be compared in an F test of a given hypothesis. This EMS column is also useful (usually in random models) to solve for components of variance as illustrated in Section 3.6.

One special case is of interest. In a two-factor factorial when there is but one observation per cell ($n = 1$), the EMS columns of Section 6.3 reduce to those in Table 6.5.

Table 6.5 EMS for One Observation per Cell

Source	EMS (Fixed)	EMS (Random)	EMS (Mixed)
A_i	$\sigma_\varepsilon^2 + b\phi_A$	$\sigma_\varepsilon^2 + \sigma_{AB}^2 + b\sigma_A^2$	$\sigma_\varepsilon^2 + \sigma_{AB}^2 + b\phi_A$
B_j	$\sigma_\varepsilon^2 + a\phi_B$	$\sigma_\varepsilon^2 + \sigma_{AB}^2 + a\sigma_B^2$	$\sigma_\varepsilon^2 + a\sigma_B^2$
AB_{ij} or ε_{ij}	$\sigma_\varepsilon^2 + \phi_{AB}$	$\sigma_\varepsilon^2 + \sigma_{AB}^2$	$\sigma_\varepsilon^2 + \sigma_{AB}^2$

A glance at these EMS values will show that there is no test for the main effects A and B in a fixed model, as interaction and error are hopelessly confounded. The only test possible is to assume that there is no interaction; then $\phi_{AB} = 0$, and the main effects are tested against the error. If a no-interaction assumption is not reasonable from information outside the experiment, the investigator should not run one observation per cell but should replicate the data in a fixed model.

For a random model both main effects can be tested whether interaction is present or not. For a mixed model there is a test for the fixed effect A but no test for the random effect B. This may not be a serious drawback, since the fixed effect is often the most important; the B effect is included chiefly for reduction of the error term. Such a situation is seen in a randomized block design where treatments are fixed, but blocks may be chosen at random.

In the discussion of the single-factor experiment it was assumed that there were equal sample sizes n for each treatment. If this is not the case, it can be shown that the expected treatment mean square is

$$\sigma_\varepsilon^2 + n_0\sigma_\tau^2 \quad \text{or} \quad \sigma_\varepsilon^2 + n_0\phi_\tau$$

and

$$n_0 = \frac{N^2 - \sum_{j=1}^{k} n_j^2}{(k-1)N}$$

where

$$N = \sum_{j=1}^{k} n_j$$

The test for treatment effect is to compare the treatment mean square with the error mean square; the use of n_0 is primarily for computing components of variance.

Problems

6.1   An experiment is run on the effects of three randomly selected operators and five fixed aluminizers on the aluminum thickness of a TV tube. Two readings are made for each operator-aluminizer combination. The following ANOVA table is compiled.

Source	df	SS	MS
Operators	2	107,540	53,770
Aluminizers	4	139,805	34,951
$O \times A$ interaction	8	84,785	10,598
Error	15	230,900	15,393
Totals	29	563,030	

Assuming complete randomization, determine the EMS column for this problem and make the indicated significance tests.

6.2   Consider a three-factor experiment where factor A is at a levels, factor B at b levels, and factor C at c levels. The experiment is to be run in a completely randomized manner with n observations for each treatment combination. Assuming factor A is run at a random levels and both B and C at fixed levels, determine the EMS column and indicate what tests would be made after the analysis.

6.3   Repeat Problem 6.2 with A and B at random levels, but C at fixed levels.

6.4   Repeat Problem 6.2 with all three factors at random levels.

6.5   Consider the completely randomized design of a four-factor experiment similar to Example 6.2. Assuming factors A and B are at fixed levels and C and D are at random levels, set up the EMS column and indicate the tests to be made.

6.6   Assuming three factors are at random levels and one is at fixed levels in Problem 6.5, work out the EMS column and the tests to be run.

6.7   A physical education experiment will be conducted to investigate the effects of four types of exercise on heart rate. Five subjects are chosen at random from a physical education class. A subject does each exercise twice. The 40 measurements are done in a completely randomized order with each subject allowed at least 10 minutes' rest between exercises.
   a. Give the model. State the parameters of its random variables.
   b. Work out the expected mean squares of all the effects. Also give the degrees of freedom for the effects.
   c. Tell how to form the F's for the tests that can be made.

6.8   Four elementary schools are chosen at random in a large city system and two methods of instruction are tried in the third, fourth, and fifth grades. *Fixed*
   a. Outline an ANOVA layout to test the effect of schools, methods, and grades on gain in reading scores assuming ten students per class.
   b. Work out the EMS column and indicate what tests are possible in this situation.

Fixed

6.9   As part of an experiment on testing viscosity four operators were chosen at random and four instruments were randomly selected from many instruments available. Each operator was to test a sample of product twice on each instrument. Results showed:

Source	df	MS
O_i	3	1498
I_j	3	1816
OI_{ij}	9	218
$\varepsilon_{k(ij)}$	16	67

Determine the EMS column and find the percentage of the total variance in these readings that is attributable to each term in the model, if it is significant.

6.10 Six different formulas or mixes of concrete are to be purchased from five competing suppliers. Tests of crushing strength are to be made on two blocks constructed according to each formula–supplier combination. One block will be poured, hardened, and tested for each combination before the second block is formed, thus making two replications of the whole experiment. Assuming mixes and suppliers fixed and replications random, set up a mathematical model for this situation, outline its ANOVA table, and show what F tests are appropriate.

6.11 Determine the EMS column for Problem 3.6 and solve for components of variance.

6.12 Derive the expression for n_0 in the EMS column of a single-factor completely randomized experiment where the n's are unequal.

6.13 Verify the numerical results of Table 6.4 using the methods of Chapter 5.

6.14 Verify the EMS column of Table 6.4 by the method of Section 6.4.

6.15 Set up F' tests for Problem 6.4 and explain how the df would be determined.

6.16 Set up F' tests for Problem 6.5 and explain how the df would be determined.

Nested and Nested-Factorial Experiments

7

7.1 Introduction

■ **EXAMPLE 7.1** In a recent in-plant training course the members of the class were assigned a final problem. Each class member was to go into the plant and set up an experiment using the techniques that had been discussed in class. One engineer wanted to study the strain readings of glass cathode supports from five different machines. Each machine had four "heads" on which the glass was formed, and she decided to take four samples from each head. She treated this experiment as a 5 × 4 factorial with four replications per cell. Complete randomization of the testing for strain readings presented no problem. Her model was

$$Y_{ijk} = \mu + M_i + H_j + MH_{ij} + \varepsilon_{k(ij)}$$

with

$$i = 1, 2, \ldots, 5 \qquad j = 1, \ldots, 4 \qquad k = 1, \ldots, 4$$

Her data and analysis appear in Table 7.1. In this model she assumed that both machines and heads were fixed, and used the 10 percent significance level. The results indicated no significant machine or head effect on strain readings, but there was a significant interaction at the 10 percent level of significance.

The question was raised as to whether the four heads were actually removed from machine A and mounted on machine B, then on C, and so on. Of course, the answer was "no" as each machine had its own four heads. Thus, machines and heads did not form a factorial experiment, as the heads on each machine were unique for that particular machine. In such a case the experiment is called a *nested experiment:* levels of one factor are nested within, or are subsamples of, levels of another factor. Such experiments are also sometimes called *hierarchical* experiments. When factors are arranged in a factorial experiment as in Chapter 5, they are often referred to as *crossed factors.*

Table 7.1 Data and ANOVA for Strain-Reading Problem

Head	Machine A	B	C	D	E
1	6	10	0	11	1
	2	9	0	0	4
	0	7	5	6	7
	8	12	5	4	9
2	13	2	10	5	6
	3	1	11	10	7
	9	1	6	8	0
	8	10	7	3	3
3	1	4	8	1	3
	10	1	5	8	0
	0	7	0	9	2
	6	9	7	4	2
4	7	0	7	0	3
	4	3	2	8	7
	7	4	5	6	4
	9	1	4	5	0

Source	df	SS	MS	EMS	F	$F_{0.90}$
M_i	4	45.08	11.27	$\sigma_\varepsilon^2 + 16\phi_M$	1.05	2.04
H_j	3	46.45	15.48	$\sigma_\varepsilon^2 + 20\phi_H$	1.45	2.18
MH_{ij}	12	236.42	19.70	$\sigma_\varepsilon^2 + 4\phi_{MH}$	1.84	1.66
$\varepsilon_{k(ij)}$	60	642.00	10.70			
Totals	79	969.95				

7.2 Nested Experiments

The preceding example can now be reanalyzed by treating it as a nested experiment, since heads are nested within machines. Such a factor may be represented in the model as $H_{j(i)}$, where j covers all levels 1, 2, . . . within the ith level of M_i. The number of levels of the nested factor need not be the same for all levels of the other factor. They are all equal in this problem, that is $j = 1, 2, 3, 4$ for all i. The errors, in turn, are nested within the levels of i and j; $\varepsilon_{k(ij)}$ and $k = 1,$ 2, 3, 4 for all i and j.

To emphasize the fact that the heads on each machine are different heads, the data layout in Table 7.2 shows heads 1, 2, 3, and 4 on machine A; heads 5, 6, 7, and 8 on machine B; and so on.

Table 7.2 Data for Strain-Reading Problem in a Nested Arrangement

Machine	A				B				C				D				E			
Head	1	2	3	4	5	6	7	8	9	10	11	12	13	14	15	16	17	18	19	20
	6	13	1	7	10	2	4	0	0	10	8	7	11	5	1	0	1	6	3	3
	2	3	10	4	9	1	1	3	0	11	5	2	0	10	8	8	4	7	0	7
	0	9	0	7	7	1	7	4	5	6	0	5	6	8	9	6	7	0	2	4
	8	8	6	9	12	10	9	1	5	7	7	4	4	3	4	5	9	3	2	0
Head totals	16	33	17	27	38	14	21	8	10	34	20	18	21	26	22	19	21	16	7	14
Machine totals	93				81				82				88				58			

As the heads that are mounted on the machine can be chosen from many possible heads, we might consider the four heads as a random sample of heads that might be used on a given machine. If such heads were selected at random for the machines, the model would be

$$Y_{ijk} = \mu + M_i + H_{j(i)} + \varepsilon_{k(ij)}$$

with

$$i = 1, \ldots, 5 \qquad j = 1, \ldots, 4 \qquad k = 1, \ldots, 4$$

This nested model has no interaction present, as the heads are not crossed with the five machines. If heads are considered random and machines fixed, the proper EMS values can be determined by the rules given in Chapter 6, as shown in Table 7.3.

Table 7.3 EMS for Nested Experiment

		5 F	4 R	4 R	
df	Source	i	j	k	EMS
4	M_i	0	4	4	$\sigma_\varepsilon^2 + 4\sigma_H^2 + 16\phi_M$
15	$H_{j(i)}$	1	1	4	$\sigma_\varepsilon^2 + 4\sigma_H^2$
60	$\varepsilon_{k(ij)}$	1	1	1	σ_ε^2

Total 79

This breakdown shows that the head effect is to be tested against the error, and the machine effect is to be tested against the heads-within-machines effect.

To analyze the data for a nested design, first determine the total sum of squares

$$SS_{total} = 6^2 + 2^2 + \cdots + 4^2 + 0^2 - \frac{(402)^2}{80} = 969.95$$

and for machines

$$SS_M = \frac{(93)^2 + (81)^2 + (82)^2 + (88)^2 + (58)^2}{16} - \frac{(402)^2}{80} = 45.08$$

To determine the sum of squares between heads within machines, consider each machine separately.

Machine A:

$$SS_H = \frac{(16)^2 + (33)^2 + (17)^2 + (27)^2}{4} - \frac{(93)^2}{16} = 50.19$$

Machine B:

$$SS_H = \frac{(38)^2 + (14)^2 + (21)^2 + (8)^2}{4} - \frac{(81)^2}{16} = 126.18$$

Machine C:

$$SS_H = \frac{(10)^2 + (34)^2 + (20)^2 + (18)^2}{4} - \frac{(82)^2}{16} = 74.75$$

Machine D:

$$SS_H = \frac{(21)^2 + (26)^2 + (22)^2 + (19)^2}{4} - \frac{(88)^2}{16} = 6.50$$

Machine E:

$$SS_H = \frac{(21)^2 + (16)^2 + (7)^2 + (14)^2}{4} - \frac{(58)^2}{16} = 25.25$$

$$\text{total } SS_H = \overline{282.87}$$

The error sum of squares by subtraction is then

$$969.95 - 45.08 - 282.87 = 642.00$$

The degrees of freedom between heads within machine A are $4 - 1 = 3$; for all five machines the degrees of freedom will be $5 \times 3 = 15$. The analysis follows in Table 7.4.

Table 7.4 ANOVA for Nested Strain-Reading Problem

Source	df	SS	MS	EMS	F	$F_{0.90}$
M_i	4	45.08	11.27	$\sigma_\varepsilon^2 + 4\sigma_H^2 + 16\phi_M$	< 1	2.36
$H_{j(i)}$	15	282.87	18.85	$\sigma_\varepsilon^2 + 4\sigma_H^2$	1.76	1.60
$\varepsilon_{k(ij)}$	60	642.00	10.70	σ_ε^2		
Totals	79	969.95				

From this analysis, machines appear to have no significant effect on strain readings, but there is a slightly significant (10 percent level) effect of heads within machines on the strain readings. Note that what the experimenter took as head effect (3 df) and interaction effect (12 df) is really heads-within-machines effect (15 df). These results might suggest a more careful adjustment between heads on the same machine. Since there are significant differences between heads within machines, the heads-within-machines sum of squares of 282.87 might well be analyzed further. Returning to the method by which this term was computed, one can look at its partitioning into five parts with 3 df per machine, and because all parts are orthogonal, further breakdown and tests of Table 7.4 can be made as in Table 7.5.

Table 7.5 Detailed ANOVA of Example 7.1

Source	df	SS		MS	F	$F_{0.90}$
M_i	4	45.08		11.27	< 1	2.36
$H_{j(i)}$	15	282.87				
$\quad H_{j(A)}$	3		50.19	16.73	1.56	2.18
$\quad H_{j(B)}$	3		126.18	42.06	3.93*	2.18
$\quad H_{j(C)}$	3		74.75	24.92	2.33*	2.18
$\quad H_{j(D)}$	3		6.50	2.17	< 1	2.18
$\quad H_{j(E)}$	3		25.25	8.42	< 1	2.18
$\varepsilon_{k(ij)}$	60	642.00		10.70		

* Significant at 10 percent level.

From Table 7.5, using a 10 percent significance level, we see a significant difference between heads on machine B and machine C. One might then examine the head means within each of these machines in a Newman–Keuls sense.

$$\text{For machine } B, \text{ means are:} \quad 2.00 \quad 3.50 \quad 5.25 \quad 9.50$$
$$\text{For heads:} \quad 8 \quad 6 \quad 7 \quad 5$$
$$s_{\bar{Y}} = \sqrt{\frac{10.70}{4}} = 1.64 \text{ with 60 df for } n_2$$

From Statistical Table E.1 (although one should have a 10 percent table)

p:	2	3	4	
	2.83	3.40	3.74	
LSR:	4.64	5.58	6.13	
So groups are	2.00	3.50	5.25	9.50

One should then examine head 5, as its mean is larger than two of the other three.

For machine C means are 2.50, 4.50, 5.00, 8.50 for heads 9, 12, 11, 10, and because the largest gap is $8.5 - 2.5 = 6$, which is less than 6.13, no significant differences are detected. This is so because the differences are significant at the 10 percent level and not at the 5 percent level on which Table E.1 is based. One might still examine the heads on machine C for lack of consistency. In a nested model it is also seen that the 5 SS between heads within each machine are "pooled," or added, to give 282.87. This assumes that these sums of squares (which are proportional to variances) within each machine are of about the same magnitude. This might be questioned, as these sums of squares are 50.19, 126.18, 74.75, 6.50, and 25.25. If these five are really different, it appears as if the greatest variability is in machine B. This should indicate the need for work on each machine with an aim toward more homogeneous strain readings between heads on each machine. This example shows the importance of recognizing the difference between a nested experiment and a factorial experiment. ∎

7.3 ANOVA Rationale

To see that the sums of squares computed in Section 7.2 were correct, consider the nested model

$$Y_{ijk} = \mu + A_i + B_{j(i)} + \varepsilon_{k(ij)}$$

or

$$Y_{ijk} \equiv \mu + (\mu_i - \mu) + (\mu_{ij} - \mu_i) + (Y_{ijk} - \mu_{ij})$$

which is an identity.

Using the best estimates of these population means from the sample data, the sample model is

$$Y_{ijk} \equiv \bar{Y}_{\ldots} + (\bar{Y}_{i\ldots} - \bar{Y}_{\ldots}) + (\bar{Y}_{ij.} - \bar{Y}_{i\ldots}) + (Y_{ijk} - \bar{Y}_{ij.})$$

with

$$i = 1, 2, \ldots, a \qquad j = 1, 2, \ldots, b \qquad k = 1, 2, \ldots, n$$

Transposing $\bar{Y}_{\ldots}$ to the left of this expression, squaring both sides, and adding over i, j, and k gives

$$\sum_i^a \sum_j^b \sum_k^n (Y_{ijk} - \bar{Y}_{\ldots})^2 = \sum_{i=1}^a nb(\bar{Y}_{i\ldots} - \bar{Y}_{\ldots})^2 + \sum_i^a \sum_j^b n(\bar{Y}_{ij.} - \bar{Y}_{i\ldots})^2$$

$$+ \sum_i^a \sum_j^b \sum_k^n (Y_{ijk} - \bar{Y}_{ij.})^2$$

as the sums of cross products equal zero. This expresses the idea that the total sum of squares is equal to the sum of squares between levels of A, plus the sums of squares between levels of B within each level of A, plus the sum of the squares of the errors. The degrees of freedom are

$$(abn - 1) \equiv (a - 1) + a(b - 1) + ab(n - 1)$$

Dividing each independent sum of squares by its corresponding degrees of freedom gives estimates of population variance as usual. For computing purposes, the sum of squares as given above should be expanded in terms of totals, with the general results shown in Table 7.6.

Table 7.6 General ANOVA for a Nested Experiment

Source	df	SS	MS
A_i	$a - 1$	$\sum\limits_{i}^{a} \dfrac{T_{i..}^2}{nb} - \dfrac{T_{...}^2}{nab}$	$\dfrac{SS_A}{a - 1}$
$B_{j(i)}$	$a(b - 1)$	$\sum\limits_{i}^{a}\sum\limits_{j}^{b} \dfrac{T_{ij.}^2}{n} - \sum\limits_{i}^{a} \dfrac{T_{i..}^2}{nb}$	$\dfrac{SS_B}{a(b - 1)}$
$\varepsilon_{k(ij)}$	$ab(n - 1)$	$\sum\limits_{i}^{a}\sum\limits_{j}^{b}\sum\limits_{k}^{n} Y_{ijk}^2 - \sum\limits_{i}^{a}\sum\limits_{j}^{b} \dfrac{T_{ij.}^2}{n}$	$\dfrac{SS_\varepsilon}{ab(n - 1)}$
Totals	$abn - 1$	$\sum\limits_{i}^{a}\sum\limits_{j}^{b}\sum\limits_{k}^{n} Y_{ijk}^2 - \dfrac{T_{...}^2}{nab}$	

This is essentially the form followed in the problem of the last section. Note that the

$$SS_{B_{j(i)}} = \sum_{i}^{a}\sum_{j}^{b} \frac{T_{ij.}^2}{n} - \sum_{i}^{a} \frac{T_{i..}^2}{nb} = \sum_{i}^{a}\left(\sum_{j}^{b} \frac{T_{ij.}^2}{n} - \frac{T_{i..}^2}{nb}\right)$$

which shows the way in which the sum of squares was calculated in the last section: by getting the sum of squares between levels of B for each level of A and then pooling over all levels of A.

7.4 Nested-Factorial Experiments

In many experiments where several factors are involved, some may be factorial or crossed with others; some may be nested within levels of the others. When both factorial and nested factors appear in the same experiment, it is known as a *nested-factorial experiment*. The analysis of such an experiment is simply an extension of the methods of Chapter 5 and this chapter. Care must be exercised, however, in computing some of the interactions. Levels of both factorial and nested factors may be either fixed or random. The methods of Chapter 6 can be used to determine the EMS values and the proper tests to be run.

■ **EXAMPLE 7.2** The nested-factorial experiment is best explained by an example. An investigator wished to improve the number of rounds per minute that could be fired from a Navy gun. He devised a new loading method, which he hoped would increase the number of rounds per minute when compared with the existing method of loading. To test this hypothesis he needed teams of men to operate the equipment. As the general physique of a man might affect the speed with which he could handle the loading of the gun, he chose teams of men in three general groupings—slight, average, and heavy or rugged men. The classification of such men was on the basis of an Armed Services classification table. He chose three teams at random to represent each of the three physique groupings. Each team was presented with the two methods of gun loading in a random order and each team used each method twice. The model for this experiment was

$$Y_{ijkm} = \mu + M_i + G_j + MG_{ij} + T_{k(j)} + MT_{ik(j)} + \varepsilon_{m(ijk)}$$

where

$$M_i = \text{methods, } i = 1, 2$$
$$G_j = \text{groups, } j = 1, 2, \ 3$$
$$T_{k(j)} = \text{teams within groups, } k = 1, 2, 3, \text{ for all } j$$
$$\varepsilon_{m(ijk)} = \text{random error, } m = 1, 2 \text{ for all } i, j, k$$

The EMS values are shown in Table 7.7, which indicates the proper F tests to run.

Table 7.7 EMS for Gun-Loading Problem

Source	2 F i	3 F j	3 R k	2 R m	EMS
M_i	0	3	3	2	$\sigma_\varepsilon^2 + 2\sigma_{MT}^2 + 18\phi_M$
G_j	2	0	3	2	$\sigma_\varepsilon^2 + 4\sigma_T^2 + 12\phi_G$
MG_{ij}	0	0	3	2	$\sigma_\varepsilon^2 + 2\sigma_{MT}^2 + 6\phi_{MG}$
$T_{k(j)}$	2	1	1	2	$\sigma_\varepsilon^2 + 4\sigma_T^2$
$MT_{ik(j)}$	0	1	1	2	$\sigma_\varepsilon^2 + 2\sigma_{MT}^2$
$\varepsilon_{m(ijk)}$	1	1	1	1	σ_ε^2

The data and analysis of this experiment appear in Table 7.8.

Table 7.8 Data and ANOVA for Gun-Loading Problem

Group	I			II			III		
Team	1	2	3	4	5	6	7	8	9
Method I	20.2	26.2	23.8	22.0	22.6	22.9	23.1	22.9	21.8
	24.1	26.9	24.9	23.5	24.6	25.0	22.9	23.7	23.5
Method II	14.2	18.0	12.5	14.1	14.0	13.7	14.1	12.2	12.7
	16.2	19.1	15.4	16.1	18.1	16.0	16.1	13.8	15.1

Source	df	SS	MS	EMS
M_i	1	651.95	651.95	$\sigma_\varepsilon^2 + 2\sigma_{MT}^2 + 18\phi_M$
G_j	2	16.05	8.02	$\sigma_\varepsilon^2 + 4\sigma_T^2 + 12\phi_G$
MG_{ij}	2	1.19	0.60	$\sigma_\varepsilon^2 + 2\sigma_{MT}^2 + 6\phi_{MG}$
$T_{k(j)}$	6	39.26	6.54	$\sigma_\varepsilon^2 + 4\sigma_T^2$
$MT_{ik(j)}$	6	10.72	1.79	$\sigma_\varepsilon^2 + 2\sigma_{MT}^2$
$\varepsilon_{m(ijk)}$	18	41.59	2.31	σ_ε^2
Totals	35	760.76		

It would be well for the reader to verify that the sums of squares of this table are correct. The only term that is somewhat different in this model than those previously handled is $MT_{ik(j)}$, that is, the interaction between methods and teams within groups. The safest way to compute this term is to compute the $M \times T$ interaction sums of squares within each of the three groups separately and then pool these sums of squares. (See Tables 7.9 through 7.11.)

Table 7.9 Data on Gun-Loading Problem for Group I

	Team			Method
	1	2	3	Totals
Method I	20.2	26.2	23.8	
	24.1	26.9	24.9	
	44.3	53.1	48.7	146.1
Method II	14.2	18.0	12.5	
	16.2	19.1	15.4	
	30.4	37.1	27.9	95.4
Team totals	74.7	90.2	76.6	241.5

Table 7.10 Data on Gun-Loading Problem for Group II

| | Team | | | Method |
	4	5	6	Totals
Method I	22.0	22.6	22.9	
	23.5	24.6	25.0	
	45.5	47.2	47.9	140.6
Method II	14.1	14.0	13.7	
	16.1	18.1	16.0	
	30.2	32.1	29.7	92.0
Team totals	75.7	79.3	77.6	232.6

Table 7.11 Data on Gun-Loading Problem for Group III

| | Team | | | Method |
	7	8	9	Totals
Method I	23.1	22.9	21.8	
	22.9	23.7	23.5	
	46.0	46.6	45.3	137.9
Method II	14.1	12.2	12.7	
	16.1	13.8	15.1	
	30.2	26.0	27.8	84.0
Team totals	76.2	72.6	73.1	221.9

To compute the $M \times T$ interaction for group I, we have

$$SS_{cell} = \frac{(44.3)^2 + (53.1)^2 + (48.7)^2 + (30.4)^2 + (37.1)^2 + (27.9)^2}{2}$$

$$- \frac{(241.5)^2}{12} = 5116.39 - 4860.21 = 256.18$$

$$SS_{method} = \frac{(146.1)^2 + (95.4)^2}{6} - 4860.21 = 214.19$$

$$SS_{team} = \frac{(74.7)^2 + (90.2)^2 + (76.6)^2}{4} - 4860.21 = 35.74$$

$$SS_{M \times T\ interaction} = 256.18 - 214.19 - 35.74 = 6.25$$

For group II:

$$SS_{cell} = \frac{(45.5)^2 + (47.2)^2 + (47.9)^2 + (30.2)^2 + (32.1)^2 + (29.7)^2}{2}$$

$$- \frac{(232.6)^2}{12} = 199.96$$

$$SS_{method} = \frac{(140.6)^2 + (92.0)^2}{6} - 4508.56 = 196.83$$

$$SS_{team} = \frac{(75.7)^2 + (79.3)^2 + (77.6)^2}{4} - 4508.56 = 1.62$$

$$SS_{M \times T \text{ interaction}} = 199.96 - 196.83 - 1.62 = 1.51$$

For group III:

$$SS_{cell} = \frac{(46.0)^2 + (46.6)^2 + (45.3)^2 + (30.2)^2 + (26.0)^2 + (27.8)^2}{2}$$

$$- \frac{(221.9)^2}{12} = 246.96$$

$$SS_{method} = \frac{(137.9)^2 + (84.0)^2}{6} - 4103.30 = 242.10$$

$$SS_{team} = \frac{(76.2)^2 + (72.6)^2 + (73.1)^2}{4} - 4103.30 = 1.90$$

$$SS_{M \times T \text{ interaction}} = 246.96 - 242.10 - 1.90 = 2.96$$

Pooling for all three groups gives

$$SS_{M \times T} = 6.25 + 1.51 + 2.96 = 10.72$$

which is recorded in Table 7.8.

The results of this experiment show a very significant method effect (the new method averaged 23.58 rounds per minute, and the old method averaged only 15.08 rounds per minute). The results also show a significant difference between teams within groups at the 5 percent significance level. This points out individual differences in the men. No other effects or interactions were significant.

A further analysis of the significant differences between teams within the three groups shows the sums of squares to be 35.74, 1.62, and 1.90, respectively. With the error mean square in Table 7.8 of 2.31 it is obvious that the significant difference between teams within the three groups is concentrated in group I. The mean number of rounds per minute for group I's three teams (across methods) are

Team	1	3	2
Mean	18.68	19.15	22.55

A Newman–Keuls will show that team 2 is exceptionally faster than the other two teams. ∎

7.5 Repeated-Measures Design and Nested-Factorial Experiments

Many statisticians who work with psychologists and educators on the design of their experiments treat repeated-measures designs as a unique topic in these fields of application of statistics. It is the purpose of this section to show that these designs are but a special case of factorial and nested-factorial experiments. Numerical examples and mathematical models are used to illustrate the correspondence between these designs.

One of the simplest of these cases is when a pretest and post-test are given to the same group of subjects after a certain time lapse during which some special instruction may have been administered.

■ **EXAMPLE 7.3**   A recent study at Purdue University was made on a measurement of physical strength (in pounds) given to seven subjects before and after a specified training period. Results are shown in Table 7.12.

Table 7.12 Pretest and Post-test Measures for Example 7.3

Subjects	Pretest	Post-test
1	100	115
2	110	125
3	90	105
4	110	130
5	125	140
6	130	140
7	105	125

Since the same seven subjects were given each test, we have two repeated measures on each subject. Winer [29, p. 266] would handle this as a between-subjects, within-subjects repeated-measures design and report results as shown in Table 7.13.

Table 7.13 ANOVA for Table 7.12

Source of Variation	df	SS		MS	F
Between subjects (S_i)	6	2084.71		347.45	
Within subjects	7	901.00		—	
Tests (T_j)	1		864.29	864.29	145
Residual	6		35.71	5.96	
Totals	13	2985.71			

This analysis implies that the subjects model is

$$Y_{ij} = \mu + S_i + \varepsilon_{j(i)}$$

with df: 6   7

And then the within-subjects' data are further broken down into tests and residual. This residual is actually the subject by test interaction so the model becomes

$$Y_{ij} = \mu + S_i + T_j + ST_{ij}$$

with df: 6   1   6

This model contains no error term since only one pretest score and one post-test score are available on each subject.

Now if this problem were considered in more general terms instead of the so-called repeated-measures format, one would consider this data layout as a two-factor factorial experiment with one observation per treatment. Subjects should be chosen at random and tests should be fixed. Using the algorithm given in Chapter 6, the resulting ANOVA layout would be as shown in Table 7.14.

Table 7.14 EMS Determination for Table 7.13

Source	df	7 R i	2 F j	1 R k	EMS
S_i	6	1	2	1	$\sigma_\varepsilon^2 + 2\sigma_S^2$
T_j	1	7	0	1	$\sigma_\varepsilon^2 + \sigma_{ST}^2 + 7\phi_T$
ST_{ij}	6	1	0	1	$\sigma_\varepsilon^2 + \sigma_{ST}^2$
$\varepsilon_{k(ij)}$	0	1	1	1	σ_ε^2 (not retrievable)

Here $k = 1$ and a separate error $\varepsilon_{k(ij)}$ is not retrievable. A glance at the EMS column indicates that the only proper F test is to compare the test mean square with the $S \times T$ interaction mean square, which was the test as shown in Table 7.13.

Of course, another approach to this example is to take differences between post- and pretest scores on each individual and test the hypothesis that the true mean of the difference is zero. Using this scheme, a t of 12.05 is found and it is well known that this t when squared equals the F of our ANOVA table:

$$t^2 = (12.05)^2 = 145 = F \qquad \blacksquare$$

■ **EXAMPLE 7.4** A slightly more complex problem involves three factors in which the subjects are nested within groups. These groups could be classes or experimental conditions and then again each subject is subjected to repeated measures. The study cited on physical strength was extended so that three groups of subjects were involved, two being subjected to special experimental training and the third acting as a control with no special training. Again, each subject was given a pre- and post-test where the measurement was velocity of a baseball throw in meters per second. The resulting data are given in Table 7.15. If these data are treated as a repeated-measures design, the results are usually presented as in Winer [29, p. 520], shown as Table 7.16.

Table 7.15 Pretest and Post-test Throwing Velocities of Three Groups of Subjects in Meters/Second

Group	Subject	Pretest	Post-test
	1	26.25	29.50
	2	24.33	27.62
	3	22.52	25.71
I	4	29.33	31.55
	5	28.90	31.35
	6	25.13	29.07
	7	29.33	31.15
	8	27.47	28.74
	9	25.19	26.11
	10	23.53	25.45
II	11	24.57	25.58
	12	26.88	27.70
	13	27.86	28.82
	14	28.09	28.99
	15	22.27	22.52
	16	21.55	21.79
	17	23.31	23.53
III	18	30.03	30.21
	19	28.17	28.65
	20	28.09	28.33
	21	27.55	27.86

If the layout of Table 7.15 is treated as a nested-factorial experiment in which subjects are nested within groups and then tests are factorial on both groups and subjects, the model can be written as follows:

$$Y_{ijk} = \mu + G_i + S_{j(i)} + T_k + GT_{ik} + TS_{kj(i)} + \varepsilon_{m(ijk)}$$

with df: 2   18   1   2   18   0

Table 7.16 ANOVA of Table 7.15

Source	df	SS	MS	EMS	F
Between subjects	20	271.05	—		
Groups (G_i)	2		28.14	14.07	$\sigma_e^2 + 2\sigma_S^2 + 14\phi_G$	1.04
Subjects within groups ($S_{j(i)}$)	18		242.91	13.50	$\sigma_e^2 + 2\sigma_S^2$	
Within subjects	21	35.73	—		
Tests (T_k)	1		21.26	21.26	$\sigma_e^2 + \sigma_{TS}^2 + 21\phi_T$	183***
$G \times T$	2		12.38	6.19	$\sigma_e^2 + \sigma_{TS}^2 + 7\phi_{GT}$	53**
$T \times S_{kj(i)}$	18		2.09	0.116	$\sigma_e^2 + \sigma_{TS}^2$	

Table 7.17 EMS Determination for Table 7.15

Source	df	F 3 i	R 7 j	F 2 k	R 1 m	EMS
G_i	2	0	7	2	1	$\sigma_e^2 + 2\sigma_S^2 + 14\phi_G$
$S_{j(i)}$	18	1	1	2	1	$\sigma_e^2 + 2\sigma_S^2$
T_k	1	3	7	0	1	$\sigma_e^2 + \sigma_{TS}^2 + 21\phi_T$
GT_{ik}	2	0	7	0	1	$\sigma_e^2 + \sigma_{TS}^2 + 7\phi_{GT}$
$TS_{kj(i)}$	18	1	1	0	1	$\sigma_e^2 + \sigma_{TS}^2$
$\varepsilon_{m(ijk)}$	0	1	1	1	1	σ_e^2 (not retrievable)

including a nonretrievable error term. Use of this model and the algorithm cited above gives Table 7.17.

This scheme generates the EMS column that was simply reported in Table 7.16, and that column indicates the proper tests to be made. The numerical results are, of course, the same. ■

The nested-factorial experiment covers many situations in addition to the repeated-measures design. The factor in the nest may be farms within townships, classes within schools, heads within machines, samples within batches, and so on. It treats all situations where three factors are involved where one is nested and another is factorial. Usually in a repeated-measures design the subjects are chosen at random and the other two factors are considered as fixed. By setting up the model and using the EMS algorithm one may have factors that are random as well as fixed. For example, classes could very well be chosen at random from a given grade level and then the students within the classes chosen at random for the experiment.

As more factors are added in an experiment, one only has to expand the mathematical model and determine the proper tests to make based on the EMS column. One does not need to think in terms of repeat measures. To illustrate further, Winer [29, p. 540] considers two cases of three-factor experiments with repeated measures. Case I has two of the factors (fixed) crossing (or repeated on) all subjects, and in Case II subjects are nested within two factors (fixed) and the third factor crosses (or is repeated on) all subjects. Both of these cases, and more complex ones, can easily be handled as nested-factorial experiments by writing the proper model. Handling them in this way does not require that the three factors all be fixed. It simply requires that the subjects be treated as another factor nested within one or more of the other factors.

Models for Case I and Case II show the relationship between the two approaches.

Case I:

$$Y_{ijkm}$$

$$= \mu + \overbrace{A_i + S_{j(i)}}^{\text{between subjects}}$$

$$+ \overbrace{B_k + AB_{ik} + BS_{kj(i)} + C_m + AC_{im} + CS_{mkj(i)} + BC_{km} + ABC_{ikm} + BCS_{mkj(i)}}^{\text{within subjects}}$$

Case II:

$$Y_{ijkm} = \mu + \overbrace{A_i + B_j + AB_{ij} + S_{k(ij)}}^{\text{between subjects}}$$

$$+ \overbrace{C_m + AC_{im} + BC_{jm} + ABC_{ijm} + CS_{mk(ij)}}^{\text{within subjects}}$$

No error terms have been included because they are not retrievable.

These examples, it is hoped, are sufficient to remove any mystery surrounding the repeated-measures designs by showing that they fit into more general models familiar to statisticians in all fields.

7.6 Summary

The summary at the end of Chapter 5 may now be extended for Part II (below).

Experiment	Design	Analysis
II. Two or more factors A. Factorial (crossed)		
	1. Completely randomized $Y_{ijk} = \mu + A_i + B_j + AB_{ij} + \varepsilon_{k(ij)}, \ldots$ for more factors	1.
B. Nested (hierarchical)	General case	ANOVA with interactions
	1. Completely randomized $Y_{ijk} = \mu + A_i + B_{j(i)} + \varepsilon_{k(ij)}$	1. Nested ANOVA
C. Nested factorial		
	1. Completely randomized $Y_{ijkm} = \mu + A_i + B_{j(i)} + C_k + AC_{ik}$ $+ BC_{jk(i)} + \varepsilon_{m(ijk)}$	1. Nested-factorial ANOVA

7.7 SAS Programs for Nested and Nested-Factorial Experiments

For nested and nested-factorial experiments, two new concepts are added to the SAS program:

(1) Nesting notation. If A is one factor and the levels of a factor B are to be nested within levels of A, we write $B(A)$ and the model is RDG= $A\ B(A)$. In more complex problems one may wish to use the upright pipe to break out the interactions, and the pipe can be inserted between A and $B(A)$ and the program will not try to compute an $A * B$ interaction because of the nested notation. For example, with a class statement including A, B, and C, a model might be RDG= $A|B(A)|C$, and the output will include A, $B(A)$, C, $A * C$, $C * B(A)$.

(2) Because nested factors are almost always run at random levels, it is necessary to determine the EMS appropriate for a given problem in order to decide what error term to use for each main effect and interaction. The

SAS program will test all effects against the error term if one exists in the problem, so the program must be instructed as to which tests are appropriate. This is done by adding one or more **TEST** commands where H (for hypothesis) = Term to be tested, and E (for error) = proper divisor for testing H. This again shows the importance of the EMS determination *before* proceeding to an analysis.

These additions are illustrated in Examples 7.1 and 7.2:

For Example 7.1:

```
DATA A;
INPUT M H RDG @@;
CARDS;
-(DATA)-
;
TITLE'EXAMPLE 7.1';
PROC PRINT;
PROC GLM;
CLASS M H;
MODEL RDG=M H(M);
TEST H=M E=H(M);
```

The print-out is on page 191.

These results agree with those of Table 7.4.

For Example 7.2:

```
DATA A;
INPUT M G T Y;
CARDS;
-(DATA)-
;
TITLE'EXAMPLE 7.2';
PROC PRINT;
PROC GLM;
CLASS M G T;
MODEL Y=M|G|T(G);
TEST H=M MxG E=M*T(G);
TEST H=G E=T(G);
```

The output is on page 192.

These results agree with those of Table 7.8.

GENERAL LINEAR MODELS PROCEDURE
DEPENDENT VARIABLE: RDG

SOURCE	DF	SUM OF SQUARES	MEAN SQUARE	F VALUE	PR > F	R-SQUARE	C.V.
MODEL	19	327.95000000	17.26052632	1.61	0.0822	0.338110	65.0962
ERROR	60	642.00000000	10.70000000				
CORRECTED TOTAL	79	969.95000000			ROOT MSE		RDG MEAN
					3.27108545		5.02500000

SOURCE	DF	TYPE I SS	F VALUE	PR > F	DF	TYPE III SS	F VALUE	PR > F
M	4	45.07500000	1.05	0.0876	4	45.07500000	1.05	0.3876
H(M)	15	282.87500000	1.76	0.0624	15	282.87500000	1.76	0.0624

TESTS OF HYPOTHESES USING THE TYPE III MS FOR H(M) AS AN ERROR TERM

SOURCE	DF	TYPE III SS	F VALUE	PR > F
M	4	45.07500000	0.60	0.6700

191

SAS
GENERAL LINEAR MODELS PROCEDURE
DEPENDENT VARIABLE: Y

SOURCE	DF	SUM OF SQUARES	MEAN SQUARE	F VALUE	PR > F	R-SQUARE	C.V.
MODEL	17	719.17000000	42.30411765	18.31	0.0001	0.945331	7.8623
ERROR	18	41.59000000	2.31055556		ROOT MSE		Y MEAN
CORRECTED TOTAL	35	760.76000000			1.52005117		19.33333333

SOURCE	DF	TYPE I SS	F VALUE	PR > F	DF	TYPE III SS	F VALUE	PR > F
M	1	651.95111111	282.16	0.0001	1	651.95111111	282.16	0.0001
G	2	16.05166667	3.47	0.0530	2	16.05166667	3.47	0.0530
M*G	2	1.18722222	0.26	0.7762	2	1.18722222	0.26	0.7762
T(G)	6	39.25833333	2.83	0.0403	6	39.25833333	2.83	0.0403
M*T(G)	6	10.72166667	0.77	0.6009	6	10.72166667	0.77	0.6009

TESTS OF HYPOTHESES USING THE TYPE III MS FOR M*T(G) AS AN ERROR TERM

SOURCE	DF	TYPE III SS	F VALUE	PR > F
M	1	651.95111111	364.84	0.0001
M*G	2	1.18722222	0.33	0.7297

TESTS OF HYPOTHESES USING THE TYPE III MS FOR T(G) AS AN ERROR TERM

SOURCE	DF	TYPE III SS	F VALUE	PR > F
G	2	16.05166667	1.23	0.3576

Problems

*Use Problems SAS
QFII 5-8.SAS*

7.1 Porosity readings on condenser paper were recorded for paper from four rolls taken at random from each of three lots. The results were as follows. Analyze these data, assuming lots are fixed and rolls random.

*Nested
or Nested
or Nested*

Lot	I				II				III			
Roll	1	2	3	4	5	6	7	8	9	10	11	12
	1.5	1.5	2.7	3.0	1.9	2.3	1.8	1.9	2.5	3.2	1.4	7.8
	1.7	1.6	1.9	2.4	1.5	2.4	2.9	3.5	2.9	5.5	1.5	5.2
	1.6	1.7	2.0	2.6	2.1	2.4	4.7	2.8	3.3	7.1	3.4	5.0

3 X 4 x 3 reps = 36 #s   35 db

7.2 In Problem 7.1 how would the results change (if they do) if the lots were chosen at random?

7.3 Set up the EMS column and indicate the proper tests to make if A is a fixed factor at five levels, if B is nested within A at four random levels for each level of A, if C is nested within B at three random levels, and if two observations are made in each "cell."

7.4 Repeat Problem 7.3 for A and B crossed or factorial and C nested within the A, B cells.

7.5 Two types of machines are used to wind coils. One type is hand operated; two machines of this type are available. The other type is power operated; two machines of this type are available. Three coils are wound on each machine from two different wiring stocks. Each coil is then measured for the outside diameter of the wire at a middle position on the coil. The results were as follows. (Units are 10^{-5} inch.)

Machine Type	Hand		Power	
Machine Number	2	3	5	8
Stock 1	3279	3527	1904	2464
	3262	3136	2166	2595
	3246	3253	2058	2303
Stock 2	3294	3440	2188	2429
	2974	3356	2105	2410
	3157	3240	2379	2685

Set up the model for this problem and determine what tests can be run.

7.6 Do a complete analysis of Problem 7.5.

7.7 If the outside diameter readings in Problem 7.5 were taken at three fixed positions on the coil (tip, middle, and end), what model would now be appropriate and what tests could be run?

7.8   An experiment is to be designed to compare faculty morale in three types of junior high school organizations: grades 6–8, grades 7–9, and grades 7–8. Two schools are randomly selected to represent each of these three types of organizations and data are to be collected from several teachers in each school. Someone suggests that the morale of male and female teachers might differ so it is agreed to choose random samples of five male and five female teachers from each school so that gender differences in morale may also be investigated.
   a. Set up a mathematical model for this experiment, outline into ANOVA, and indicate what tests can be made.
   b. Make any recommendations that seem reasonable based on your answers to part (a), assuming that the data have not yet been collected.

7.9   In an attempt to study the effectiveness of two corn varieties (V_1 and V_2), two corn-producing counties (C) in Iowa were chosen at random. Four farms (F) were then randomly selected within each county. Seeds from both varieties were sent to these farms and planted in random plots. When harvested, the number of bushels of corn per acre was recorded for each variety on each farm.
   a. Write a mathematical model for this situation.
   b. Determine the EMS for varieties in this experiment.
   c. Assuming only counties (C) showed significantly different average yields, and $MS_C = 130$ and $MS_F = 10$ (farms or error), find the percentage of the variance in this experiment that can be attributed to county differences.

7.10   An educator proposes a new teaching method and wishes to compare the achievement of students using his method with that of students using a traditional method. Twenty students are randomly placed into two groups of 10 students per group. Tests are given to all 20 students at the beginning of a semester, at the end of the semester, and 10 weeks after the end of the semester. The educator wishes to determine whether there is a difference in the average achievement between the two methods at each of the three time periods.
   a. Write a mathematical model for this situation.
   b. Set up an ANOVA table and show the F tests that can be made.
   c. If the educator's method is really better than the traditional method, what would you expect if you graphed the results at each time period? Show by sketch.
   d. If method A—the new method—gave a mean achievement score across all time periods of 60.0, explain in some detail how you would set confidence limits on this method mean.

7.11   In Example 7.4 in the text the ANOVA of Table 7.16 suggests that further analysis is needed. Using the data of Table 7.15 make further analyses and state your conclusions. (A graph might also be helpful.)

7.12   In studying the intellectual attitude toward science as measured on the Bratt attitude scale, a science educator gave a pretest and post-test to an experimental group exposed to a new curriculum and a control group that was not so exposed. There were 15 students in each group. Set up a mathematical model for this situation and determine what tests are appropriate.

7.13 In Problem 7.12 a strong interaction was found between the groups and the tests with means as follows: pretest control = 57.00, pretest experimental = 54.96, post-test control = 54.66, post-test experimental = 65.66. Assuming the sum of squares for the error term used to test for this interaction was 492.61, make further tests and state some conclusions based on this information.

7.14 A researcher wished to test the figural fluency of three groups of Egyptian students: those taught by the Purdue Creative Thinking Program (PCTP) in a restricted atmosphere, those taught by PCTP in a permissive atmosphere, and a control group taught in a traditional manner with no creativity training. Four classes were randomly assigned to each of the three methods. Before training began, all pupils were administered a Group Embedded Figures Test and classified as either field-dependent or field-independent students. After six weeks of training, results on the averages of students within these classes on this one subtest (figural fluency) were found to be

Source	df	SS	MS	F	Prob.
Methods (M_i)		124.60			
Classes within method ($C_{j(i)}$)		35.10			
Cognitive style (F_k)		2.76			
$M \times F$ interaction (MF_{ik})		25.76			
$F \times C$ interaction ($FC_{kj(i)}$)		20.97			

Complete the table and state your conclusions.

7.15 Fifty-four randomly selected students are assigned to one of three curricula (C) of science lessons. Over a period of a semester six lessons (L) were presented to all students in these three groups according to their prescribed curricula. The first two curricula were experimental A—taught question-asking skills with evaluation feedback, and experimental B—taught with question-asking skills without the evaluation feedback. The third curriculum was used as a control with no instruction in question-asking skill or feedback. Results on the proportion of high-level questions asked on a test gave

Source	df	SS	MS	F	Prob.
C_i	2	0.556	0.278	4.14	0.0211
$S_{k(i)}$	51	3.417	0.067		
L_j	5	0.250	0.050	2.25	0.0493
CL_{ij}	10	0.430	0.043	1.92	0.0431
$LS_{jk(i)}$	255	5.610	0.022		

Discuss these results and explain how the F tests were made.

7.16 For Problem 7.15 it was known in advance that the experimenter wished to compare the two experimental curricula results with the control, and also to compare the two experimental curricula results with each other. Set up proper a priori tests for this purpose and test the results based on the following means and the table given in Problem 7.15.

Mean proportion for curriculum A:	0.2224
Mean proportion for curriculum B:	0.1929
Mean proportion for curriculum C:	0.1236

7.17 In a study made of the characteristics associated with guidance competence versus counseling competence, 144 students were divided into 9 groups of 16 each. These nine groups represented all combinations of three levels of guidance ranking (high, medium, low) and three levels of counseling ranking (high, medium, low). All subjects were then given nine subtests. Assuming the rankings as two fixed factors, the subtests as fixed, and the subjects within the nine groups as random, set up a data layout and mathematical model for this experiment.

7.18 Determine the ANOVA layout and the EMS column for Problem 7.17 and indicate the proper tests that can be made.

7.19 Three days of sampling in which each sample was subjected to two types of size graders gave the following results, coded by subtracting 4 percent moisture and multiplying by 10.

Day	1		2		3	
Grader	A	B	A	B	A	B
Sample 1	4	11	5	11	0	6
2	6	7	17	13	−1	−2
3	6	10	8	15	2	5
4	13	11	3	14	8	2
5	7	10	14	20	8	6
6	7	11	11	19	4	10
7	14	16	6	11	5	18
8	12	10	11	17	10	13
9	9	12	16	4	16	17
10	6	9	−1	9	8	15
11	8	13	3	14	7	11

Assuming graders fixed, days random, and samples within days random, set up a mathematical model for this experiment and determine the EMS column.

7.20 Complete the ANOVA for Problem 7.19 and comment on the results.

7.21 In a filling process two random runs are made. For each run six hoppers are used, but they may not be the same hoppers on the second run. Assume that hoppers are a random sample of possible hoppers. Data are taken from the bottom, middle, and top of each hopper in order to check on a possible position effect on filling. Two observations are made within positions and runs on all hoppers. Set up a mathematical model for this problem and the associated EMS column.

7.22 The data below are from the experiment described in Problem 7.21.

	Run 1						Run 2				
Hopper	Bottom		Middle		Top		Bottom		Middle		Top	
1	19	0	31	25	25	13	6	6	0	6	0	0
2	13	0	19	38	31	25	0	13	0	31	6	0
3	13	0	31	19	13	0	28	6	25	19	13	28
4	13	0	19	31	25	13	28	0	6	0	28	25
5	19	19	19	38	19	25	16	13	13	19	0	0
6	6	0	0	0	0	0	13	6	6	19	0	19

Do a complete analysis of these data.

7.23 Some research on the abrasive resistance of filled epoxy plastics was carried out by molding blocks of plastic using five fillers: iron oxide, iron filings, copper, alumina, and graphite. Two concentration ratios were used for the ratio of filler to epoxy resin: $\frac{1}{2}$:1 and 1:1. Three sample blocks were made up of each of the ten combinations above. These blocks were then subjected to a reciprocating motion in which gritcloth was used as the abrasive material in all tests. After 10,000 cycles, each sample block was measured in three places at each of three fixed positions on the plates: I, II, and III. These blocks had been measured before cycling so that the difference in thickness was used as the measured variable. Measurements were made to the nearest 0.0001 inch. Assuming complete randomization of the order of testing of the 30 blocks, set up a mathematical model for this situation and set up the ANOVA table with an EMS column. We are interested in the effect of concentration ratio, fillers, and positions on the block. The three observations at each position will be considered as error, and there may well be differences between the average of the three sample blocks within each treatment combination.

7.24 Data for Problem 7.23 were found to be as shown in the tables on pages 198–99. Complete an ANOVA for these data and state your conclusions.

7.25 From the results of Problem 7.24, what filler–concentration combination would you recommend if you were interested in a minimum amount of wear? The 1:1 ratio is cheaper than $\frac{1}{2}$:1.

Concentration	Alumina						Graphite					
	$\frac{1}{2}:1$			$1:1$			$\frac{1}{2}:1$			$1:1$		
Position	I	II	III	I	II	III	I	II	III	I	II	III
Sample 1	7.0	5.4	6.4	7.1	5.0	6.4	8.8	5.5	9.0	18.0	14.4	18.5
	7.3	5.1	5.7	6.9	5.7	6.6	8.3	5.9	11.0	17.6	12.4	19.2
	6.6	5.0	7.7	6.5	4.0	6.2	6.9	4.7	11.3	18.4	12.6	19.4
	20.9	15.5	19.8	20.5	14.7	19.2	24.0	16.1	31.3	54.0	39.4	57.1
Sample 2	6.8	5.1	4.6	6.8	4.0	7.5	7.7	6.8	10.1	15.6	11.6	17.6
	6.9	5.5	5.8	7.2	5.2	7.8	6.5	7.0	11.2	14.6	12.8	18.7
	6.2	4.8	5.8	4.8	4.4	6.0	5.9	7.2	10.8	15.1	13.4	19.3
	19.9	15.4	16.2	18.8	13.6	21.3	20.1	21.0	32.1	45.3	37.8	55.6
Sample 3	7.3	6.3	5.1	6.2	4.4	5.7	7.2	7.0	7.7	13.3	11.7	15.5
	7.6	5.3	6.5	6.6	4.8	6.6	6.8	6.9	9.3	13.8	13.4	18.0
	5.6	4.8	7.1	4.4	3.8	5.6	6.2	5.2	9.8	13.6	12.1	19.6
	20.5	16.4	18.7	17.2	13.0	17.9	20.2	19.1	26.8	40.7	37.2	53.1
Totals	163.3			156.2			210.7			420.2		

	Iron Filings						Iron Oxide						Copper					
Concentration	$\frac{1}{2}:1$			$1:1$			$\frac{1}{2}:1$			$1:1$			$\frac{1}{2}:1$			$1:1$		
Position	I	II	III	I	II	III	I	II	III	I	II	III	I	II	III	I	II	III
Sample 1	2.1	1.1	1.3	3.6	1.6	2.3	3.0	1.1	3.2	1.8	1.3	2.4	2.7	1.4	2.8	2.8	1.5	2.4
	2.1	1.1	1.7	3.8	1.0	2.5	3.8	1.8	4.1	2.1	1.0	1.9	2.9	2.2	3.8	2.6	1.1	2.1
	1.0	0.9	1.7	2.9	1.6	2.4	3.0	1.1	3.3	1.9	1.6	1.8	3.0	1.8	3.4	1.9	1.4	2.0
	5.2	3.1	4.7	10.3	4.2	7.2	9.8	4.0	10.6	5.8	3.9	6.1	8.6	5.4	10.0	7.3	4.0	6.5
Sample 2	1.7	1.1	1.7	2.1	1.1	1.7	3.1	1.6	2.2	2.1	1.0	1.5	2.5	1.8	2.4	2.1	1.0	1.5
	1.8	1.0	2.0	2.6	1.1	1.0	3.0	1.7	3.0	2.0	0.8	2.0	3.0	2.6	3.3	2.5	1.0	1.6
	1.3	0.8	2.0	1.6	0.9	1.4	2.7	1.2	2.7	2.3	1.0	2.0	2.0	2.1	2.4	1.5	1.2	1.7
	4.8	2.9	5.7	6.3	3.1	4.1	8.8	4.5	7.9	6.4	2.8	5.5	7.5	6.5	8.1	6.1	3.2	4.8
Sample 3	3.2	0.8	1.4	2.3	1.1	1.8	2.1	1.5	2.0	1.6	1.2	1.6	3.3	2.1	3.4	2.6	1.3	2.6
	2.8	0.8	2.0	2.6	1.2	2.0	2.9	1.9	2.5	1.6	1.1	1.4	3.2	2.0	4.1	2.9	1.5	3.3
	2.0	0.5	1.9	2.1	0.7	2.4	1.6	2.5	2.4	1.9	1.2	2.1	3.2	2.0	3.8	2.8	1.2	2.7
	8.0	2.1	5.3	7.0	3.0	6.2	6.6	5.9	6.9	5.1	3.5	5.1	9.7	6.1	11.3	8.3	4.0	8.6
Totals	41.8			51.4			65.0			44.2			73.2			52.8		

7.26 Glass lenses were produced on a production line that required 10 handling stations. The company was concerned about scratched-lens damage, which might occur at any one of these 10 stations. Two trays of lenses were randomly selected at each station. The number of defective lenses found in samples of 64 out of 576 lenses from each of two positions (inner and outer sections) of each tray were reported. Write a mathematical model for this experiment to check whether the stations differ, whether the two positions in the trays differ, or whether the trays within the stations differ in the number of defective lenses found in the samples of 64.

7.27 For the model in Problem 7.26, determine the appropriate tests that can be run by examining the EMS column for this model.

7.28 Data from Problem 7.26 are given as follows:

	LENS DAMAGE					**LENS DAMAGE**			
OBS	POS	Station	Tray	RDG	OBS	POS	Station	Tray	RDG
1	1	1	1	0	21	2	1	1	0
2	1	1	2	1	22	2	1	2	0
3	1	2	3	3	23	2	2	3	1
4	1	2	4	1	24	2	2	4	0
5	1	3	5	0	25	2	3	5	0
6	1	3	6	3	26	2	3	6	1
7	1	4	7	0	27	2	4	7	0
8	1	4	8	0	28	2	4	8	1
9	1	5	9	0	29	2	5	9	1
10	1	5	10	0	30	2	5	10	0
11	1	6	11	0	31	2	6	11	0
12	1	6	12	2	32	2	6	12	0
13	1	7	13	1	33	2	7	13	0
14	1	7	14	1	34	2	7	14	0
15	1	8	15	0	35	2	8	15	0
16	1	8	16	4	36	2	8	16	2
17	1	9	17	1	37	2	9	17	4
18	1	9	18	0	38	2	9	18	1
19	1	10	19	0	39	2	10	19	0
20	1	10	20	0	40	2	10	20	2

Do a complete analysis of these data and state your conclusions.

Experiments of Two or More Factors— Restrictions on Randomization

8

8.1 Introduction

In the discussion of factorial and nested experiments it was assumed that the whole experiment was performed in a completely randomized manner. In practice, however, it may not be feasible to run the entire experiment several times in one day, or by one experimenter, and so forth. It then becomes necessary to restrict this complete randomization and block the experiment in the same manner that a single-factor experiment was blocked in Chapter 4. Instead of running several replications of the experiment all at one time, it may be possible to run one complete replication on one day, a second complete replication on another day, a third replication on a third day, and so on. In this case each replication is a block, and the design is a randomized block design with a complete factorial or nested experiment randomized within each block.

Occasionally a second restriction on randomization is necessary, in which case a Latin square design might be used, with the treatments in the square representing a complete factorial or nested experiment.

8.2 Factorial Experiment in a Randomized Block Design

Just as the factorial arrangements were extracted from the treatment effect in Equations (5.1) and (5.2), so can these arrangements be extracted from the treatment effect in the randomized complete block design of Chapter 4 (Section 4.2). For the completely randomized design, you will recall that

$$Y_{ij} = \mu + \tau_{ij} + \varepsilon_{k(ij)}$$

can be subdivided into

$$Y_{ijk} = \mu + A_i + B_j + AB_{ij} + \varepsilon_{k(ij)}$$

where $A_i + B_j + AB_{ij} = \tau_{ij}$ of the first model.

When a complete experiment is replicated several times and R_k represents the blocks or replications, the randomized block model is

$$Y_{ijkm} = \mu + R_k + \tau_{ij} + \varepsilon_{m(ijk)}$$

where R_k represents the blocks or replications and τ_{ij} represents the treatments. If the treatments are formed from a two-factor factorial experiment, then

$$\tau_{ij} = A_i + B_j + AB_{ij}$$

and the complete model is

$$Y_{ijk} = \mu + R_k + A_i + B_j + AB_{ij} + \varepsilon_{ijk} \qquad (8.1)$$

Here it is assumed that there is no interaction between replications (blocks) and treatments, which is the usual assumption underlying a randomized block design. Any such interaction is confounded in the error term ε_{ijk}.

To get a more concrete picture of this type of design, consider three levels of factor A, two levels of B, and four replications. The treatment combinations may be written as A_1B_1, A_1B_2, A_2B_1, A_2B_2, A_3B_1, and A_3B_2. Such a design assumes that all six of these treatment combinations can be run on a given day, if days are the replications or blocks. A layout in which these six treatment combinations are randomized within each replication might be

Replication I:	A_1B_2	A_3B_1	A_3B_2	A_2B_1	A_1B_1	A_2B_2
Replication II:	A_2B_2	A_1B_1	A_3B_2	A_2B_1	A_1B_2	A_3B_1
Replication III:	A_2B_1	A_3B_2	A_1B_2	A_3B_1	A_2B_2	A_1B_1
Replication IV:	A_1B_1	A_3B_1	A_3B_2	A_2B_1	A_1B_2	A_2B_2

An analysis breakdown would be that shown in Table 8.1.

Table 8.1 Two-Factor Experiment in a Randomized Block Design

Source	df	
Replications R_k	3	
Treatments τ_{ij}	5	
A_i		2
B_j		1
AB_{ij}		2
Error ε_{ijk}	15	
Total	23	

To see how the interaction of replications and treatments is used as error, consider the expanded model and the corresponding EMS terms exhibited in Table 8.2. Here the factors are fixed and replications are considered as random.

Table 8.2 EMS Terms for Two-Factor Experiment in a Randomized Block

Source	df	3 F i	2 F j	4 R k	1 R m	EMS
R_k	3	3	2	1	1	$\sigma_\varepsilon^2 + 6\sigma_R^2$
A_i	2	0	2	4	1	$\sigma_\varepsilon^2 + 2\sigma_{RA}^2 + 8\phi_A$
RA_{ik}	6	0	2	1	1	$\sigma_\varepsilon^2 + 2\sigma_{RA}^2$
B_j	1	3	0	4	1	$\sigma_\varepsilon^2 + 3\sigma_{RB}^2 + 12\phi_B$
RB_{jk}	3	3	0	1	1	$\sigma_\varepsilon^2 + 3\sigma_{RB}^2$
AB_{ij}	2	0	0	4	1	$\sigma_\varepsilon^2 + \sigma_{RAB}^2 + 4\phi_{AB}$
RAB_{ijk}	6	0	0	1	1	$\sigma_\varepsilon^2 + \sigma_{RAB}^2$
$\varepsilon_{m(ijk)}$	0	1	1	1	1	σ_ε^2 (not retrievable)
Total	23					

Since the degrees of freedom are quite low for the tests indicated above, and since there is no separate estimate of error variance, it is customary to assume that $\sigma_{RA}^2 = \sigma_{RB}^2 = \sigma_{RAB}^2 = 0$ and to pool these three terms to serve as the error variance. The reduced model is then as shown in Table 8.3. Here all effects are tested against the 15-df error term. Although we usually pool the interactions of replications with treatment effects for the error term, this is not always necessary. It will depend upon the degrees of freedom available for testing various hypotheses in Table 8.2, and also upon whether or not any repeat measurements can be taken within a replication. If the latter is possible, a separate estimate of σ_ε^2 is available and all tests in Table 8.2 may be made.

Table 8.3 EMS Terms for Two-Factor Experiment, Randomized Block Design, Reduced Model

Source	df	EMS
R_k	3	$\sigma_\varepsilon^2 + 6\sigma_R^2$
A_i	2	$\sigma_\varepsilon^2 + 8\phi_A$
B_j	1	$\sigma_\varepsilon^2 + 12\phi_B$
AB_{ij}	2	$\sigma_\varepsilon^2 + 4\phi_{AB}$
ε_{ijk}	15	σ_ε^2
Total	23	

■ **EXAMPLE 8.1**   At Purdue a researcher was interested in determining the effect of both nozzle types and operators on the rate of fluid flow in cubic centimeters through these nozzles. The researcher considered three fixed nozzle types and five randomly chosen operators. Each of these 15 combinations was to be run in a random order on each of three days. These three days will be considered as three replications of the complete 3×5 factorial experiment. After subtracting 96.0 cm^3 from each reading and multiplying all readings by 10, the data layout and observed readings (in cubic centimeters) are as recorded in Table 8.4.

Table 8.4 Nozzle Example Data

Operator	1			2			3			4			5		
Nozzle	A	B	C	A	B	C	A	B	C	A	B	C	A	B	C
Replication															
I	6	13	10	26	4	−35	11	17	11	21	−5	12	25	15	−4
II	6	6	10	12	4	0	4	10	−10	14	2	−2	18	8	10
III	−15	13	−11	5	11	−14	4	17	−17	7	−5	−16	25	1	24

The model for this example is

$$Y_{ijk} = \mu + R_k + N_i + O_j + NO_{ij} + \varepsilon_{ijk}$$

with

$$k = 1, 2, 3 \qquad i = 1, 2, 3 \qquad j = 1, 2, \dots, 5$$

The R_k levels are random, O_j levels are random, and N_i levels are fixed. If m repeat measurements are considered in the model, but $m = 1$, the EMS values are those appearing in Table 8.5.

Table 8.5 EMS for Nozzle Example

Source	df	3 F i	5 R j	3 R k	1 R m	EMS
R_k	2	3	5	1	1	$\sigma_\varepsilon^2 + 15\sigma_R^2$
N_i	2	0	5	3	1	$\sigma_\varepsilon^2 + 3\sigma_{NO}^2 + 15\phi_N$
O_j	4	3	1	3	1	$\sigma_\varepsilon^2 + 9\sigma_O^2$
NO_{ij}	8	0	1	3	1	$\sigma_\varepsilon^2 + 3\sigma_{NO}^2$
$\varepsilon_{m(ijk)}$	28	1	1	1	1	σ_ε^2
Total	44					

These EMS values indicate that all main effects and the interaction can be tested with reasonable precision (df). The sums of squares are computed in the usual manner from the subtotals in Table 8.6.

Table 8.6 Subtotals in Nozzle Example

Replication	Total	Nozzle	Operator 1	2	3	4	5	Nozzle Total
I	127	A	−3	43	19	42	68	169
II	92	B	32	19	44	−8	24	111
III	29	C	9	−49	−16	−6	30	−32
Totals	248		38	13	47	28	122	248

Using these subtotals, the usual sums of squares can easily be computed and the results displayed as in Table 8.7.

Table 8.7 ANOVA for Nozzle Example

Source	df	SS	MS	EMS	F	$F_{0.05}$
R_k	2	328.84	164.42	$\sigma_\varepsilon^2 + 15\sigma_R^2$	1.70	3.34
N_i	2	1426.97	713.48	$\sigma_\varepsilon^2 + 3\sigma_{NO}^2 + 15\phi_N$	3.13	4.46
O_j	4	798.79	199.70	$\sigma_\varepsilon^2 + 9\sigma_O^2$	2.06	2.71
NO_{ij}	8	1821.48	227.68	$\sigma_\varepsilon^2 + 3\sigma_{NO}^2$	2.35*	2.29
ε_{ijk}	28	2709.16	96.76	σ_ε^2		
Totals	44	7085.24				

The results of tests of hypotheses on this model show only a significant nozzle × operator interaction at the 5 percent level of significance. The plot in Figure 8.1 of all totals for the right-hand data of Table 8.6 shows this interaction graphically.

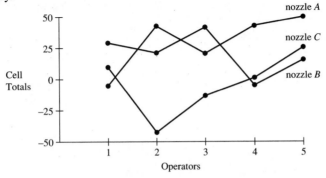

Figure 8.1 Plot of nozzle × operator interaction in nozzle example.

If desirable, the model of this example could be expanded to include an estimate of replication by operator and replication by nozzle interactions using the three-way interaction ($N \times O \times$ Replication) as the error. The results of such a breakdown are recorded in Table 8.8.

Table 8.8 ANOVA for Nozzle Example with Replication Interactions

Source	df	SS	MS	EMS	F
R_k	2	328.84	164.42	$\sigma_\varepsilon^2 + 3\sigma_{RO}^2 + 15\sigma_R^2$	1.86
N_i	2	1426.97	713.48	$\sigma_\varepsilon^2 + 3\sigma_{NO}^2 + 5\sigma_{NR}^2 + 15\phi_N$	*
NR_{ik}	4	272.23	68.06	$\sigma_\varepsilon^2 + 5\sigma_{NR}^2$	< 1
O_j	4	798.79	199.70	$\sigma_\varepsilon^2 + 3\sigma_{RO}^2 + 9\sigma_O^2$	2.26
RO_{jk}	8	705.60	88.20	$\sigma_\varepsilon^2 + 3\sigma_{RO}^2$	< 1
NO_{ij}	8	1821.48	227.68	$\sigma_\varepsilon^2 + 3\sigma_{NO}^2$	2.10
ε_{ijk} or					
NRO_{ijk}	16	1731.33	108.21	σ_ε^2	
Totals	44	7085.24			

Replication	Operator						Nozzle			
	1	2	3	4	5	Total	A	B	C	Total
I	29	−5	39	28	36	127	89	44	−6	127
II	22	16	4	14	36	92	54	30	8	92
III	−13	2	4	−14	50	29	26	37	−34	29
Totals	38	13	47	28	122	248	169	111	−32	248

The F test values are shown in Table 8.8 except for the test on the nozzle labeled with an asterisk. Because there is no direct test, an F' test is used as described in Chapter 6. Here

$$MS = MS_{NR} + MS_{NO} - MS_{NRO}$$

Here the coefficients are 1, 1, and −1, and the corresponding mean squares and degrees of freedom are

$$MS_{NR} = 68.06 \quad \text{with } v_1 = 4 \text{ df}$$
$$MS_{NO} = 227.68 \quad \text{with } v_2 = 8 \text{ df}$$
$$MS_{NRO} = 108.21 \quad \text{with } v_3 = 16 \text{ df}$$

The F' test is then

$$F' = \frac{MS_N}{MS_{NR} + MS_{NO} - MS_{NRO}}$$

$$= \frac{713.48}{68.06 + 227.68 - 108.21}$$

$$= \frac{713.48}{187.53} = 3.80$$

and

$$v = \frac{(187.53)^2}{(1)^2[(68.06)^2 / 4] + (1)^2[(227.68)^2 / 8] + (-1)^2[(108.21)^2 / 16]} = 4.2$$

With 2 and 4 df the 5 percent F value is 6.94, and with 2 and 5 df it is 5.79; hence the $F' = 3.80$ is not significant at the 5 percent level. Hence all F tests indicate no significant effects. The $N \times O$ interaction does not show significance as it did in Table 8.7 because the test is less sensitive due to the reduction in degrees of freedom from 28 to 16 in the error mean square. Because both the $N \times R$ and $R \times O$ interaction F tests are less than one, we are probably justified in assuming that σ_{NR}^2 and σ_{RO}^2 are zero and in pooling these with σ_{NRO}^2 for a 28-df error mean square as in Table 8.7. ∎

8.3 Factorial Experiment in a Latin Square Design

If there are two restrictions on the randomization, and an equal number of restriction levels and treatment levels are used, Chapter 4 has shown that a Latin square design may be used. The model for this design is

$$Y_{ijk} = \mu + R_i + \tau_j + \gamma_k + \varepsilon_{ijk}$$

where τ_j represents the treatments, and both R_i and γ_k represent restrictions on the randomization. If a 4×4 Latin square is used for a given experiment, such as the one in Section 4.5, the four treatments A, B, C, and D could represent the four treatment combinations of a 2×2 factorial (Table 8.9). Here

$$\tau_j = A_m + B_q + AB_{mq}$$

Table 8.9 4×4 Latin Square (Same as Table 4.12)

Position	Car			
	I	II	III	IV
1	C	D	A	B
2	B	C	D	A
3	A	B	C	D
4	D	A	B	C

with its 3 df. Thus a factorial experiment ($A \times B$) is run in a Latin square design. For the problem of Section 4.5, instead of four tire brands, the four treatments could consist of two brands (A_1 and A_2) and two types (B_1 and B_2). The types might be black- and white-walled tires. The complete model would then be

$$Y_{ikmq} = \mu + R_i + \gamma_k + A_m + B_q + AB_{mq} + \varepsilon_{ikmq}$$

and the degree-of-freedom breakdown would be that shown in Table 8.10.

Table 8.10 Degree-of-Freedom
Breakdown for Table 8.9

Source	df	
R_i	3	
γ_k	3	
A_m	1	
B_q	1	} 3 df for τ_j
AB_{mq}	1	
ε_{ikmq}	6	
Total	15	

This idea could be extended to running a factorial in a Graeco–Latin square, and so on.

8.4 Remarks

In the preceding sections factorial experiments were considered when the design of the experiment was a randomized block or a Latin square. It is also possible to run a nested experiment or a nested-factorial experiment in a randomized block or Latin square design. In the case of a nested experiment, the treatment effect might be broken down as

$$\tau_m = A_i + B_{j(i)}$$

and when this experiment has one restriction on the randomization, the model is

$$Y_{ijk} = \mu + R_k + \underbrace{A_i + B_{j(i)}}_{\tau_m} + \varepsilon_{ijk}$$

If a nested factorial is repeated on several different days, the model would be

$$Y_{ijkm} = \mu + R_k + \underbrace{A_i + B_{j(i)} + C_m + AC_{im} + BC_{mj(i)}}_{\tau_m} + \varepsilon_{ijkm}$$

where the whole nested factorial is run in a randomized block design. The analysis of such designs follows from the methods of Chapter 7.

8.5 Summary

The summary at the end of Chapter 7 may now be extended.

Experiment	Design	Analysis
II. Two or more factors A. Factorial (crossed)		
	1. Completely randomized $Y_{ijk} = \mu + A_i + B_j$ $\quad\quad + AB_{ij} + \varepsilon_{k(ij)}, \ldots$ for more factors a. General case	1. a. ANOVA with interactions
	2. Randomized block a. Complete $Y_{ijk} = \mu + R_k + A_i$ $\quad\quad + B_j + AB_{ij} + \varepsilon_{ijk}$	2. a. Factorial ANOVA with replications R_k
	3. Latin square a. Complete $Y_{ijkm} = \mu + R_k + \gamma_m + A_i$ $\quad\quad + B_j + AB_{ij} + \varepsilon_{ijkm}$	3. a. Factorial ANOVA with replications and positions
B. Nested (hierarchical)		
	1. Completely randomized $Y_{ijk} = \mu + A_i + B_{j(i)} + \varepsilon_{k(ij)}$	1. Nested ANOVA
	2. Randomized block a. Complete $Y_{ijk} = \mu + R_k + A_i$ $\quad\quad + B_{j(i)} + \varepsilon_{ijk}$	2. Nested ANOVA with blocks R_k a. Nested ANOVA with blocks R_k
	3. Latin square a. Complete $Y_{ijkm} = \mu + R_k + \gamma_m$ $\quad\quad + A_i + B_{j(i)} + \varepsilon_{ijkm}$	3. a. Nested ANOVA with blocks and positions
C. Nested factorial		
	1. Completely randomized $Y_{ijkm} = \mu + A_i$ $\quad\quad + B_{j(i)} + C_k + AC_{ik}$ $\quad\quad + BC_{kj(i)} + \varepsilon_{m(ijk)}$	1. Nested-factorial ANOVA

Table continued on page 210.

Experiment	Design	Analysis
	2. Randomized block	2.
	a. Complete	
	$Y_{ijkm} = \mu + R_k + A_i$	a. ANOVA with
	$+ B_{j(i)} + C_m + AC_{im}$	blocks R_k
	$+ BC_{mj(i)} + \varepsilon_{ijkm}$	
	3. Latin square	3.
	a. Complete	
	$Y_{ijkmq} = \mu + R_k + \gamma_m + A_i$	a. ANOVA with blocks
	$+ B_{j(i)} + C_q + AC_{iq}$	and positions
	$+ BC_{qj(i)} + \varepsilon_{ijkmq}$	

8.6 SAS Programs

Example 8.1 can be handled with a SAS program for both the reduced model and the full model of Section 8.2. The following program gives the essential steps:

```
DATA A;
INPUT REP OPERATOR NOZZLE $ FLOWRATE @@;
CARDS;
  1  1  A   6  1  1  B  13  1  1  C  10  1  2  A   26
              etc. thru
  3  5  C  24
;
TITLE 'EXAMPLE OF A FACTORIAL IN A RANDOMIZED BLOCK DESIGN';
TITLE2 'REDUCED MODEL ANALYSIS';
PROC GLM;
CLASS REP OPERATOR NOZZLE;
MODEL FLOWRATE = REP OPERATOR|NOZZLE;
TEST H = NOZZLE E = OPERATOR*NOZZLE;
PROC GLM;
TITLE2'FULL MODEL ANALYSIS';
TITLE3 'NO DIRECT TEST FOR NOZZLE';
CLASS REP OPERATOR NOZZLE;
MODEL FLOWRATE = REP|OPERATOR|NOZZLE;
TEST H = OPERATOR E = REP*OPERATOR;
TEST = OPERATOR*NOZZLE E = REP*OPERATOR*NOZZLE;
RUN;
```

and the printout is on pages 211–212.

These results agree with those in Tables 8.7 and 8.8.

EXAMPLE OF A FACTORIAL IN A RANDOMIZED BLOCK DESIGN
REDUCED MODEL ANALYSIS
GENERAL LINEAR MODELS PROCEDURE
CLASS LEVEL INFORMATION

CLASS	LEVELS	VALUES
REP	3	1 2 3
OPERATOR	5	1 2 3 4 5
NOZZLE	3	1 2 3

NUMBER OF OBSERVATIONS IN DATA SET = 45

EXAMPLE OF A FACTORIAL IN A RANDOMIZED BLOCK DESIGN
REDUCED MODEL ANALYSIS
GENERAL LINEAR MODELS PROCEDURE

DEPENDENT VARIABLE: FLOWRATE

SOURCE	DF	SUM OF SQUARES	MEAN SQUARE	F VALUE	PR > F	R-SQUARE	C.V.
MODEL	16	4376.08888889	273.50555556	2.83	0.0078	0.617634	178.4838
ERROR	28	2709.15555556	96.75555556			ROOT MSE	FLOWRATE MEAN
CORRECTED TOTAL	44	7085.24444444				9.83644019	5.51111111

SOURCE	DF	TYPE I SS	F VALUE	PR > F	TYPE III SS	F VALUE	PR > F
REP	2	328.84444444	1.70	0.2011	328.84444444	1.70	0.2011
NOZZLE	2	1426.97777778	7.37	0.0027	1426.97777778	7.37	0.0027
OPERATOR	4	798.80000000	2.06	0.1124	798.80000000	2.06	0.1124
OPERATOR*NOZZLE	8	1821.46666667	2.35	0.0448	1821.46666667	2.35	0.0448

TESTS OF HYPOTHESES USING THE TYPE III MS FOR OPERATOR*NOZZLE AS AN ERROR TERM

SOURCE	DF	TYPE III SS	F VALUE	PR > F
NOZZLE	2	1426.97777778	3.13	0.0989

EXAMPLE OF A FACTORIAL IN A RANDOMIZED BLOCK DESIGN
FULL MODEL ANALYSIS
NO DIRECT TEST FOR NOZZLE
GENERAL LINEAR MODELS PROCEDURE
CLASS LEVEL INFORMATION

CLASS	LEVELS	VALUES
REP	3	1 2 3
OPERATOR	5	1 2 3 4 5
NOZZLE	3	1 2 3

NUMBER OF OBSERVATIONS IN DATA SET = 45

EXAMPLE OF A FACTORIAL IN A RANDOMIZED BLOCK DESIGN
FULL MODEL ANALYSIS
NO DIRECT TEST FOR NOZZLE
GENERAL LINEAR MODELS PROCEDURE

DEPENDENT VARIABLE: FLOWRATE

SOURCE	DF	SUM OF SQUARES	MEAN SQUARE	F VALUE	PR > F	R-SQUARE	C.V.
MODEL	44	7085.24444444	161.02828283	.	.	1.000000	0.0000
ERROR	0	0.00000000	0.00000000				
CORRECTED TOTAL	44	7085.24444444			ROOT MSE		FLOWRATE MEAN
					0.00000000		5.5111111

SOURCE	DF	TYPE I SS	F VALUE	PR > F	DF	TYPE III SS	F VALUE	PR > F
REP	2	328.84444444	.	.	2	328.84444444	.	.
NOZZLE	2	1426.97777778	.	.	2	1426.97777778	.	.
REP*NOZZLE	4	272.22222222	.	.	4	272.22222222	.	.
OPERATOR	4	798.80000000	.	.	4	798.80000000	.	.
REP*OPERATOR	8	705.60000000	.	.	8	705.60000000	.	.
OPERATOR*NOZZLE	8	1821.46666667	.	.	8	1821.46666667	.	.
REP*OPERATOR*NOZZLE	16	1731.33333333	.	.	16	1731.33333333	.	.

TESTS OF HYPOTHESES USING THE TYPE III MS FOR REP*OPERATOR AS AN ERROR TERM

SOURCE	DF	TYPE III SS	F VALUE	PR > F
OPERATOR	4	798.80000000	2.26	0.1512

TESTS OF HYPOTHESES USING THE TYPE III MS FOR REP*OPERATOR*NOZZLE AS AN ERROR TERM

SOURCE	DF	TYPE III SS	F VALUE	PR > F
OPERATOR*NOZZLE	8	1821.46666667	2.10	0.0977

Problems

8.1 Data on the glass rating of tubes taken from two fixed stations and three shifts are recorded each week for three weeks. Considering the weeks as blocks, the six treatment combinations (two stations by three shifts) were tested in a random order each week, with results as follows:

		Station 1			Station 2		
Shift		1	2	3	1	2	3
Week 1		3	3	3	6	3	6
		6	4	6	8	9	8
		6	7	7	11	11	13
Week 2		14	8	11	4	15	4
		16	8	12	6	15	7
		19	9	17	7	17	10
Week 3		2	2	2	2	2	10
		3	3	4	5	4	12
		6	4	6	7	6	13

Analyze these data as a factorial run in a randomized block design.

8.2 In the example of Section 4.5 (Table 4.13) consider tire brands A, B, C, D as four combinations of two factors, ply of tires and type of tread, where

$$A = P_1 T_1 \qquad B = P_1 T_2 \qquad C = P_2 T_1 \qquad D = P_2 T_2$$

representing the four combinations of two ply values and two tread types. Analyze the data for this revised design—a 2^2 factorial run in a Latin square design.

8.3 If in Problem 7.5 the stock factor is replaced by "days," considered random, and the rest of the experiment is performed in a random manner on each of the two days, use the same numerical results and reanalyze with this restriction.

8.4 Again, in Problem 7.5 assume that the whole experiment (24 readings) was repeated on five randomly selected days. Set up an outline of this experiment, including the EMS column, df, and so on.

8.5 Three complete replications were run in a study of aluminum thickness. Samples were taken from five machines and two types of slug—rivet and staple—in each replicate. Considering replicates random and machines and type of slug fixed, set up a model for this problem and determine what tests can be made. Comment on the adequacy of the various tests. Data are as follows:

		Machine				
Replication	Slug	1	2	3	4	5
I	Rivet	175	95	180	170	155
	Staple	165	165	175	185	130
II	Rivet	190	185	180	200	190
	Staple	170	160	175	165	190
III	Rivet	185	165	175	195	200
	Staple	190	160	200	185	200

8.6 Analyze the data of Problem 8.5 using a computer program and state your conclusions.

8.7 Data were collected for three consecutive years on the number of days after planting that the first flowering of a plant occurred. Each year five varieties of plants were planted at six different stations. Considering the repeat years random and varieties and stations fixed, outline the ANOVA for this problem.

8.8 The data for Problem 8.7 are given below.

	Variety				
Station/Year	Hazel	Coltsfoot	Anemone	Blackthorn	Mustard
Broadchalke					
1932	57	67	95	102	123
1933	46	72	90	88	101
1934	28	66	89	109	113
Total	131	205	274	299	337
Bratton					
1932	26	44	92	96	93
1933	38	68	89	89	110
1934	20	64	106	106	115
Total	84	176	287	291	318
Lenham					
1932	48	61	78	99	113
1933	35	60	89	87	109
1934	48	75	95	113	111
Total	131	196	262	299	333

Table continued on page 215.

Station/Year	Variety				
	Hazel	Coltsfoot	Anemone	Blackthorn	Mustard
Dorstone					
1932	50	68	85	117	124
1933	37	65	74	93	102
1934	19	61	80	107	118
Total	106	194	239	317	344
Coaley					
1932	23	74	105	103	120
1933	36	47	85	90	101
1934	18	69	85	105	111
Total	77	190	275	298	332
Ipswich					
1932	39	57	91	102	112
1933	39	61	82	93	104
1934	43	61	98	98	112
Total	121	179	271	293	328

Do a complete ANOVA on these data and discuss the results.

8.9 Puffed cereal was packaged from a special machine with six heads, and the filling could come from any one of three positions—top, middle, or bottom of a hopper. Data were collected on a measure of filling capacity from each of these 18 possible combinations of head and position. Measurements were taken on each of two cycles within each of the combinations above. These cycles within treatments can be considered random. The whole experiment was repeated in a second run at a later date. Assuming runs are random, set up an analysis of variance table for this problem and also make up a data layout table for the collection of the data.

8.10 Data for Problem 8.9 are given below for two runs, three positions (B, M, T), six heads (H), and two cycles (C) with each combination:

	Run 1						Run 2					
	B		M		T		B		M		T	
H	C_1	C_2	C_1	C_2	C_1	C_2	C_1	C_2	C_1	C_2	C_1	C_2
1	19	0	31	25	25	13	6	6	0	6	0	0
2	13	0	19	38	31	25	0	13	0	31	6	0
3	13	0	31	19	13	0	28	6	25	19	13	28
4	13	0	19	31	25	13	28	0	6	0	28	25
5	19	19	19	38	19	25	16	13	13	19	0	0
6	6	0	0	0	0	0	13	6	6	19	0	19

Do a complete analysis of these data. Explain whether or not you could combine terms in the table as done in the text example.

8.11 Discuss the similarities and differences between a nested experiment with two factors and a randomized block design with one factor.

8.12 Assuming the nested experiment involving machines and heads in Section 7.2 was completely repeated on three subsequent randomly selected days, outline its mathematical model and ANOVA table complete with the EMS column.

8.13 For Problem 8.12 show a complete model with *all* interactions and the reduced model. Make suggestions regarding how this experiment might be made less costly and still give about the same information.

8.14 For the nested-factorial example in Section 7.4 on gun loading set up the model assuming the whole experiment is repeated on four more randomly selected days.

8.15 Examine both the full and reduced models for Problem 8.14 and comment.

2^f Factorial Experiments

9

9.1 Introduction

In Chapter 5 factorial experiments were considered and a general method for their analysis was given. There are a few special cases that are of considerable interest in future designs. One of these is the case of f factors in which each factor is at just two levels. These levels might be two extremes of temperature, two extremes of pressure, two time values, two machines, and so on. Although this may seem like a rather trivial case since only two levels are involved, it is, nevertheless, very useful for at least two reasons: to introduce notation and concepts useful when more involved designs are discussed and to illustrate what main effects and interactions there really are in this simple case. It is also true that in practice many experiments are run at just two levels of each factor. The ceramic tool-cutting example of Chapters 1 and 5 is a $2 \times 2 \times 2$, or 2^3 factorial, with four observations per cell. Throughout this chapter the two levels will be considered as fixed levels. This is quite reasonable because the two levels are chosen at points near the extremes, rather than at random.

9.2 2^2 Factorial

The simplest case to consider is one in which two factors are of interest and each factor is set at just two levels. This is a $2 \times 2 = 2^2$ factorial, and the design will be considered as completely randomized. The example in Section 5.1 is of this type. The factors are temperature and altitude, and each is set at two levels: temperature at 25°C and 55°C and altitude at 0 K and 3 K. This gives four treatment combinations, displayed in Figure 5.3 where the response variable is the current flow in milliamperes. To generalize a bit, consider temperature as factor A and altitude as factor B. The model for this completely randomized design would be

$$Y_{ij} = \mu + A_i + B_j + AB_{ij} + \varepsilon_{ij} \qquad (9.1)$$

where $i = 1, 2$ and $j = 1, 2$ in this case. Unless there is some replication, of course, no assessment of interaction can be made independent of error. From the data of Figure 5.3 when both temperature A and altitude B are at their low levels, the response is 210 mA. This may be designated by subscripts on AB as follows: $A_0B_0 = 210$ mA. Following this notation, A_1B_0 means A at its high level and B at its low level, or temperature 55°C and altitude 0 K. The response is $A_1B_0 = 240$ mA. Likewise, $A_0B_1 = 180$ mA and $A_1B_1 = 200$ mA. Since a 2^f experiment is encountered so often in the literature, most authors have adopted another notation for these treatment combinations. For this new notation, just the subscripts on AB are used as exponents on the small letters ab. If both factors are at their low levels, $a^0b^0 = (1)$, and (1) represents the response of both factors at their low level. $a^0b^1 = b$ represents B at its high level and A at its low level, $a^1b^0 = a$ represents the high level of A and low level of B, and $a^1b^1 = ab$ represents the response when both factors are at their high levels. This notation can easily be extended to more factors, provided only two levels of each factor are involved.

For the example of Figure 5.3 the treatment combinations can be represented by the vertices of a square as in Figure 9.1. In Figure 9.1 the low and high levels of factors A and B are represented by 0 and 1, respectively, on the A and B axes. The intersection of these levels in the plane of the figure shows the four treatment combinations. For example: $00 = (1)$ represents both factors at their low levels, $10 = a$ represents A high, B low, and so on, $01 = b$, $11 = ab$. These expressions are only symbolic and are to be considered as merely a mnemonic device to simplify the design and its analysis. The normal order for writing these treatment combinations is (1), a, b, ab. Note that (1) is written first, then the high level of each factor with the low level of the other (a, b), and the fourth term is the algebraic product of the second and third (ab). When a third factor

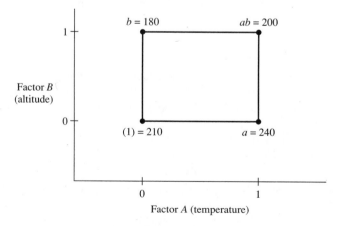

Figure 9.1 2^2 factorial experiment.

is introduced, it is placed at the end of this sequence and then multiplied by all of its predecessors. For example, if factor C is also present at two levels, the treatment combinations are

$$(1), a, b, ab, c, ac, bc, abc$$

which can be represented as the vertices of a cube.

Returning to Figure 9.1 we find that the *effect of a factor* is defined as the change of response produced by a change in the level of that factor. At the low level of B, the effect of A is then $240 - 210$ or $a - (1)$, whereas the effect of A at the high level of B is $200 - 180$ or $ab - b$. The average effect of A is then

$$A = \frac{1}{2}[a - (1) + ab - b]$$

or

$$A = \frac{1}{2}[-(1) + a - b + ab]$$

Note that the coefficients on this A effect are all $+1$ when A is at its high level in the treatment combinations $(+a, +ab)$ and the coefficients are all -1 when A is at its low level as in (1) and b. Note also that

$$2A = -(1) + a - b + ab$$

is a contrast as defined in Section 3.4 (the sum of its coefficients is $-1 + 1 - 1 + 1 = 0$). This concept will be useful, as the sum of squares due to this contrast or effect can easily be determined.

The average effect of B, based on low level of A $[180 - 210 = b - (1)]$ and high level of A $(200 - 240 = ab - a)$, is

$$B = \frac{1}{2}[b - (1) + ab - a]$$

or

$$B = \frac{1}{2}[-(1) - a + b + ab]$$

and again it is seen that the same four responses are used, but $+1$ coefficients are on the treatment combinations for the high level of B and -1 coefficients for the low level of B. Also, $2B$ is a contrast.

To determine the effect of the interaction between A and B, note that at the high level of B the A effect is $ab - b$, and at the low level of B the A effect is $a - (1)$. If these two effects differ, there is an interaction between A and B. Thus the interaction is the average *difference* between these two differences

$$AB = \frac{1}{2}\{(ab - b) - [a - (1)]\}$$

$$= \frac{1}{2}[ab - b - a + (1)]$$

or

$$= \frac{1}{2}[(1) - a - b + ab]$$

Here again the same four treatment combinations are used with a different combination of coefficients. Note that $2AB$ is also a contrast, and all contrasts $2A$, $2B$, $2AB$ are orthogonal to each other. Summarizing in normal order gives

$$2A = -(1) + a - b + ab$$
$$2B = -(1) - a + b + ab$$
$$2AB = +(1) - a - b + ab$$

Note that the interaction effect takes the responses on one diagonal of the square with $+1$ coefficients and the responses on the other diagonal with -1 coefficients. Note also that the coefficients for the interaction effect can be found by multiplying the corresponding coefficients of the two main effects. As the only coefficients used are $+1$'s and -1's, the proper coefficients on the treatment combinations for each main effect and interaction can be determined from Table 9.1.

Table 9.1 Coefficients for Effects in a 2^2 Factorial Experiment

Treatment Combination	Effect		
	A	B	AB
(1)	$-$	$-$	$+$
a	$+$	$-$	$-$
b	$-$	$+$	$-$
ab	$+$	$+$	$+$

From Table 9.1 the orthogonality of the effects is easily seen, as well as the generation of the interaction coefficients from the main effect coefficients.

Another approach to the interaction between A and B would be to consider that at the high level of A the effect of B is $ab - a$, and at the low level of A the effect of B is $b - (1)$, so that the average difference is

$$AB = \frac{1}{2}\{(ab - a) - [b - (1)]\}$$

$$= \frac{1}{2}[ab - a - b + (1)]$$

$$= \frac{1}{2}[+(1) - a - b + ab]$$

which is the same expression as given before.

For the response data of Figure 9.1 the effects are

$$A = \frac{1}{2}(-210 + 240 - 180 + 200) = 25 \text{ mA}$$

$$B = \frac{1}{2}(-210 - 240 + 180 + 200) = -35 \text{ mA}$$

$$AB = \frac{1}{2}(+210 - 240 - 180 + 200) = -5 \text{ mA}$$

Because $2A$, $2B$, and $2AB$ are contrasts, the sum of squares due to a contrast is

$$SS_{C_m} = \frac{(\text{contrast})^2}{n \sum c_{jm}^2}$$

where n is the number of observations in each total (here $n = 1$). Also $\sum c_{jm}^2 = 1 + 1 + 1 + 1 = 4$ (or 2^2). From this definition

$$SS_A = \frac{[2(25)]^2}{4} = 625$$

$$SS_B = \frac{[2(-35)]^2}{4} = 1225$$

$$SS_{AB} = \frac{[2(-5)]^2}{4} = 25$$

$$SS_{\text{total}} = 1875$$

Because each effect and the interaction has but 1 df and there is no measure of error as only one observation was taken in each cell, this ANOVA is trivial, although it does show another approach to analysis based on effects. The simple ANOVA table would be that shown by Table 9.2.

Table 9.2 ANOVA for a 2^2 Factorial with No Replication

Source	df	SS	MS
A_i	1	625	625
B_j	1	1225	1225
Error or AB_{ij}	1	25	25
Totals	3	1875	

If the general methods of Chapter 5 are used on the data in Figure 9.1

(coded by subtracting 200), the results are those in Table 9.3,

$$SS_{total} = (10)^2 + (-20)^2 + (40)^2 + (0)^2 - \frac{(30)^2}{4} = 1875$$

$$SS_A = \frac{(-10)^2 + (40)^2}{2} - \frac{(30)^2}{4} = 850 - 225 = 625$$

$$SS_B = \frac{(50)^2 + (-20)^2}{2} - \frac{(30)^2}{4} = 1450 - 225 = 1225$$

$$SS_{AB} \text{ (by subtraction)} = 1875 - 625 - 1225 = 25$$

which are the same sums of squares as given in Table 9.2. Even though no separate measure of error is available, a glance at the mean squares of Table 9.2 shows that both main effects are large compared with the interaction effect.

Table 9.3 Coded Data of Figure 9.1

Factor B	Factor A		Totals
	0	1	
0	+10	+40	+50
1	-20	0	-20
Totals	-10	+40	+30

The results of this section can easily be extended to cases where there are n replications in the cells, by using cell totals for the responses at (1), a, b, and ab and adjusting the sum of squares for the effects accordingly (see Example 9.1).

■ **EXAMPLE 9.1** An example of a 2^2 factorial experiment with two replications per cell is considered here, using hypothetical responses to illustrate the principles of the previous section. Consider the data of Table 9.4.

Table 9.4 2^2 Factorial with Two Replications

Factor B	Factor A		Totals
	0	1	
0	4 6	2 −2	
	10	0	+10
1	3 7	−4 −6	
	10	−10	0
Totals	+20	−10	+10

Using the general methods of Chapter 5, the sums of squares are

$$SS_{total} = 4^2 + 6^2 + 3^2 + 7^2 + 2^2 + (-2)^2 + (-4)^2 + (-6)^2 - \frac{(10)^2}{8}$$

$$= 170 - 12.5 = 157.5$$

$$SS_A = \frac{(20)^2 + (-10)^2}{4} - 12.5 = 112.5$$

$$SS_B = \frac{(10)^2 + (0)^2}{4} - 12.5 = 12.5$$

$$SS_{A \times B \text{ interaction}} = \frac{(10)^2 + (10)^2 + (0)^2 + (-10)^2}{2} - 12.5 - 112.5 - 12.5$$

$$= 12.5$$

$$SS_{error} = 157.5 - 112.5 - 12.5 - 12.5 = 20.0$$

These results could be displayed in an ANOVA table. Using the methods of Section 9.2, we get, with the treatment combinations, the total response

$$(1) = 10 \qquad a = 0 \qquad b = 10 \qquad ab = -10$$

and the contrasts

$$4A = -10 + 0 - 10 + (-10) = -30$$
$$4B = -10 - 0 + 10 + (-10) = -10$$
$$4AB = +10 - 0 - 10 + (-10) = -10$$

The coefficient 4 used with each response represents the two individual responses at each level. In general the coefficient of these effects is $n \cdot 2^{f-1}$ where n is the number of replications and f the number of factors. Here $f = 2$ and $n = 2$. The sums of squares for these three contrasts are

$$SS_A = \frac{(4A)^2}{n \cdot 2^2} = \frac{(-30)^2}{2 \cdot 4} = \frac{900}{8} = 112.5$$

$$SS_B = \frac{(4B)^2}{n \cdot 2^2} = \frac{(-10)^2}{2 \cdot 4} = \frac{100}{8} = 12.5$$

$$SS_{AB} = \frac{(4AB)^2}{n \cdot 2^2} = \frac{(-10)^2}{2 \cdot 4} = \frac{100}{8} = 12.5$$

because

$$2^2 = \sum_{i=1}^{4} c_{jm}^2 = 1 + 1 + 1 + 1 = 4$$

These results are the same as given by the general method. The total sum of squares must be calculated as usual in order to get the error sum of squares by subtraction.

For this special case of a 2^f factorial experiment, Yates [31] developed a rather simple scheme for computing these contrasts. The method can best be illustrated on Example 9.1, using Table 9.5.

Table 9.5 Yates Method on a 2^2 Factorial

Treatment Combination	Response	(1)	(2)	SS
(1)	10	10	10 = total	12.5
a	0	0	−30 = 4A	112.5
b	10	−10	−10 = 4B	12.5
ab	−10	−20	−10 = 4AB	12.5

In Table 9.5, list all treatment combinations in the first column. Place the total response to each of these treatment combinations in the second column. For the third column, labeled (1), add the adjacent responses in pairs, for example, $10 + 0 = 10$ for the first two and $10 - 10 = 0$ for the next two; this completes half of column (1). For the second half, subtract the responses in pairs, always subtracting the first from the second, for example, $0 - 10 = -10$; $-10 - (10) = -20$. This completes column (1). Column (2) is determined in the same manner as column (1) using the column (1) results: $10 + 0 = 10$; $-10 - 20 = -30$; $0 - 10 = -10$; $-20 - (-10) = -10$. Proceed in this same manner until the fth column is reached: (1), (2), (3), . . . , (f). In this case $f = 2$, so there are just two columns. The values in column f are the contrasts, where the first entry is the grand total of all readings; the one corresponding to a is $n \cdot 2^{f-1} \cdot A$, b is $n \cdot 2^{f-1} \cdot B$, and ab is $n \cdot 2^{f-1} \cdot AB$.

When the results of the last column [column (f)] are squared and divided by $n \cdot 2^f$, the results are the sums of squares as shown above. The first sum of squares

$(\text{total})^2 / 8$ is the correction term for the grand mean given in Chapter 5. The Yates method reduces the analysis to simply adding and subtracting numbers. It is very useful provided the experiment is a 2^f factorial with n observations per treatment. ■

As proof for the Yates method on a 2^2 factorial go through the steps above using the treatment combination symbols for the responses as in Table 9.6. It is obvious that the last column does give the proper treatment contrasts.

Table 9.6 Yates Method on a 2^2 in General; One Observation per Cell

Treatment Combination	(1)	(2)
(1)	$(1) + a$	$(1) + a + b + ab = \text{total}$
a	$b + ab$	$-(1) + a - b + ab = 2A$
b	$a - (1)$	$-(1) - a + b + ab = 2B$
ab	$ab - b$	$(1) - a - b + ab = 2AB$

9.3 2^3 Factorial

Considering a third factor C, also at two levels, the experiment will be a $2 \times 2 \times 2$ or 2^3 factorial, again run in a completely randomized manner. The treatment combinations are now (1), a, b, ab, c, ac, bc, abc, and they may be represented as vertices of a cube as in Figure 9.2.

With these $8 = 2^3$ observations, or $8n$ if there are replications, the main effects and each interaction may be expressed by using the proper coefficients $(-1$ or $+1)$ on these eight responses. For the effect of factor A, consider all

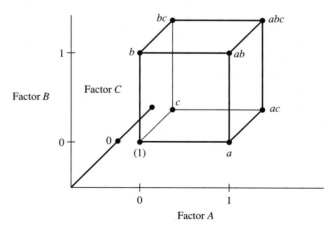

Figure 9.2 2^3 factorial arrangements.

responses in the right-hand plane (a, ab, ac, abc) with plus signs and all in the left-hand plane [(1), b, bc, c] with minus signs as these will show the effect of increasing the level of A. Or,

$$4A = -(1) + a - b + ab - c + ac - bc + abc$$

For factor B consider responses in the lower and higher planes of the cube

$$4B = -(1) - a + b + ab - c - ac + bc + abc$$

The AB interaction is determined by the difference in the A effect from level 0 of B to level 1 of B, regardless of C

$$\text{at } B_0: \quad 2A \text{ effect } a + ac - c - (1)$$
$$\text{at } B_1: \quad 2A \text{ effect } abc + ab - b - bc$$

The difference in these is the interaction

$$4AB = [abc + ab - b - bc] - [a + ac - c - (1)]$$
$$4AB = abc + ab - b - bc - a - ac + c + (1)$$

or

$$4AB = +(1) - a - b + ab + c - ac - bc + abc$$

which gives the same signs as the products of corresponding signs on $4A$ and $4B$.

For factor C compare the responses in the back plane of the cube with those in the front plane of the cube

$$4C = c + bc + ac + abc - (1) - b - a - ab$$

or

$$4C = -(1) - a - b - ab + c + ac + bc + abc$$

The AC and BC interactions can be determined as AB was, and it can be seen that the resulting interaction effects are

$$4AC = +(1) - a + b - ab - c + ac - bc + abc$$
$$4BC = +(1) + a - b - ab - c - ac + bc + abc$$

To determine the ABC interaction, consider the BC interaction at level 0 of A versus the BC interaction at level 1 of A. Any difference in these is an ABC interaction

$$2BC \text{ interaction at } A_0: \quad +(1) - b - c + bc$$
$$2BC \text{ interaction at } A_1: \quad +a - ab - ac + abc$$

Their difference is

$$4ABC = (a - ab - ac + abc) - [(1) - b - c + bc]$$

or

$$4ABC = -(1) + a + b - ab + c - ac - bc + abc$$

which can also be obtained from multiplying the coefficients of A and BC, or B and AC, or C and AB. These others can also be used to get the ABC interaction, but the results are the same. Summarizing gives us Table 9.7. Table 9.7 illustrates once again the orthogonality of the effects and can easily be extended to four, five, and more factors if each factor is at two levels only. In a 2^3 factorial experiment, the sum of squares is given by

$$SS_{contrast} = \frac{(contrast)^2}{n \cdot 2^3} = \frac{(contrast)^2}{8n}$$

and again the Yates method leads to an easy computation of the contrasts.

Table 9.7 Coefficients for Effects in a 2^3 Factorial Experiment

Treatment Combination	Total	A	B	AB	C	AC	BC	ABC
(1)	+	−	−	+	−	+	+	−
a	+	+	−	−	−	−	+	+
b	+	−	+	−	−	+	−	+
ab	+	+	+	+	−	−	−	−
c	+	−	−	+	+	−	−	+
ac	+	+	−	−	+	+	−	−
bc	+	−	+	−	+	−	+	−
abc	+	+	+	+	+	+	+	+

■ **EXAMPLE 9.2** In Chapters 1 and 5 the problem on power requirements for cutting with ceramic tools was analyzed in detail. It is readily seen that this is a $2 \times 2 \times 2 = 2^3$ factorial with four replications per cell. This problem could be analyzed by the special methods of this chapter. From Table 5.10 on coded data the treatment combinations might be summarized as in Table 9.8.

Table 9.8 Ceramic Tool Data in 2^3 Form

	Tool Type			
	1		2	
	Bevel 15°	Angle 30°	Bevel 15°	Angle 30°
Type of Cut				
Continuous	(1)	b	a	ab
	2	15	−5	13
Interrupted	c	bc	ac	abc
	−12	−2	−17	−7

Tool type = factor A, bevel = factor B, cut = factor C.

In Table 9.8 the totals of the four observations for each treatment combination have been entered in their corresponding cells. By the Yates method we get the entries in Table 9.9.

Table 9.9 Yates Method on Ceramic Tool Data

Treatment Combination	Response	(1)	(2)	(3)	SS
(1)	2	−3	25	−13 = total	5.28
a	−5	28	−38	−19 = 16A	11.28
b	15	−29	−9	51 = 16B	81.28
ab	13	−9	−10	5 = 16AB	0.78
c	−12	−7	31	−63 = 16C	124.03
ac	−17	−2	20	−1 = 16AC	0.03
bc	−2	−5	5	−11 = 16BC	3.78
abc	−7	−5	0	−5 = 16ABC	0.78

The sum of squares is seen to be in substantial agreement with those of Table 5.10. We must resort to the individual readings, however, to determine the total sum of squares and then the error sum of squares. This was done in Table 5.10 and the interpretation of this problem is given in Chapter 5. The purpose of repeating the problem here was merely to show the use of the Yates method on a 2^3 factorial experiment. ■

9.4 2^f Remarks

The methods shown for 2^2 and 2^3 factorials may easily be extended to 2^f factorials where f factors are each considered at two levels. The contrasts are determined from the responses to the treatment combinations by associating plus signs with high levels of a factor and minus signs with low levels; the contrasts for interactions are found by multiplication of corresponding coefficients.

The general relationships for 2^f factorials with n observations per treatment are

$$\text{contrast} = n \cdot 2^{f-1} \text{ (effect)}$$

or

$$\text{effect} = \frac{1}{n2^{f-1}} \text{ (contrast)}$$

and

$$SS_{\text{contrast}} = \frac{(\text{contrast})^2}{n2^f}$$

A general ANOVA would be as in Table 9.10 but not in "normal" order.

Table 9.10 ANOVA for a 2^f Factorial with n Replications

Source		df	
Main effects	A	1	$\left.\vphantom{\begin{matrix}1\\1\\1\end{matrix}}\right\} f$
	B	1	
	C	1	
	$\vdots$	$\vdots$	
Two-factor interactions	AB	1	$\left.\vphantom{\begin{matrix}1\\1\\1\end{matrix}}\right\} C(f,2) = \dfrac{f(f-1)}{2}$
	AC	1	
	BC	1	
	$\vdots$	$\vdots$	
Three-factor interactions	ABC	1	$\left.\vphantom{\begin{matrix}1\\1\\1\end{matrix}}\right\} C(f,3) = \dfrac{f(f-1)(f-2)}{6}$
	ABD	1	
	BCD	1	
	$\vdots$	$\vdots$	
Four-factor interactions, and so on			
Sum of all treatment combinations		$2^f - 1$	
Residual or error		$2^f \cdot (n-1)$	
Total		$n \cdot 2^f - 1$	

9.5 Summary

Experiment	Design	Analysis
I. Single factor		
	1. Completely randomized $Y_{ij} = \mu + \tau_j + \varepsilon_{ij}$	1. One-way ANOVA
	2. Randomized block $Y_{ij} = \mu + \beta_i + \tau_j + \varepsilon_{ij}$	2.
	a. Complete	a. Two-way ANOVA
	3. Latin square $Y_{ijk} = \mu + \beta_i + \tau_j$ $\quad + \gamma_k + \varepsilon_{ijk}$	
	a. Complete	a. Three-way ANOVA
	4. Graeco-Latin square $Y_{ijkm} = \mu + \beta_i + \tau_j$ $\quad + \gamma_k + \omega_m + \varepsilon_{ijkm}$	4. Four-way ANOVA

Table continued on page 230.

Experiment	Design	Analysis
II. Two or more factors A. Factorial (crossed)		
	1. Completely randomized $Y_{ijk} = \mu + A_i + B_j$ $+ AB_{ij} + \varepsilon_{k(ij)} \cdots$ for more factors	1.
	a. General case	a. ANOVA with interactions
	b. 2^f case	b. Yates method or general ANOVA; use (1), a, b, ab, ...

Problems

9.1   For the 2^2 factorial with three observations per cell given below (hypothetical data), do an analysis by the method of Chapter 5.

	Factor A	
Factor B	A_1	A_2
B_1	0	4
	2	6
	1	2
B_2	-1	-1
	-3	-3
	1	-7

9.2   Redo Problem 9.1 by the Yates method and compare results.

9.3   In an experiment on chemical yield three factors were studied, each at two levels. The experiment was completely randomized and the factors were known only as A, B, and C. The results are listed below. Analyze by the methods of Chapter 5.

	A_1				A_2			
	B_1		B_2		B_1		B_2	
C_1	C_2	C_1	C_2	C_1	C_2	C_1	C_2
1595	1745	1835	1838	1573	2184	1700	1717
1578	1689	1823	1614	1592	1538	1815	1806

9.4 Analyze Problem 9.3 by the Yates method and compare your results.

9.5 Plot any results from Problem 9.3 that might be meaningful from a management point of view.

9.6 The results of Problem 9.3 lead to another experiment with four factors each at two levels, with the following data.

A_1							
B_1				B_2			
C_1		C_2		C_1		C_2	
D_1	D_2	D_1	D_2	D_1	D_2	D_1	D_2
1985	2156	1694	2184	1765	1923	1806	1957
1592	2032	1712	1921	1700	2007	1758	1717

A_2							
B_1				B_2			
C_1		C_2		C_1		C_2	
D_1	D_2	D_1	D_2	D_1	D_2	D_1	D_2
1595	1578	2243	1745	1835	1863	1614	1917
2067	1733	1745	1818	1823	1910	1838	1922

Analyze these data by the general methods of Chapter 5.

9.7 Analyze Problem 9.6 using the Yates method.

9.8 Plot from Problem 9.6 any results you think are meaningful.

9.9 Consider a four-factor experiment with each factor at two levels.
   a. Write out the treatment combinations in this experiment in normal order.
   b. Assuming there is only one observation per treatment, outline an ANOVA for this example and describe and justify what you would use for an error term to test for significance of various factors and interactions.
   c. Explain how you would determine the AD interaction from the responses to the treatment combinations given in part (a).

9.10 An experimenter is interested in the effects of five factors—A, B, C, D, and E—each at two levels on some response variable Y. Assuming that data are available for each of the possible treatment combinations in this experiment, answer the following.
   a. How would you determine the effect of factor C and its sum of squares based on data taken at each of the treatment combinations?
   b. Set up an ANOVA table for this problem assuming n observations for each treatment combination.
   c. If $n = 2$ for part (b), at what numerical value of F would you reject the hypothesis of no ACE interaction?

   d. How would you propose testing for ACE interaction and what F would be necessary for rejection in this problem if $n = 1$?

9.11 Consider an experiment with five factors A, B, C, D, and E, each at two levels.
   a. How many treatment combinations are there ?
   b. Outline an ANOVA table to test all main effects, all two-way and all three-way interactions, if there is only one observation per treatment combination.
   c. Explain how you would find the proper signs on the treatment combinations for the ACE interaction.
   d. An experimenter comments that "since all factors are at only two levels, it is unnecessary to run a Newman–Keuls test as one only has to look and see which level gives the greater mean response." Explain situations in which the comment would be true for your results and situations in which it would not be true.

9.12 Three factors are studied for their effect on the horsepower necessary to remove a cubic inch of metal per minute. The factors are feed rate at two levels, tool condition at two levels, and tool type at two levels. For each treatment combination three observations are taken and all 24 observations of horsepower are made in a completely randomized order. The data appear below.

Tool Type	Feed Rate			
	0.011		0.015	
	Tool Condition		Tool Condition	
	Same	New	Same	New
Utility	0.576	0.548	0.514	0.498
	0.576	0.555	0.515	0.519
	0.565	0.540	0.518	0.504
Precision	0.526	0.547	0.494	0.521
	0.542	0.524	0.504	0.480
	0.548	0.525	0.530	0.494

   Do a complete ANOVA on these data as a 2^f experiment and state your conclusions.

9.13 If minimum horsepower is desired for the experiment given in Problem 9.12, what are your recommendations as to the best set of factor conditions on the basis of your results in Problem 9.12?

9.14 Any factor whose levels are powers of 2 such as 4, 8, 16, . . . , may be considered as pseudo factors in the form 2^2, 2^3, 2^4, ... The Yates method can than be used if all factor levels are 2 or powers of 2 and the resulting sums of squares for the pseudo factors can be added to give the sums of squares for the actual factors and appropriate interactions. Try out this scheme on the data of Problem 5.21.

9.15 A systematic test was made to determine the effects on the coil breakdown voltage of the following six variables, each at two levels as indicated.
   a. Firing furnace: Number 1 or 3
   b. Firing temperature: 1650°C or 1700°C

c. Gas humidification: Yes or no
d. Coil outside diameter: Large or small
e. Artificial chipping: Yes or no
f. Sleeve: Number 1 or 2

Assuming complete randomization, devise a data sheet for this experiment, outline its ANOVA, and explain your error term if only one coil is to be used for each of the treatment combinations.

9.16 If seven factors are to be studied at two levels each, outline the ANOVA for this experiment in a table similar to Table 9.10 and, assuming $n = 1$, suggest a reasonable error term.

9.17 In a steel mill an experiment was designed to study the effect of dropout temperature, back-zone temperature, and type of atmosphere on heat slivers in samples of steel. This response variable is a mean surface factor obtained for each type of sliver. Results are

	Dropout Temperature			
	2145°		2165°	
	Back-Zone Temperature			
Atmosphere	2040–2140°	2060–2160°	2040–2140°	2060–2160°
Oxidizing	7.8	4.3	4.5	8.2
Reducing	6.8	5.9	11.0	6.0

Assuming complete randomization, analyze these results. To get an error term for testing, three of the data points were repeated to give an SS_{error} of 1.11 based on 3 degrees of freedom.

9.18 Further analyze the results of Problem 9.17 and make any recommendations you can for minimizing heat slivers.

9.19 At the same time that data were collected on heat slivers in Problem 9.17, data were also collected on grinder slivers. The results were found to be as follows:

	Dropout Temperature			
	2145°		2165°	
	Back-Zone Temperature			
Atmosphere	2040–2140°	2060–2160°	2040–2140°	2060–2160°
Oxidizing	0.0	4.3	9.8	4.0
Reducing	0.0	4.4	10.1	3.9

For these data the $SS_{error} = 13.22$ based on three repeated measurements. Completely analyze these data.

9.20 Sketch rough graphs for the following situations in which each factor is at two levels only:

a. Main effects A and B both significant, AB interaction not significant.

b. Main effect A significant, B not significant, and AB significant.

c. Both main effects not significant but interaction significant.

9.21 For a 2^3 factorial can you sketch a graph (or graphs) to show no significant two-factor interactions, but a significant three-factor interaction?

9.22 The effect of chip lot (A or B) and temperature (cold, room) on the failure rate of integrated circuits in parts per million (ppm) was studied at one plant. Two lots were taken at each of the four treatment conditions, and tested with results as follows:

		Chip Lot	
		A	B
	Cold	3,548	6,704
Temperature		3,932	6,656
	Room	10,107	34,653
		9,562	31,097

Use any method you like to analyze these data and state your conclusions.

9.23 The company wished to determine whether it could eliminate the cold test. How would you answer this query in light of the results in Problem 9.22?

9.24 Are you bothered by the results in Problem 9.22? If so, what do you think may have happened in this experiment?

9.25 A steel company wished to determine which of three factors might be affecting the rating of carbon edge on some coils. The factors that it manipulated were type of anneal (shelf or regular), stack position (A or B), and coil position (north or south edge). Carbon edge ratings were determined on four coils in each of the eight treatments. Analyze the data below and state your conclusions.

		Type of Anneal			
		Regular Position		Shelf Position	
		A	B	A	B
	North	0	0	0	0
		0	0	0	0
		81	0	8	0
Edge		36	28	36	0
	South	81	0	81	0
		108	0	36	0
		72	0	36	0
		108	0	36	0

3^f Factorial Experiments

10

10.1 Introduction

As 2^f factorial experiments represent an interesting special case of factorial experimentation, so also do 3^f factorial experiments. The 3^f factorials consider f factors at three levels; thus there are 2 df between the levels of each of these factors. The 3^f factorials play an important role in more complicated design problems, which will be discussed in subsequent chapters. For this chapter, it will be assumed that the design is a completely randomized design and that the levels of the factors considered are fixed levels. Such levels may be either qualitative or quantitative.

For a three-level factor, one can partition the 2 degrees of freedom into two orthogonal contrasts, as shown in Chapter 3, each with 1 degree of freedom.

If an experiment can be designed such that the levels of an effect are quantitative and equispaced, two orthogonal contrasts can be found that will have physical meaning in terms of the equispaced variable, say X.

Figure 10.1 shows three equispaced levels of a factor X; namely X_1, X_2, and X_3; and the corresponding responses of a Y variable at these three points. The Y totals are shown as $T_{.1}, T_{.2}$, and $T_{.3}$. In Figure 10.1, if X increases from X_1 to X_3 the increase in response totals goes from $T_{.1}$ to $T_{.3}$, or $T_{.3} - T_{.1}$ becomes a linear contrast if there really is a straight-line response between X_1 and X_3.

The linear effect from X_1 to X_2 is $T_{.2} - T_{.1}$ and from X_2 to X_3 is $T_{.3} - T_{.2}$. The cumulative linear effect is then the sum of these, or $T_{.2} - T_{.1} + T_{.3} - T_{.2} = T_{.3} - T_{.1}$. If, however, the response is not linear, the linear effect from X_1 to X_2 is different from the linear effect from X_2 to X_3, and this difference produces a curvature in the response or a quadratic effect. This is then given by $(T_{.3} - T_{.2}) - (T_{.2} - T_{.1}) = T_{.3} - 2T_{.2} + T_{.1}$, which is another contrast called the quadratic effect. Note then that the linear and quadratic contrasts are

$$C_L = T_{.3} \qquad - T_{.1}$$
$$C_Q = T_{.3} + 2T_{.2} - T_{.1}$$

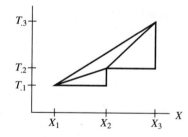

Figure 10.1 Linear and quadratic contrasts.

and these are orthogonal, allowing one to "pull out" a linear and quadratic effect from every quantitative, equispaced three-level factor.

10.2 3^2 Factorial

If just two factors are crossed in an experiment and each of the two is set at three levels, there are $3 \times 3 = 9$ treatment combinations. Because each factor is at three levels, the notation of Chapter 9 will no longer suffice. There is now a low, intermediate, and high level for each factor, which may be designated as 0, 1, and 2. A model for this arrangement would be

$$Y_{ij} = \mu + A_i + B_j + AB_{ij} + \varepsilon_{ij}$$

where $i = 1, 2, 3$, $j = 1, 2, 3$, and the error term is confounded with the AB interaction unless there are some replications in the nine cells, in which case,

$$Y_{ijk} = \mu + A_i + B_j + AB_{ij} + \varepsilon_{k(ij)}$$

and $k = 1, 2, \ldots, n$ for n replications.

To introduce some notation for treatment combinations when three levels are involved, consider the data layout in Figure 10.2, in which two digits are used to describe each of the nine treatment combinations. The first digit indicates the level of factor A, and the second digit the level of factor B. Thus, 12 means A at its intermediate level and B at its highest level. This notation can easily be extended to more factors and as many levels as are necessary. It could have been used for 2^f factorials as 00, 10, 01, and 11, corresponding respectively to (1), a, b, and ab. The only reason for not using this digital notation on 2^f factorials is that so much of the literature includes this (1), a, b, . . . notation. By proper choice of coefficients on these treatment combinations, the linear and quadratic effects of both A and B can be determined, as well as their interactions, such as $A_L \times B_L$, $A_L \times B_Q$, $A_Q \times B_L$, and $A_Q \times B_Q$. The methods of analysis will be illustrated in a simple hypothetical example.

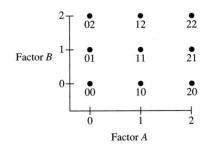

Figure 10.2 3^2 data layout.

■ **EXAMPLE 10.1** Suppose the responses in Table 10.1 were recorded for the treatment combinations indicated in the upper-left-hand corner of each cell.

Table 10.1 3^2 Factorial with Responses and Totals

	Factor A			
Factor B	0	1	2	$T_{.j}$
0	00 ⠀⠀⠀ 1	10 ⠀⠀⠀ -2	20 ⠀⠀⠀ 3	2
1	01 ⠀⠀⠀ 0	11 ⠀⠀⠀ 4	21 ⠀⠀⠀ 1	5
2	02 ⠀⠀⠀ 2	12 ⠀⠀⠀ -1	22 ⠀⠀⠀ 2	3
$T_{.i}$	3	1	6	10

Analyzing the data of Table 10.1 by the general methods of Chapter 5 gives

$$SS_{total} = 1^2 + 0^2 + 2^2 + \cdots + 2^2 - \frac{(10)^2}{9} = 28.89$$

$$SS_A = \frac{3^2 + 1^2 + 6^2}{3} - \frac{(10)^2}{9} = 4.22$$

$$SS_B = \frac{2^2 + 5^2 + 3^2}{3} - \frac{(10)^2}{3} = 1.56$$

$$SS_{error} = 28.89 - 4.22 - 1.56 = 23.11$$

and the ANOVA is given in Table 10.2.

Table 10.2 ANOVA for 3^2 Factorial of Table 10.1

Source	df	SS	MS
A_i	2	4.22	2.11
B_j	2	1.56	0.78
AB_{ij}	4	23.11	5.77
Totals	8	28.89	

A further breakdown of this analysis is now possible by recalling that coefficients of $-1, 0, +1$ applied to the responses at low, intermediate, and high levels of a factor will measure its linear effect, whereas coefficients of $+1, -2, +1$ applied to these same responses will measure the quadratic effect of this factor. As in the case of 2^f factorials, products of coefficients will give the proper coefficients for various interactions. This can best be shown by Table 10.3, which indicates the coefficients for each effect to be used with the nine treatment combinations.

Table 10.3 Coefficients for a 3^2 Factorial with Quantitative Levels

Factor	00	01	02	10	11	12	20	21	22	$\sum c_i^2$
A_L	-1	-1	-1	0	0	0	$+1$	$+1$	$+1$	6
A_Q	$+1$	$+1$	$+1$	-2	-2	-2	$+1$	$+1$	$+1$	18
B_L	-1	0	$+1$	-1	0	$+1$	-1	0	$+1$	6
B_Q	$+1$	-2	$+1$	$+1$	-2	$+1$	$+1$	-2	$+1$	18
$A_L B_L$	$+1$	0	-1	0	0	0	-1	0	$+1$	4
$A_L B_Q$	-1	$+2$	-1	0	0	0	$+1$	-2	$+1$	12
$A_Q B_L$	-1	0	$+1$	$+2$	0	-2	-1	0	$+1$	12
$A_Q B_Q$	$+1$	-2	$+1$	-2	$+4$	-2	$+1$	-2	$+1$	36
Y_{ij}	1	0	2	-2	4	-1	3	1	2	

From Table 10.3 it can be seen that A_L compares all highest levels of $A(+1)$ with all lowest levels of $A(-1)$. A_Q compares the extreme levels with twice the intermediate levels. Both of these effects are taken across *all* levels of B. Now B_L compares the highest versus the lowest level of B at the 0 level of A, then at level 1 of A, then at level 2 of A, reading from left to right across B_L. Similarly, B_Q compares the extreme levels of B with twice the intermediate level at all three levels of A. The coefficients for interaction are found by multiplying corresponding main-effect coefficients. An examination of these coefficients in the light of what interactions there are should make the coefficients seem plausible. The sums of squares of the coefficients are given at the right of Table 10.3.

Applying these coefficients to the responses for each treatment combination gives

$$A_L = -1(1) - 1(0) - 1(2) + 0(-2)$$
$$+ 0(4) + 0(-1) + 1(3) + 1(1) + 1(2) = 3$$

$$A_Q = +1(1) + 1(0) + 1(2) - 2(-2)$$
$$- 2(4) - 2(-1) + 1(3) + 1(1) + 1(2) = 7$$

$$B_L = -1(1) + 0(0) + 1(2) - 1(-2)$$
$$+ 0(4) + 1(-1) - 1(3) + 0(1) + 1(2) = 1$$

$$B_Q = +1(1) - 2(0) + 1(2) + 1(-2)$$
$$- 2(4) + 1(-1) + 1(3) - 2(1) + 1(2) = -5$$

$$A_L B_L = +1(1) + 0(0) - 1(2) + 0(-2)$$
$$+ 0(4) + 0(-1) - 1(3) + 0(1) + 1(2) = -2$$

$$A_L B_Q = -1(1) + 2(0) - 1(2) + 0(-2)$$
$$+ 0(4) + 0(-1) + 1(3) - 2(1) + 1(2) = 0$$

$$A_Q B_L = -1(1) + 0(0) + 1(2) + 2(-2)$$
$$+ 0(4) - 2(-1) - 1(3) + 0(1) + 1(2) = -2$$

$$A_Q B_Q = +1(1) - 2(0) + 1(2) - 2(-2)$$
$$+ 4(4) - 2(-1) + 1(3) - 2(1) + 1(2) = 28$$

The corresponding sums of squares become

$$SS_{A_L} = \frac{(3)^2}{6} = 1.50 \qquad SS_{A_L B_L} = \frac{(-2)^2}{4} = 1.00$$

$$SS_{A_Q} = \frac{(7)^2}{18} = 2.72 \qquad SS_{A_L B_Q} = \frac{0^2}{12} = 0.00$$

$$SS_{B_L} = \frac{(1)^2}{6} = 0.17 \qquad SS_{A_Q B_L} = \frac{(-2)^2}{12} = 0.33$$

$$SS_{B_Q} = \frac{(-5)^2}{18} = 1.39 \qquad SS_{A_Q B_Q} = \frac{(28)^2}{36} = 21.78$$

Summarizing, we obtain Table 10.4.

Table 10.4 ANOVA Breakdown for 3^2 Factorial

Source	df	SS	
A_i	2	4.22	
$\quad A_L$	1		1.50
$\quad A_Q$	1		2.72
B_j	2	1.56	
$\quad B_L$	1		0.17
$\quad B_Q$	1		1.39
AB_{ij}	4	23.11	
$\quad A_L B_L$	1		1.00
$\quad A_L B_Q$	1		0.00
$\quad A_Q B_L$	1		0.33
$\quad A_Q B_Q$	1		21.78
Totals	8	28.89	

The results of this analysis will not be tested, as there is no separate measure of error, and the interaction effect is obviously large compared with other effects. As the numbers used here are purely hypothetical, the purpose has been only to show how such data can be analyzed and how the notation can be used. ■

It may be instructive to examine graphically the meaning of this high $A_Q B_Q$ interaction because it dwarfs all main effects and other interactions in the example.

If the $A_L B_L$ interaction is graphed using only the extreme levels of each factor, the results are as shown in Figure 10.3. Although the lines do cross, there is a very small interaction. Note also that the average change in response for the lowest level of A to the highest level of A (shown by ×) is very small, indicating a very small A_L effect as the data show.

3^f Factorial Experiments

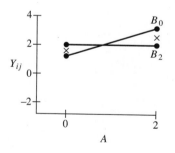

Figure 10.3 $A_L B_L$ interaction.

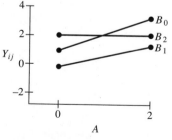

Figure 10.4 $A_L B_Q$ interaction.

Plotting $A_L B_Q$ means using all three levels of B, giving Figure 10.4, which again shows little interaction.

For $A_Q B_L$ the results are as shown in Figure 10.5, which once again shows little interaction. This graph does indicate that the quadratic effect of A is more pronounced than its linear effect. This is borne out by the data.

Replotting Figure 10.4 with B as abscissa gives Figure 10.6, which shows the same lack of serious interaction as in Figure 10.4, but does show a slight quadratic bend in factor B over the linear effect.

So far no startling results have been seen. Now plot the middle level of B on Figure 10.5 to obtain Figure 10.7. Note the way this "curve" reverses its trend compared with the other two. This shows that not until both factors are considered at all three of their levels does this $A_Q B_Q$ interaction show up. It can also be seen by adding the middle level of A on Figure 10.6 (see Figure 10.8).

Before leaving this problem, reconsider the data of Table 10.1. Add the data by diagonals rather than by rows or columns. First consider the diagonals downward from left to right where the main diagonal is $1 + 4 + 2 = 7$, the next one to the right is $-2 + 1 + 2 = 1$, and the last $+2$ is found by repeating the table again to the right of the present one as in Table 10.5.

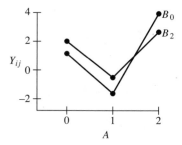

Figure 10.5 $A_Q B_L$ interaction.

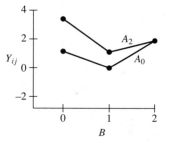

Figure 10.6 $A_L B_Q$ interaction again.

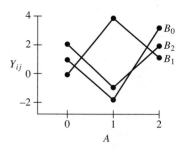

Figure 10.7 $A_Q B_Q$ interaction.

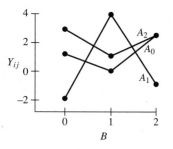

Figure 10.8 $A_Q B_Q$ interaction again.

Table 10.5 Diagonal Computations

Factor B	Factor A \0	\1	\2	Factor A 0	1	2
0	1	−2	3	1	−2	3
1	0	4	1	0	4	1
2	2	−1	2	2	−1	2

Similarly, the next downward diagonal gives $3 + 0 + (-1) = 2$. The sum of squares between these three diagonal terms is then

$$\frac{(7)^2 + (1)^2 + (2)^2}{3} - \frac{(10)^2}{9} = 6.89$$

If the diagonals are now considered downward and to the left, their totals are

$$3 + 4 + 2 = 9$$
$$1 + 1 - 1 = 1$$
$$-2 + 0 + 2 = 0$$

and their sum of squares is

$$\frac{(9)^2 + (1)^2 + (0)^2}{3} - \frac{(10)^2}{9} = 16.22$$

These two somewhat artificial sums of squares of 6.89 and 16.22 are seen to add up to the interaction sum of squares

$$6.89 + 16.22 = 23.11$$

These two components of interaction have no physical significance, but simply illustrate another way to extract two orthogonal components of interaction. Testing each of these separately for significance has no meaning, but this arbitrary breakdown is very useful in more complex designs. Some authors refer to these two components as the I and J components of interaction

$$I(AB) = 6.89 \quad 2 \text{ df}$$
$$J(AB) = 16.22 \quad 2 \text{ df}$$
$$\text{total } A \times B = 23.11 \quad 4 \text{ df}$$

Each such component carries 2 df. These are sometimes referred to as the AB^2 and AB components of $A \times B$ interaction. In this notation, effects can be multiplied together using a modulus of 3, since this is a 3^f factorial. A *modulus* of 3 means that the resultant number is equal to the remainder when the number in the usual base of 10 is divided by 3. Thus $4 = 1$ in modulus 3, as 1 is the remainder when 4 is divided by 3. The following associations also hold:

$$\text{numbers: } 0 \quad 3 = 0 \quad 6 = 0 \quad 9 = 0$$
$$1 \quad 4 = 1 \quad 7 = 1 \quad 10 = 1$$
$$2 \quad 5 = 2 \quad 8 = 2 \quad 11 = 2$$
$$\vdots$$

When using the form $A^p B^q$, it is postulated that the only exponent allowed on the first letter in the expression is a 1. To make it a 1, the expression can be squared and reduced, modulus 3. For example,

$$A^2 B = (A^2 B)^2 = A^4 B^2 = AB^2$$

Hence AB and AB^2 are the only components of the $A \times B$ interaction with 2 df each. Here the two types of notation are related as follows:

$$I(AB) = AB^2$$
$$J(AB) = AB$$

To summarize this simple experiment, all effects can be expressed with 2 df each, as in Table 10.6.

Table 10.6 3^2 Factorial by 2-df Analysis

Source	df	SS
A_i	2	4.22
B_j	2	1.56
$I(AB) = AB^2$	2	6.89
$J(AB) = AB$	2	16.22
Totals	8	28.89

It will be found very useful to break such an experiment down into 2-df effects when more complex designs are considered. Here, this breakdown is presented merely to show another way to partition the interaction effect.

Let us consider a practical example.

■ **EXAMPLE 10.2** The effect of two factors—prechamber volume ratio (V) and injection timing (T)—on the parts per million noxious gas emitted from an engine was to be studied. The volume ratio was set at three equispaced levels and the time was also set at three equispaced levels. Two engines were built for each of the nine treatment combinations and the amount of gas emitted was recorded for the 18 engines as shown in Table 10.7. The ANOVA for Table 10.7, assuming a completely randomized design, is given in Table 10.8.

Table 10.7 Gas Emission Data of Example 10.2

Time	Volume (V)		
(T)	Low	Medium	High
Short	6.27	8.08	7.34
	5.43	8.04	7.87
Medium	6.94	7.48	8.61
	6.51	7.52	8.32
Long	7.22	8.65	9.02
	7.05	8.97	9.07

Table 10.8 ANOVA for Gas Emission Data

Source	df	SS	MS	F	Prob.
Volume (V_i)	2	11.44	5.72	71.5	0.000
Time (T_j)	2	4.17	2.08	26.0	0.000
$V \times T$ interaction (VT_{ij})	4	1.39	0.35	4.4	0.031
Error	9	0.70	0.08		

The ANOVA of Table 10.8 shows strong main effects and an interaction significant at the 5 percent significance level. As both factors are quantitative and equispaced, we can treat these data as we did the hypothetical data of Table 10.3 to give Table 10.9.

Table 10.9 Quantitative Level Breakdown of Example 10.2

Factor	00	01	02	10	11	12	20	21	22	Contrast
V_L	−1	−1	−1	0	0	0	+1	+1	+1	10.81
V_Q	+1	+1	+1	−2	−2	−2	+1	+1	+1	−7.83
T_L	−1	0	+1	−1	0	+1	−1	0	+1	6.95
T_Q	+1	−2	+1	+1	−2	+1	+1	−2	+1	2.25
$V_L T_L$	+1	0	−1	0	0	0	−1	0	+1	0.31
$V_L T_Q$	−1	+2	−1	0	0	0	+1	−2	+1	0.37
$V_Q T_L$	−1	0	+1	+2	0	−2	−1	0	+1	2.45
$V_Q T_Q$	+1	−2	+1	−2	+4	−2	+1	−2	+1	−8.97
$T_{ij}.$	11.70	13.45	14.27	16.12	15.00	17.62	15.21	16.93	18.09	

Upon taking each contrast and squaring and dividing by 2 (as $n = 2$) and the sum of the squares of the coefficients, one gets Table 10.10.

Table 10.10 ANOVA Breakdown for Example 10.2

Source	df	SS	MS	F	Prob.
V_i	2	11.44	5.72		
V_L	1		9.74	9.74	121.75***	0.000
V_Q	1		1.70	1.70	21.25**	0.001
T_j	2	4.17	2.08		
T_L	1		4.03	4.03	50.38***	0.000
T_Q	1		0.14	0.14	1.75	0.219
VT_{ij}	4	1.38	0.35		
$V_L T_L$	1		0.01	0.01	< 1	
$V_L T_Q$	1		0.00	0.00	< 1	
$V_Q T_L$	1		0.25	0.25	3.12	0.111
$V_Q T_Q$	1		1.12	1.12	14.00**	0.005
Error	9	0.70	0.08		

From Table 10.10 we see that volume ratio produced both a strong linear and quadratic effect on the amount of gas emitted. The timing shows a strong linear effect, and a significant interaction is shown as a quadratic by quadratic interaction. One must then plot all nine treatment means to see the significant effects as shown in Figure 10.9. Note that without the T_2 line in Figure 10.9 there would be little obvious interaction. Careful study of this illustration should confirm the reasonableness of the results given in Table 10.10.

As an exercise let us also pull out the VT and VT^2 components of interaction from the totals on the diagonals of Table 10.7,

$$SS_{VT} = \frac{(44.48)^2 + (47.66)^2 + (46.25)^2}{6} - \frac{(138.39)^2}{18} = 0.85$$

$$SS_{VT^2} = \frac{(44.79)^2 + (47.32)^2 + (46.28)^2}{6} - \frac{(138.39)^2}{18} = 0.54$$

which is approximately (1.39 versus 1.38) the total interaction.

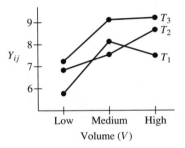

Figure 10.9 Treatment means of Example 10.2.

Since the chief concern in this example is to reduce the amount of noxious gas to a minimum one could examine all nine treatment means in a Newman–Keuls sense and look for conditions that might result in a minimum mean.
The means are

$$5.85, \ 6.72, \ 7.13, \ 7.50, \ 7.60, \ 8.06, \ 8.47, \ 8.81, \ 9.04$$

$$S_{\bar{Y}_{ij}} = \sqrt{\frac{0.08}{2}} = 0.2$$

From Statistical Table E.1 for $n_2 = 9$ df:

$p =$	2	3	4	5	6	7	8	9
	3.20	3.95	4.42	4.76	5.02	5.24	5.43	5.60

and LSRs $= 0.64, 0.79, 0.88, 0.95, 1.00, 1.05, 1.09, 1.12$.

From these the lines can be drawn as shown above. Clearly, the minimum emission is when one has low prechamber volume ratio and the shortest injection time as the mean of 5.85 ppm is significantly lower than any other of the nine means. ■

10.3 3^3 Factorial

If an experimenter has three factors, each at three levels, or a $3 \times 3 \times 3 = 3^3$ factorial, there are several ways to break down the effects of factors A, B, and C and their associated interactions. If the order of experimentation is completely randomized, the model for such an experiment is

$$Y_{ijk} = \mu + A_i + B_j + AB_{ij} + C_k + AC_{ik} + BC_{jk} + ABC_{ijk} + \varepsilon_{ijk}$$

with the last two terms confounded unless there is replication within the cells. In this model $i = 1, 2, 3, j = 1, 2, 3$, and $k = 1, 2, 3$, making 27 treatment combinations. These 27 treatment combinations may be as shown in Table 10.11.

Association of the proper coefficients on these 27 treatment combinations would allow the Table 10.12 breakdown of an ANOVA if all effects were set at quantitative levels.

In an actual problem these three-way interactions would be hard to explain, and quite often the ABC interaction is left with its 8 df for use as an error term to test the main effects A, B, C, and the two-way interactions.

Another possible partitioning of these effects is in terms of 2-df effects using I and J components on AB, AC, and BC interactions. These could be designated as AB, AB^2, AC, AC^2, and BC, BC^2, each with 2 df. However, the three-way interaction with its 8 df may need a further breakdown. Sometimes ABC is broken

Table 10.11 3^3 Factorial Treatment Combinations

		Factor A		
Factor B	**Factor C**	0	1	2
0	0	000	100	200
	1	001	101	201
	2	002	102	202
1	0	010	110	210
	1	011	111	211
	2	012	112	212
2	0	020	120	220
	1	021	121	221
	2	022	122	222

Table 10.12 3^3 Factorial Analysis for Linear and Quadratic Effects

Source	df	Source	df
A_i	2	$A_Q C_L$	1
A_L	1	$A_Q C_Q$	1
A_Q	1	BC_{jk}	4
B_j	2	$B_L C_L$	1
B_L	1	$B_L C_Q$	1
B_Q	1	$B_Q C_L$	1
AB_{ij}	4	$B_Q C_Q$	1
$A_L B_L$	1	ABC_{ijk}	8
$A_L B_Q$	1	$A_L B_L C_L$	1
$A_Q B_L$	1	$A_L B_L C_Q$	1
$A_Q B_Q$	1	$A_L B_Q C_L$	1
C_k	2	$A_L B_Q C_Q$	1
C_L	1	$A_Q B_L C_L$	1
C_Q	1	$A_Q B_L C_Q$	1
AC_{ik}	4	$A_Q B_Q C_L$	1
$A_L C_L$	1	$A_Q B_Q C_Q$	1
$A_L C_Q$	1		
		Total	26

Table 10.13 3^3 Factorial in 2-df Analyses

Source	df	
A	2	
B	2	
AB	2	$\left.\vphantom{\begin{matrix}2\\2\end{matrix}}\right\}4$
AB^2	2	
C	2	
AC	2	$\left.\vphantom{\begin{matrix}2\\2\end{matrix}}\right\}4$
AC^2	2	
BC	2	$\left.\vphantom{\begin{matrix}2\\2\end{matrix}}\right\}4$
BC^2	2	
ABC	2	
ABC^2	2	$\left.\vphantom{\begin{matrix}2\\2\\2\\2\end{matrix}}\right\}8$
AB^2C	2	
AB^2C^2	2	
Total	26	

into four 2-df components called $X(ABC)$, $Y(ABC)$, $Z(ABC)$, and $W(ABC)$, or, using the notation of the last section, AB^2C, ABC^2, ABC, and AB^2C^2. Here again no first letter is squared, and $A^2BC = (A^2BC)^2 = A^4B^2C^2 = AB^2C^2$ modulus 3. Such a partitioning would yield Table 10.13.

■ **EXAMPLE 10.3**   A problem involving the effect of three factors, each at three levels, was proposed by Professor Burr of Purdue University. Here the measured variable was yield and the factors that might affect this response were days, operators, and concentrations of solvent. Three days, three operators, and three concentrations were chosen. Days and operators were qualitative effects, concentrations were quantitative and set at 0.5, 1.0, 2.0. Although these are not equispaced, the logarithms of these three levels are equispaced, and the logarithms can then be used if a curve fitting is warranted. For the purposes of this chapter, all levels of all factors will be considered as fixed and the design will be considered as completely randomized. It was decided to take three replications of each of the $3^3 = 27$ treatment combinations. The data, after coding by subtracting 20.0, are as presented in Table 10.14.

Table 10.14 Example Data on 3^3 Factorial with Three Replications

	Day D_i								
	5/14			5/15			5/16		
				Operator O_j					
Concentration C_k	A	B	C	A	B	C	A	B	C
0.5	1.0	0.2	0.2	1.0	1.0	1.2	1.7	0.2	0.5
	1.2	0.5	0.0	0.0	0.0	0.0	1.2	0.7	1.0
	1.7	0.7	−0.3	0.5	0.0	0.5	1.2	1.0	1.7
1.0	5.0	3.2	3.5	4.0	3.2	3.7	4.5	3.7	3.7
	4.7	3.7	3.5	3.5	3.0	4.0	5.0	4.0	4.5
	4.2	3.5	3.2	3.5	4.0	4.2	4.7	4.2	3.7
2.0	7.5	6.0	7.2	6.5	5.2	7.0	6.7	7.5	6.2
	6.5	6.2	6.5	6.0	5.7	6.7	7.5	6.0	6.5
	7.7	6.2	6.7	6.2	6.5	6.8	7.0	6.0	7.0

   If these data are analyzed on a purely qualitative basis, the methods of Chapter 5 can be used. The resulting ANOVA is shown in Table 10.15.
   The model for this example is merely $Y_{ijkm} = \mu$ plus the sum of the terms in the Source column in Table 10.15. From this analysis the concentration effect is tremendous, and the days, operators, and day × operator interaction are all significant at the 1 percent level of significance.
   Because concentrations are at quantitative levels, the linear and quadratic effects of concentrations may be computed, as well as the interactions between linear effect of concentration and days, quadratic effect of concentration and days, linear effect of concentration and operators, and quadratic effect of concentration and operators. It is not usually worthwhile to extract three-way interaction

Table 10.15 First ANOVA for Example 10.3

Source	df	SS	MS
D_i	2	3.48	1.74**
O_j	2	6.14	3.07**
DO_{ij}	4	4.07	1.02**
C_k	2	468.99	234.49***
DC_{ik}	4	0.59	0.15
OC_{jk}	4	0.89	0.22
DOC_{ijk}	8	1.09	0.14
$\varepsilon_{m(ijk)}$	54	9.98	0.18
Totals	80	495.23	

in this way. To calculate these quantitative effects it is usually helpful to construct some two-way tables for the interactions that are being computed. Two of these are shown as Table 10.16(a) and Table 10.16(b).

Table 10.16 Cell Totals for $D \times C$ and $O \times C$ Interactions

	(a) Concentration				(b) Concentration			
Day	0.5	1.0	2.0	Operator	0.5	1.0	2.0	
5/14	5.2	34.5	60.5	A	9.5	39.1	61.6	
5/15	4.2	33.1	56.6	B	4.3	32.5	55.3	
5/16	9.2	38.0	60.4	C	4.8	34.0	60.6	
Totals	18.6	105.6	177.5	301.7	Totals	18.6	105.6	177.5	301.7

From Table 10.16(a), applying the linear and quadratic coefficients to the concentration totals, we have

Sums of Squares

$$C_L = -1(18.6) + 0(105.6) + 1(177.5) = 158.9$$

$$SS_{C_L} = \frac{(158.9)^2}{27(2)} = 467.58$$

$$C_Q = +1(18.6) - 2(105.6) + 1(177.5) = -15.1$$

$$SS_{C_Q} = \frac{(-15.1)^2}{27(6)} = 1.41$$

$$SS_C = 468.99$$

For the $D \times C$ interactions, consider each level of days separately. At

$$5/14: \quad C_L = -1(5.2) + 0(34.5) + 1(60.5) = 55.3$$
$$5/15: \quad C_L = -1(4.2) + 0(33.1) + 1(56.6) = 52.4$$
$$5/16: \quad C_L = -1(9.2) + 0(38.0) + 1(60.4) = 51.2$$

The $D \times C_L$ SS$_{\text{interaction}}$ is then

$$\frac{(55.3)^2 + (52.4)^2 + (51.2)^2}{9(2)} - \frac{(158.9)^2}{27(2)} = 0.49$$

For quadratic effects, at

$$5/14: \quad C_Q = +1(5.2) - 2(34.5) + 1(60.5) = -3.3$$
$$5/15: \quad C_Q = +1(4.2) - 2(33.1) + 1(56.6) = -5.4$$
$$5/16: \quad C_Q = +1(9.2) - 2(38.0) + 1(60.4) = -6.4$$

The $D \times C_Q$ SS$_{\text{interaction}}$ is then

$$\frac{(-3.3)^2 + (-5.4)^2 + (-6.4)^2}{9(6)} - \frac{(-15.1)^2}{27(6)} = 0.09$$

and

$$\text{SS}_{D \times C} = \text{SS}_{D \times C_L} + \text{SS}_{D \times C_Q} = 0.49 + 0.09 = 0.58$$

If the same procedure is now applied to the data of Table 10.16(b), we have

$$\text{SS}_{O \times C_L} = \frac{(52.1)^2 + (51.0)^2 + (55.8)^2}{9(2)} - \frac{(158.9)^2}{27(2)} = 0.70$$

$$\text{SS}_{O \times C_Q} = \frac{(-7.1)^2 + (-5.4)^2 + (-2.6)^2}{9(6)} - \frac{(-15.1)^2}{27(6)} = 0.19$$

and

$$\text{SS}_{O \times C} = \text{SS}_{O \times C_L} + \text{SS}_{O \times C_Q} = 0.70 + 0.19 = 0.89$$

The resulting ANOVA can now be shown in Table 10.17.

This second analysis shows that the linear effect and the quadratic effect of concentration are extremely significant. Two plots of Figure 10.9 may help in picturing what is really happening in this experiment. Figure 10.10(a) shows the effect of operators, days, and $D \times O$ interaction. Figure 10.10(b) indicates that the linear effect of concentration far outweighs the quadratic effect and there is no significant interaction. If a straight line or three straight lines were fit to these data, the logs of the concentrations would be used, as the logs are equispaced.

Table 10.17 Second ANOVA for Example 10.3

Source	df	SS	MS
D_i	2	3.48	1.74**
O_j	2	6.14	3.07**
DO_{ij}	4	4.07	1.02**
C_L	1	467.58	467.58***
C_Q	1	1.41	1.41**
$D \times C_L$	2	0.49	0.24
$D \times C_Q$	2	0.09	0.04
$O \times C_L$	2	0.70	0.35
$O \times C_Q$	2	0.19	0.09
DOC_{ijk}	8	1.09	0.14
$\varepsilon_{m(ijk)}$	54	9.98	0.18
Totals	80	495.22	

Although this would usually conclude the analysis of this problem, each interaction will be broken down into its diagonal, or I and J, components in order to illustrate the technique. To compute the two diagonal components of the two-factor interactions, the two parts of Table 10.16 can be used, along with a similar table for the $D \times O$ cells (see Table 10.18). From Table 10.18 the diagonal components of the $D \times O$ interaction are

$$I(D \times O)$$
$$= \frac{(39.5 + 28.6 + 34.8)^2 + (30.2 + 34.1 + 39.5)^2 + (30.5 + 33.3 + 31.2)^2}{27}$$
$$- \frac{(301.7)^2}{81} = 1.74$$

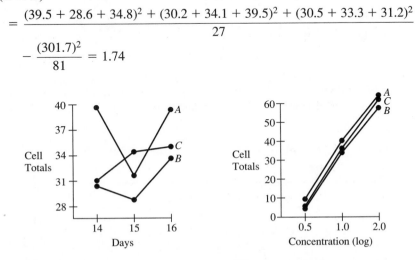

(a)   (b)

Figure 10.10 Plots of 3^3 example of Table 10.14.

Table 10.18 Cell Totals for $D \times O$ Interactions

		Operator	
Day	A	B	C
5/14	39.5	30.2	30.5
5/15	31.2	28.6	34.1
5/16	39.5	33.3	34.8

Call it DO^2.

$$J(D \times O) = \frac{(30.5 + 28.6 + 39.5)^2 + (96.2)^2 + (106.9)^2}{27} - \frac{(301.7)^2}{81} = 2.33$$

Call it DO. These total $1.74 + 2.33 = 4.07$, the $D \times O$ interaction sum of squares. Applying the same technique to the two parts of Table 10.16 gives

$$I(DC) = DC^2 = \frac{(98.7)^2 + (102.7)^2 + (100.3)^2}{27} - \frac{(301.7)^2}{81} = 0.30$$

$$J(DC) = DC = \frac{(102.8)^2 + (99.8)^2 + (99.1)^2}{27} - \frac{(301.7)^2}{81} = 0.29$$

$$D \times C = 0.59$$

$$I(OC) = OC^2 = \frac{(102.6)^2 + (99.9)^2 + (99.2)^2}{27} - \frac{(301.7)^2}{81} = 0.24$$

$$J(OC) = OC = \frac{(98.9)^2 + (104.0)^2 + (98.8)^2}{27} - \frac{(301.7)^2}{81} = 0.65$$

$$O \times C = 0.89$$

To break down the 8 df of the $D \times O \times C$ interaction, form an $O \times D$ table showing each of the three levels of concentration C as in Table 10.19.

Table 10.19 Cell Totals for $D \times O$ Interaction at Each Level of Concentration

O_i	D_i at C_1			D_i at C_2			D_i at C_3		
A	3.9	1.5	4.1	13.9	11.0	14.2	21.7	18.7	21.2
B	1.4	1.0	1.9	10.4	10.2	11.9	18.4	17.4	19.5
C	−0.1	1.7	3.2	10.2	11.9	11.9	20.4	20.5	19.7

For each of these concentration levels, find the I and J effect totals, for example, at C_1,

$$I \text{ components are } 8.1, 3.3, 7.2$$

$$J \text{ components are } 5.0, 6.1, 7.5$$

Now form a table with these I and J components at each level of C (see Table 10.20).

Table 10.20 Diagonal Totals for Each Level of Concentration

C_k	$I(DO)$			$J(DO)$		
	i_0	i_1	i_2	j_0	j_1	j_2
0.5	8.1	3.3	7.2	5.0	6.1	7.5
1.0	36.0	33.1	36.5	34.6	33.3	37.7
2.0	58.8	58.6	60.1	59.0	56.8	61.7

Treat each half of Table 10.20 as a simple interaction and compute the I and J components. Thus,

$$DO^2C^2 = I[C \times I(DO)] = \frac{(101.3)^2 + (98.6)^2 + (101.8)^2}{27} - \frac{(301.7)^2}{81} = 0.22$$

$$DO^2C = J[C \times I(DO)] = \frac{(99.1)^2 + (99.4)^2 + (103.2)^2}{27} - \frac{(301.7)^2}{81} = 0.39$$

$$DOC^2 = I[C \times J(DO)] = \frac{(100.0)^2 + (102.8)^2 + (98.9)^2}{27} - \frac{(301.7)^2}{81} = 0.30$$

$$DOC = J[C \times J(DO)] = \frac{(99.8)^2 + (102.4)^2 + (99.5)^2}{27} - \frac{(301.7)^2}{81} = 0.19$$

$$\text{total } D \times O \times C = 1.10$$

compared with 1.09 in Table 10.17. This last breakdown into four parts could also have been accomplished by considering the $C \times O$ interaction at three levels of D_i, or the $C \times D$ interaction at three levels of O_j. The resulting analysis is summarized in Table 10.21. This analysis is in substantial agreement with Tables 10.15 and 10.17. No new tests would be performed on the data in Table 10.21 because they represent only an arbitrary breakdown of the interactions into 2-df components. The purpose of such a breakdown is discussed in subsequent chapters. For testing hypotheses on interaction, these components are added together again.

Table 10.21 Third ANOVA for 3^3
Experiment

Source	df	SS
D_i	2	3.48
O_j	2	6.14
DO	2	2.33
DO^2	2	1.74
C_k	2	468.99
DC	2	0.29
DC^2	2	0.30
OC	2	0.65
OC^2	2	0.24
DOC	2	0.19
DOC^2	2	0.30
DO^2C	2	0.39
DO^2C^2	2	0.22
$\varepsilon_{m(ijk)}$	54	9.98
Totals	80	495.24

■

10.4 Summary

The summary at the end of Chapter 9 may now be extended for Part II.

Experiment	Design	Analysis
II. Two or more factors A. Factorial (crossed)		
	1. Completely randomized $Y_{ijk} = \mu + A_i + B_j + AB_{ij}$ $\qquad + \varepsilon_{k(ij)}, \ldots$ for more factors	1.
	a. General case	a. ANOVA with interactions
	b. 2^f case	b. Yates method or general ANOVA; use (1), a, b, ab, ...
	c. 3^f case	c. General ANOVA; use 00, 10, 20, 01, 11, ... and $A \times B = AB + AB^2$, ... for interaction

10.5 SAS Programs

Of course, SAS programs may be used for the analyses of 3^f factorials provided one can set up the proper coefficients for the linear and quadratic contrasts as seen in Chapter 3. Conversely, because these contrasts are used when the factor is quantitative and equispaced, it may be best to use a regression program to give the equation of "best fit" in such cases. This will be discussed in a later chapter.

Problems

10.1 Pull-off force in pounds on glued boxes at three temperatures and three humidities with two observations per treatment combination in a completely randomized experiment gives the table below. Do a complete analysis of this problem by the general methods of Chapter 5.

	Temperature *A*		
Humidity *B*	Cold	Ambient	Hot
50%	0.8	1.5	2.5
	2.8	3.2	4.2
70%	1.0	1.6	1.8
	1.6	1.8	1.0
90%	2.0	1.5	2.5
	2.2	0.8	4.0

10.2 Assuming the temperatures in Problem 10.1 are equispaced, extract linear and quadratic effects of both temperature and humidity as well as all components of interaction.

10.3 From Problem 10.1 extract the AB and AB^2 components of interaction.

10.4 Develop a "Yates method" for this 3^2 experiment and check the results with those above.

10.5 When the data were collected in Example 10.2 on ppm of noxious gas, data were also recorded on percent smoke emitted by the 18 engines. Results were as given below. Do an ANOVA by the methods of Chapter 5.

Time	**Volume (*V*)**		
(*T*)	Low	Medium	High
Short	0.3	0.1	0.4
	0.4	0.4	0.4
Medium	0.1	0.1	0.4
	0.4	0.2	1.2
Long	0.8	0.7	2.5
	2.0	1.6	3.6

10.6   From Problem 10.5 extract linear and quadratic components of both main effects and the interaction.

10.7   From Problem 10.5 extract the VT and the VT^2 components of interaction.

10.8   Plot any results that stand out in Problem 10.6.

10.9   Run further tests to recommend the best volume and time combinations for minimizing smoke in Problem 10.5.

10.10   Compare the results obtained in Problem 10.8 with the results of Example 10.2 in the text and make some overall recommendations about these engine designs.

10.11   A behavior variable on concrete pavements was measured for three surface thicknesses: 3 inches, 4 inches, and 5 inches; three base thicknesses: 0 inch, 3 inches, and 6 inches; and three subbase thicknesses: 4 inches, 8 inches, and 12 inches. Two observations were made under each of the 27 pavement conditions and complete randomization performed.

Do a complete ANOVA of this experiment by the methods of Chapter 5.

	Surface Thickness (inches)								
	3			**4**			**5**		
Subbase Thickness (inches)	Base Thickness (inches)			Base Thickness (inches)			Base Thickness (inches)		
	0	3	6	0	3	6	0	3	6
4	2.8	4.3	5.7	4.1	5.4	6.7	6.0	6.3	7.1
	2.6	4.5	5.3	4.4	5.5	6.9	6.2	6.5	6.9
8	4.1	5.7	6.9	5.3	6.5	7.7	6.1	7.2	8.1
	4.4	5.8	7.1	5.1	6.7	7.4	5.8	7.1	8.4
12	5.5	7.0	8.1	6.5	7.7	8.8	7.0	8.0	9.1
	5.3	6.8	8.3	6.7	7.5	9.1	7.2	8.3	9.0

10.12   Since all three factors are quantitative and equispaced, determine linear and quadratic effects for each factor and all interaction breakdowns. Test for significance in Problem 10.11.

10.13   Break down the interactions of Problem 10.11 into 2-df components such as AB, AB^2, ABC, AB^2C, and so on.

10.14   Use a Yates method to solve Problem 10.11 and check the results.

10.15   Plot any significant results of Problem 10.11.

10.16   The following data are on the wet-film thickness (in mils) of lacquer. The factors studied were type of resin (two types), gate-blade setting in mils (three settings), and weight fraction of nonvolatile material in the lacquer (three fractions).

Gate Setting	Resin Type					
	1			**2**		
	Weight Fraction			Weight Fraction		
	0.20	0.25	0.30	0.20	0.25	0.30
2	1.6	1.5	1.5	1.5	1.4	1.6
	1.5	1.3	1.3	1.4	1.3	1.4
4	2.7	2.5	2.4	2.4	2.6	2.2
	2.7	2.5	2.3	2.3	2.4	2.1
6	4.0	3.6	3.5	4.0	3.7	3.4
	3.9	3.8	3.4	4.0	3.6	3.3

Do an ANOVA of the above data and pull out any quantitative effects in terms of their components.

Factorial Experiment— Split-Plot Design

11

11.1 Introduction

In many experiments in which a factorial arrangement is desired, it may not be possible to randomize completely the order of experimentation. In Chapter 8 restrictions on randomization were considered, and both randomized block and Latin square designs were discussed. There are still many practical situations in which it is not at all feasible to even randomize within a block. Under certain conditions these restrictions will lead to a split-plot design. An example will show why such designs are quite common.

■ **EXAMPLE 11.1** The data in Table 11.1 were compiled on the effect of oven temperature T and baking time B on the life Y of an electrical component [30].

Table 11.1 Electrical Component Life-Test Data

Baking Time B (min)	Oven Temperature T (°F)			
	580	600	620	640
5	217	158	229	223
	188	126	160	201
	162	122	167	182
10	233	138	186	227
	201	130	170	181
	170	185	181	201
15	175	152	155	156
	195	147	161	172
	213	180	182	199

Looking only at this table of data, we might think of this as a 4×3 factorial with three replications per cell and proceed to the analysis in Table 11.2. Here only the temperature has a significant effect on the life of the component at the 1 percent significance level.

Table 11.2 ANOVA for Electrical Component Life-Test Data as a Factorial

Source	df	SS	MS	EMS
T_i	3	12,494	4165	$\sigma_\varepsilon^2 + 9\phi_T$
B_j	2	566	283	$\sigma_\varepsilon^2 + 12\phi_B$
BT_{ij}	6	2601	434	$\sigma_\varepsilon^2 + 3\phi_{TB}$
$\varepsilon_{k(ij)}$	24	13,670	570	σ_ε^2
Totals	35	29,331		

We might go on now and seek linear, quadratic, and cubic effects of temperature; but before proceeding, a few questions on design should be raised. What was the order of experimentation? The data analysis above assumes a completely randomized design. This means that one of the four temperatures was chosen at random and the oven was heated to this temperature, then a baking time was chosen at random and an electrical component was inserted in the oven and baked for the time selected. After this run, the whole procedure was repeated until the data were compiled. Now, was the experiment really conducted in this manner? Of course the answer is "no." Once an oven is heated to temperature, all nine components are inserted; three are baked for 5 minutes, three for 10 minutes, and three for 15 minutes. We would argue that this is the only practical way to run the experiment. Complete randomization is too impractical as well as too expensive. Fortunately, this restriction on the complete randomization can be handled in a design called a *split-plot* design.

The four temperature levels are called *plots*. They could be called blocks, but in a previous chapter blocks and replications were used for a complete rerun of the whole experiment. (The word "plots" has been inherited from agricultural applications.) In such a setup, temperature—a main effect—is confounded with these plots. If conditions change from one oven temperature to another, these changes will show up as temperature differences. Thus, in such a design, a main effect is confounded with plots. This is necessary because it is often the most practical way to order the experiment. Now once a temperature has been set up by choosing one of the four temperatures at random, three components can be placed in the oven and one component can be baked for 5 minutes, another for 10 minutes, and the third one for 15 minutes. The specific component that is to be baked for 5, 10, and 15 minutes is again decided by a random selection. These three baking-time levels may be thought of as a splitting of the plot into three parts, one part for each baking time. This defines the three parts of a main plot that are called split plots. Note here that only three components were placed in the oven, not nine. The temperature is then changed to another level and three more components are placed in the oven for 5, 10, and 15 minutes baking time. This same procedure is followed for all four temperatures; after this the whole experiment may be replicated. These replications may be run several days after the initial experiment; in fact, it is often desirable to collect data from two or three replications and then decide whether more replications are necessary.

The experiment conducted above was actually run as a split-plot experiment and laid out as shown in Table 11.3. In Table 11.3 the lines indicate where randomization restrictions have occurred. First a replication or block is run (double lines), then within a replication an oven temperature is chosen at random (single vertical line), and then within that temperature level baking time is randomized. Thus a set of observations with no lines separating them (such as 158, 138, 152 for temperature 600°, replication I) indicates random order of these three observations. This notion of using straight lines to show randomization restrictions was suggested in Chapter 4, Section 4.2. If data are presented as in Table 11.1, there is no way to tell how they were collected. Were they completely randomized—all 36 readings? Or were there randomization restrictions as shown in Table 11.3?

Table 11.3 Split-Plot Layout for Electrical Component Life-Test Data

Replication R	Baking Time B (min)	Oven Temperature T (°F)			
		580	600	620	640
I	5	217	158	229	223
	10	233	138	186	227
	15	175	152	155	156
II	5	188	126	160	201
	10	201	130	170	181
	15	195	147	161	172
III	5	162	122	167	182
	10	170	185	181	201
	15	213	180	182	199

Here temperatures are confounded with plots and the $R \times T$ cells are called whole plots. Inside a whole plot, the baking times are applied to one-third of the material. These plots associated with the three baking times are called split plots. Since one main effect is confounded with plots and the other main effect is not, it is usually desirable to place in the split the main effect we are most concerned about testing, as this factor is not confounded.

We might think that factor B (baking time) is nested with the main plots but this is not the case, since the same levels of B are used in all plots. A model for this experiment would be

$$Y_{ijk} = \mu + \underbrace{R_i + T_j + RT_{ij}}_{\text{whole plot}} + \underbrace{B_k + RB_{ik} + TB_{jk} + RTB_{ijk}}_{\text{split plot}}$$

The first three variable terms in this model represent the whole plot, and the RT interaction is often referred to as the *whole-plot error*. The usual assumption is that this interaction does not exist, that this term is really an estimate of the error within the main plot. The last four terms represent the split plot, and the RTB interaction is referred to as the *split-plot error*. Sometimes the RB term is also considered nonexistent and is combined with RTB as an error term. A separate error term might be obtained if it were feasible to repeat some observations within the split plot. The proper EMS values can be found by considering m repeat measurements where $m = 1$ (Table 11.4). Note that in Table 11.4 the degrees of freedom in the whole plot add to 11 as there are 12 whole plots (3 replications $\times$ 4 temperatures) and the degrees of freedom in the split plot add to 24 as there are 2 df within each whole plot and hence 2×12 within all whole plots.

Table 11.4 EMS for Split-Plot Electrical Component Life-Test Data

| | | | 3 | 4 | 3 | 1 | |
| | | | R | F | F | R | |
	Source	df	i	j	k	m	EMS
Whole plot	R_i	2	1	4	3	1	$\sigma_\varepsilon^2 + 12\sigma_R^2$
	T_j	3	3	0	3	1	$\sigma_\varepsilon^2 + 3\sigma_{RT}^2 + 9\phi_T$
	RT_{ij}	6	1	0	3	1	$\sigma_\varepsilon^2 + 3\sigma_{RT}^2$
Split plot	B_k	2	3	4	0	1	$\sigma_\varepsilon^2 + 4\sigma_{RB}^2 + 12\phi_B$
	RB_{ik}	4	1	4	0	1	$\sigma_\varepsilon^2 + 4\sigma_{RB}^2$
	TB_{jk}	6	3	0	0	1	$\sigma_\varepsilon^2 + \sigma_{RTB}^2 + 3\phi_{TB}$
	RTB_{ijk}	12	1	0	0	1	$\sigma_\varepsilon^2 + \sigma_{RTB}^2$
	$\varepsilon_{m(ijk)}$	—	1	1	1	1	σ_ε^2 (not retrievable)
	Total	35					

Because the error mean square cannot be isolated in this experiment, $\sigma_\varepsilon^2 + \sigma_{RTB}^2$ is taken as the split-plot error, and $\sigma_\varepsilon^2 + 3\sigma_{RT}^2$ is taken as the whole-plot error. The main effects and interaction of interest TB can be tested, as seen from the EMS column, although no exact tests exist for the replication effect nor for interactions involving the replications. This is not a serious disadvantage for this design, since tests on replication effects are not of interest but are isolated only to reduce the error variance. The analysis of the data from Table 11.5 follows the methods given in Chapter 5. The results are shown in the table.

Table 11.5 ANOVA for Split-Plot Electrical Component Life-Test Data

Source	df	SS	MS	EMS
R_i	2	1963	982	$\sigma_\varepsilon^2 + 12\sigma_R^2$
T_j	3	12,494	4165	$\sigma_\varepsilon^2 + 3\sigma_{RT}^2 + 9\phi_T$
RT_{ij}	6	1774	296	$\sigma_\varepsilon^2 + 3\sigma_{RT}^2$
B_k	2	566	283	$\sigma_\varepsilon^2 + 4\sigma_{RB}^2 + 12\phi_B$
RB_{ik}	4	7021	1755	$\sigma_\varepsilon^2 + 4\sigma_{RB}^2$
TB_{jk}	6	2600	434	$\sigma_\varepsilon^2 + \sigma_{RTB}^2 + 3\phi_{TB}$
RTB_{ijk}	12	2912	243	$\sigma_\varepsilon^2 + \sigma_{RTB}^2$
Totals	35	29,331		

Testing the hypothesis of no temperature effect gives

$$F_{3,6} = \frac{4165}{296} = 14.1 \quad \text{(significant at the 1 percent level)}$$

Testing for baking time gives

$$F_{2,4} = \frac{283}{1755} < 1 \quad \text{(not significant)}$$

Testing for TB interaction gives

$$F_{6,12} = \frac{434}{243} = 1.79 \quad \text{(not significant at the 5 percent level)}$$

No exact tests are available for testing the replication effect or replication interaction with other factors, but these effects are present only to reduce the experimental error in this split-plot design.

The results of this split-plot analysis are not too different from the results using the incorrect method of Table 11.2, but this split-plot design shows the need for careful consideration of the method of randomization before starting an experiment. This split-plot design represents a restriction on the randomization over a complete randomization in a factorial experiment.

To emphasize a bit more the difference between a completely randomized 4×3 factorial design and a split-plot 4×3 factorial design, one might let the numbers 1 through 4 on one die (red) represent temperature and the numbers 1 through 3 on another die (green) represent baking times and then use the two dice to effect the proper randomization. In the case of complete randomization, both dice are tossed and the first six results are as indicated below.

Baking Time	Oven Temperature			
	580°	600°	620°	640°
5	3		4	
10		1		5
15	6		2	

The numbers 1, 2, . . . , 6 indicate the order of the experiment: first set temperature at 600, bake for 10 minutes; then set temperature at 620, bake for 15 minutes; and so on. One notes that the order is scattered throughout the 12 treatment combinations. In the case of the split-plot design a temperature is randomly chosen (roll the red die) and then the baking times are chosen at random within that temperature (roll the green die). One such randomization is shown for the first six combinations as follows:

Baking Time	Oven Temperature			
	580°	600°	620°	640°
5	3		4	
10	1		6	
15	2		5	

In this latter case it should seem more obvious that temperature is confounded with the order in which the experiment is run, as three treatments are run at 580° first, then three are run at 620° second, and so on. It is hoped that by replicating the whole experiment several times any effect of order confounded with temperature would average out. Nevertheless, in a split-plot design a main effect is confounded.

Some statisticians recommend introducing a restriction error into a model wherever such restrictions occur; see Anderson and McLean [1]. No such term has been introduced here, and so it is strongly recommended that F tests be run only within the whole plot or within the split plot and that mean squares in the whole plot not be compared with mean squares in the split plot regardless of the EMS column.

Because this type of design is encountered often in industrial experiments, another example will be considered. ■

■ **EXAMPLE 11.2** A defense-related organization was to study the pull-off force necessary to separate boxes of chaff from a tape to which they are affixed. These boxes are made of a cardboard material, approximately 3 by 3 by 1 inches, and are mounted on a strip of cloth tape that has an adhesive backing. The tape is 2 inches wide, and the boxes are placed 7 inches center to center on the strip. There are 75 boxes mounted on each strip.

The tape is pulled from the box at a 90° angle as it is wound onto a drum. During this separation process the portion of the tape still adhering to the box carries the box onto a platform. The box trips a microswitch, which energizes a plunger. The plunger then kicks the box out of the machine.

After this problem was discussed with the plant engineers, several factors were listed that might affect this pull-off force. The most important factors were temperature and humidity. It was agreed to use three fixed temperature levels, −55°C, 25°C, and 55°C, and three fixed humidity levels, 50 percent, 70 percent, and 90 percent. These gave nine basic treatment combinations. Because there might be differences in pull-off force as a result of the strip selected, it was decided to choose five different strips at random for use in this experiment. There might also be differences within a strip, so two boxes were chosen at random and cut from each strip for the test.

The test in the laboratory was accomplished by hand-holding the package, attaching a spring scale to the strip by means of a hole previously punched in the strip, and pulling the tape from the package in a direction perpendicular to the package.

In discussing the design of this experiment it seemed best to set the climatic condition (a combination of temperature and humidity) at random from one of the $3 \times 3 = 9$ conditions, and then test two boxes from each of the five strips while these conditions were maintained. Then another of the nine conditions is set, and the results again determined on two boxes from each of the five strips. This is then a restriction on the randomization, and the resulting design is a split-plot design. It was agreed to replicate the complete experiment four times. A layout for this experiment is shown in Table 11.6.

In Table 11.6, each replication is a repeat of replication I, but a new order of randomizing the nine atmospheric conditions is taken in each replication.

Table 11.6 Split-Plot Design of Chaff Experiment

	Strip S	Box	\multicolumn{3}{c}{−55}	\multicolumn{3}{c}{25}	\multicolumn{3}{c}{55}
			50	70	90	50	70	90	50	70	90
Replication I	1	1									
		2									
R	2	3									
		4									
	3	5									
		6									
	4	7									
		8									
	5	9									
		10									

The table above has column groups: **Temperature T (°C)** spanning the values −55, 25, 55, and within each, **Humidity H (percent)** with sub-columns 50, 70, 90.

Replication II (repeat as above)
Replication III (repeat as above)
Replication IV (repeat as above)

In this design atmospheric conditions and replications form the whole plot and strips are in the split plot.

The model for this design and its associated EMS relations are set up in Table 11.7.

From the EMS column tests can be made on the effects of replications, replications by temperature interaction, replications by humidity interaction, replications by temperature by humidity interaction, strips, and strips by all other factor interactions. Unfortunately no exact tests are available for the factors of chief importance: temperature, humidity, and temperature by humidity interaction. Of course, we could first test the hypotheses that can be tested, and if some of these are not significant at a reasonably high level (say 25 percent), assume they are nonexistent and then remove these terms from the EMS column. For example, if the TS interaction can be assumed as zero ($\sigma_{TS}^2 = 0$), the temperature mean square can be tested against the RT interaction mean square with 2 and 6 df. If this is not a reasonable assumption, a pseudo-F (F') test can be used as discussed in Chapter 6.

Unfortunately, data were available only for the first replication of this experiment, so its complete analysis cannot be given. The method of analysis is the

Table 11.7 EMS for Chaff Experiment

	Source	df	**4** **R** **i**	**3** **F** **j**	**3** **F** **k**	**5** **R** **m**	**2** **R** **q**	EMS
Whole plot	R_i	3	1	3	3	5	2	$\sigma_\varepsilon^2 + 18\sigma_{RS}^2 + 90\sigma_R^2$
	T_j	2	4	0	3	5	2	$\sigma_\varepsilon^2 + 6\sigma_{RTS}^2 + 24\sigma_{TS}^2 + 30\sigma_{RT}^2 + 120\phi_T$
	RT_{ij}	6	1	0	3	5	2	$\sigma_\varepsilon^2 + 6\sigma_{RTS}^2 + 30\sigma_{RT}^2$
	H_k	2	4	3	0	5	2	$\sigma_\varepsilon^2 + 6\sigma_{RHS}^2 + 24\sigma_{HS}^2 + 30\sigma_{RH}^2 + 120\phi_H$
	RH_{ik}	6	1	3	0	5	2	$\sigma_\varepsilon^2 + 6\sigma_{RHS}^2 + 30\sigma_{RH}^2$
	TH_{jk}	4	4	0	0	5	2	$\sigma_\varepsilon^2 + 2\sigma_{RTHS}^2 + 8\sigma_{THS}^2 + 10\sigma_{RTH}^2 + 40\phi_{TH}$
	RTH_{ijk}	12	1	0	0	5	2	$\sigma_\varepsilon^2 + 2\sigma_{RTHS}^2 + 10\sigma_{RTH}^2$
Split plot	S_m	4	4	3	3	1	2	$\sigma_\varepsilon^2 + 18\sigma_{RS}^2 + 72\sigma_S^2$
	RS_{im}	12	1	3	3	1	2	$\sigma_\varepsilon^2 + 18\sigma_{RS}^2$
	TS_{jm}	8	4	0	3	1	2	$\sigma_\varepsilon^2 + 6\sigma_{RTS}^2 + 24\sigma_{TS}^2$
	RTS_{ijm}	24	1	0	3	1	2	$\sigma_\varepsilon^2 + 6\sigma_{RTS}^2$
	HS_{km}	8	4	3	0	1	2	$\sigma_\varepsilon^2 + 6\sigma_{RHS}^2 + 24\sigma_{HS}^2$
	RHS_{ikm}	24	1	3	0	1	2	$\sigma_\varepsilon^2 + 6\sigma_{RHS}^2$
	THS_{jkm}	16	4	0	0	1	2	$\sigma_\varepsilon^2 + 2\sigma_{RTHS}^2 + 8\sigma_{THS}^2$
	$RTHS_{ijkm}$	48	1	0	0	1	2	$\sigma_\varepsilon^2 + 2\sigma_{RTHS}^2$
	$\varepsilon_{q(ijkm)}$	180	1	1	1	1	1	σ_ε^2
	Total	359						

same as given in Chapter 5, even though this experiment is rather complicated. It is presented here only to show another actual example of a split-plot design. ■

11.2 A Split-Split-Plot Design

■ **EXAMPLE 11.3** In a study of the cure rate index on some samples of rubber, three laboratories, three temperatures, and three types of mix were involved. Material for the three mixes was sent to one of the three laboratories where they ran the experiment on the three mixes at the three temperatures (145°C, 155°C, and 165°C). However, once a temperature was set, all three mixes were subjected to that temperature and then another temperature was set and again all three mixes were involved, and finally the third temperature was set. Material was also sent to the second and third laboratories and similar experimental procedures were performed. There are therefore two restrictions on randomization since the laboratory is chosen, then the temperature is chosen, and then mixes can be randomized at that particular temperature and laboratory.

By complete replication of the whole experiment to achieve four replications, the three laboratories and four replications form the whole plots. Then the temperature levels may be randomized at each laboratory and in each replication to form a split plot. Then at each temperature–laboratory–replication combination, the three mixes are randomly applied, forming what is called a *split-split-plot,* indicating that two main effects (laboratory and temperature) are confounded with blocks.

Table 11.8 shows one possible order of the first nine experiments in one replication of (a) a completely randomized design, (b) a split-plot design, and (c) a split-split-plot design. The resultant data will usually not reveal how the experiment was ordered, but the use of lines in the data given in Table 11.9 should help indicate the restrictive order of randomization.

Table 11.8 Order of First Nine Rubber Cure Rate Index Experiments

	Temperature (°C)									
	145			**155**			**165**			
	Mix			Mix			Mix			
Laboratory	*A*	*B*	*C*	*A*	*B*	*C*	*A*	*B*	*C*	
1			3				4			(a)
2	8	1			5		7		9	Completely randomized
3			2						6	
1										(b)
2	4	6	1	7	9	3	2	8	5	Split plot
3										
1										(c)
2	8	9	7	3	1	2	4	6	5	Split-split-plot
3										

The model and its EMS values are given in Table 11.10.

The EMS column indicates that F tests can be made on all three fixed effects and their interactions without a separate error term. Because the effect of replication is often considered of no great interest and the interaction of replication with the other factors is often assumed to be nonexistent, this appears to be a feasible experiment. If examination of the EMS column should indicate that some F tests have too few degrees of freedom in their denominator (some statisticians say less than 6), the experimenter might consider increasing

Table 11.9 Rubber Cure Rate Index Data

Replication	Laboratory	Temperature (°C)								
		145			155			165		
		Mix			Mix			Mix		
		A	*B*	*C*	*A*	*B*	*C*	*A*	*B*	*C*
I	1	18.6	14.5	21.1	9.5	7.8	11.2	5.4	5.2	6.3
	2	20.2	18.4	22.5	11.4	10.8	13.3	6.8	6.0	7.7
	3	19.7	16.3	22.7	9.3	9.1	11.3	6.7	5.7	6.6
II	1	17.0	15.8	20.8	9.4	8.3	10.0	5.3	4.9	6.4
	2	20.1	18.1	22.7	11.5	11.1	14.0	6.9	6.1	8.0
	3	18.3	16.7	21.9	10.2	9.2	11.0	6.0	5.5	6.5
III	1	18.7	16.5	21.8	9.5	8.9	11.5	5.7	4.3	5.8
	2	19.4	16.5	21.5	11.4	9.5	12.0	6.0	5.0	6.6
	3	16.8	14.4	19.3	9.8	8.0	10.9	5.0	4.6	5.9
IV	1	18.7	17.6	21.0	10.0	9.1	11.1	5.3	5.2	5.6
	2	20.0	16.7	21.3	11.5	9.7	11.5	5.7	5.2	6.3
	3	17.1	15.2	19.3	9.5	9.0	11.4	4.8	5.4	5.8

the number of replications to increase the precision of the test. In the above experiment, if five replications were used, the whole plot error RL would have 8 df instead of 6, the split-plot error RLT would have 16 instead of 12, and so on.

Another technique that is sometimes used to increase the degrees of freedom is to pool all interactions with replications into the error term for that section of the table. Here one might pool RT and RLT and RM, RLM and RTM with $RLTM$ if additional degrees of freedom are believed necessary. One way to handle experiments with several replications is to stop after two or three replications, compute the results, and see if significance has been achieved or that the F's are large even if not significant. Then add another replication, and so on, increasing the precision (and the cost) of the experiment in the hope of detecting significant effects if they are there.

The numerical results of the rubber cure rate index are given in Table 11.11.

Here laboratory effect is significant at the 5 percent level, and temperature, mixes, and temperature–mix interaction are all highly significant (< 0.001). ∎

Table 11.10 EMS for Rubber Cure Rate Index Experiment

	Source	df	4 R i	3 F j	3 F k	3 F m	1 R q	EMS
Whole plot	R_i	3	1	3	3	3	1	$\sigma_\varepsilon^2 + 27\sigma_R^2$
	L_j	2	4	0	3	3	1	$\sigma_\varepsilon^2 + 9\sigma_{RL}^2 + 36\phi_L$
	RL_{ij}	6	1	0	3	3	1	$\sigma_\varepsilon^2 + 9\sigma_{RL}^2$
Split plot	T_k	2	4	3	0	3	1	$\sigma_\varepsilon^2 + 9\sigma_{RT}^2 + 36\phi_T$
	RT_{ik}	6	1	3	0	3	1	$\sigma_\varepsilon^2 + 9\sigma_{RT}^2$
	LT_{jk}	4	4	0	0	3	1	$\sigma_\varepsilon^2 + 3\sigma_{RLT}^2 + 12\phi_{LT}$
	RLT_{ijk}	12	1	0	0	3	1	$\sigma_\varepsilon^2 + 3\sigma_{RLT}^2$
Split-split-plot	M_m	2	4	3	3	0	1	$\sigma_\varepsilon^2 + 9\sigma_{RM}^2 + 36\phi_M$
	RM_{im}	6	1	3	3	0	1	$\sigma_\varepsilon^2 + 9\sigma_{RM}^2$
	LM_{jm}	4	4	0	3	0	1	$\sigma_\varepsilon^2 + 3\sigma_{RLM}^2 + 12\phi_{LM}$
	RLM_{ijm}	12	1	0	3	0	1	$\sigma_\varepsilon^2 + 3\sigma_{RLM}^2$
	TM_{km}	4	4	3	0	0	1	$\sigma_\varepsilon^2 + 3\sigma_{RTM}^2 + 12\phi_{TM}$
	RTM_{ikm}	12	1	3	0	0	1	$\sigma_\varepsilon^2 + 3\sigma_{RTM}^2$
	LTM_{jkm}	8	4	0	0	0	1	$\sigma_\varepsilon^2 + \sigma_{RLTM}^2 + 4\phi_{LTM}$
	$RLTM_{ijkm}$	24	1	0	0	0	1	$\sigma_\varepsilon^2 + \sigma_{RLTM}^2$
	$\varepsilon_{q(ijkm)}$	0	1	1	1	1	1	σ_ε^2 (not retrievable)
Total		107						

Table 11.11 ANOVA for Rubber Cure Rate Index Experiment

Source	df	SS	MS
R	3	9.41	3.14
L	2	40.66	20.33*
RL	6	16.11	2.68
T	2	3119.51	1559.76***
RT	6	2.07	0.34
LT	4	4.94	1.24
RLT	12	7.81	0.65
M	2	145.71	72.86***
RM	6	3.29	0.55
LM	4	0.35	0.09
RLM	12	2.93	0.24
TM	4	43.69	10.92***
RTM	12	2.22	0.18
LTM	8	1.07	0.14
RLTM	24	4.97	0.21
Totals	107	3404.74	

11.3 Summary

The summary of Chapter 8 may now be extended.

Experiment	Design	Analysis

II. Two or
   more factors
A. Factorial
   (crossed)		

	1. Completely randomized	1.
	2. Randomized block	2.
	a. Complete	a. Factorial ANOVA
	$$Y_{ijk} = \mu + R_k + A_i$$ $$+ B_j + AB_{ij} + \varepsilon_{ijk}$$	

	b. Incomplete, confounding:
	   i. Main effect—split plot
		b.
		   i. Split plot
		ANOVA

	$$Y_{ijk} = \mu + \underbrace{R_i + A_j + RA_{ij}}_{\text{whole plot}}$$ $$+ \underbrace{B_k + RB_{ik} + AB_{jk} + RAB_{ijk}}_{\text{split plot}}$$	

11.4 SAS Programs

The SAS programs used on factorial experiments (Chapter 5) and cases where special hypotheses need to be stated and tested (Chapters 7 and 8) can be used on the split-plot designs. Examples 11.1 and 11.3 illustrate such programs.

For Example 11.1

```
DATA A;
INPUT REPL TIME TEMP RDG;
CARDS;
—(DATA)—
;
TITLE'EXAMPLE 11.1';
PROC PRINT;
PROC GLM;
CLASS REPL TIME TEMP;
MODEL RDG= REPL|TEMP|TIME;
```

```
TEST H=TEMP E=REPL*TEMP;
TEST H= TIME E=REPL*TIME;
TEST H=TEMP*TIME E=REPL*TEMP*TIME;
```

The printout is on page 272.

For Example 11.3

```
DATA A;
INPUT REPL LAB TEMP MIX RDG;
CARDS;
-(DATA)-
;
TITLE'EXAMPLE 11.3';
PROC PRINT;
PROC GLM;
CLASS REPL LAB TEMP MIX;
MODEL RDG=REPL|LAB|TEMP|MIX;
TEST H=LAB E=REPL*LAB;
TEST H=TEMP E=REPL*TEMP;
TEST H= LAB*TEMP E=REPL*LAB*TEMP;
TEST H=MIX E=REPL*MIX;
TEST H=LAB*MIX E=REPL*LAB*MIX;
TEST H=TEMP*MIX E=REPL*TEMP*MIX;
TEST H=LAB*TEMP*MIX E=REPL*LAB*TEMP*MIX;
```

The printout is on pages 273–274.

These results agree with those of Tables 11.5 and 11.11 and the tests that follow.

GENERAL LINEAR MODELS PROCEDURE

DEPENDENT VARIABLE: RDG

SOURCE	DF	SUM OF SQUARES	MEAN SQUARE	F VALUE	PR > F	R-SQUARE	C.V.
MODEL	35	29330.97222222	838.02777778	99999.99	0.0000	1.000000	0.0000
ERROR	0	0.00000000	0.00000000			ROOT MSE	RDG MEAN
CORRECTED TOTAL	35	29330.97222222				0.00000000	178.47222222

SOURCE	DF	TYPE I SS	F VALUE	PR > F	DF	TYPE III SS	F VALUE	PR > F
REPL	2	1962.72222222	.	.	2	1962.72222222	.	.
TEMP	3	12494.30555556	.	.	3	12494.30555556	.	.
REP*TEMP	6	1773.94444444	.	.	6	1773.94444444	.	.
TIME	2	566.22222222	.	.	2	566.22222222	.	.
REPL*TIME	4	7021.27777778	.	.	4	7021.27777778	.	.
TIME*TEMP	6	2600.44444444	.	.	6	2600.44444444	.	.
REPL*TIME*TEMPL	12	2912.05555556	.	.	12	2912.05555556	.	.

TESTS OF HYPOTHESES USING THE TYPE III MS FOR REPL*TEMP AS AN ERROR TERM

SOURCE	DF	TYPE III SS	F VALUE	PR > F
TEMP	3	12494.30555556	14.09	0.0040

TESTS OF HYPOTHESES USING THE TYPE III MS FOR REPL*TIME AS AN ERROR TERM

SOURCE	DF	TYPE III SS	F VALUE	PR > F
TIME	2	566.22222222	0.16	0.8563

TESTS OF HYPOTHESES USING THE TYPE III MS FOR REPL*TIME*TEMP AS AN ERROR TERM

SOURCE	DF	TYPE III SS	F VALUE	PR > F
TIME*TEMP	6	2600.44444444	1.79	0.1848

GENERAL LINEAR MODELS PROCEDURE

DEPENDENT VARIABLE: RDG

SOURCE	DF	SUM OF SQUARES	MEAN SQUARE	F VALUE	PR > F	R-SQUARE	C.V.
MODEL	107	3404.74324074	31.82003029	99999.99	0.0000	1.000000	0.0000
ERROR	0	0.00000000	0.00000000			ROOT MSE	RDG MEAN
CORRECTED TOTAL	107	3404.74324074			0.00000000	11.66574074	

SOURCE	DF	TYPE I SS	F VALUE	PR > F		TYPE III SS	F VALUE	PR > F
REPL	3	9.41435185	.	.		9.41435185	.	.
LAB	2	40.66351852	.	.		40.66351852	.	.
REPL*LAB	6	16.10981481	.	.		16.10981481	.	.
TEMP	2	3119.50907407	.	.		3119.50907407	.	.
REPL*TEMP	6	2.06648148	.	.		2.06648148	.	.
LAB*TEMP	4	4.93648148	.	.		4.93648148	.	.
REPL*LAB*TEMP	12	7.81685185	.	.		7.81685185	.	.
MIX	2	145.71796296	.	.		145.71796296	.	.
REPL*MIX	6	3.27759259	.	.		3.27759259	.	.
LAB*MIX	4	0.33925926	.	.		0.33925926	.	.
REPL*LAB*MIX	12	2.94074074	.	.		2.94074074	.	.
TEMP*MIX	4	43.68703704	.	.		43.68703704	.	.
REPL*TEMP*MIX	12	2.23074074	.	.		2.23074074	.	.
LAB*TEMP*MIX	8	1.07740741	.	.		1.07740741	.	.
REPL*LAB*TEMP*MIX	24	4.95592593	.	.		4.95592593	.	.

TESTS OF HYPOTHESES USING THE TYPE III MS FOR REPL*LAB AS AN ERROR TERM

SOURCE	DF	TYPE III SS	F VALUE	PR > F
LAB	2	40.66351352	7.57	0.0228

(Continues next page)

(Continued)

TESTS OF HYPOTHESES USING THE TYPE III MS FOR REPL*TEMP AS AN ERROR TERM

SOURCE	DF	TYPE III SS	F VALUE	PR > F
TEMP	2	3119.50907407	4528.73	0.0001

TESTS OF HYPOTHESES USING THE TYPE III MS FOR REPL*LAB*TEMP AS AN ERROR TERM

SOURCE	DF	TYPE III SS	F VALUE	PR > F
LAB*TEMP	4	4.93648148	1.89	0.1762

GENERAL LINEAR MODELS PROCEDURE

DEPENDENT VARIABLE: RDG

TESTS OF HYPOTHESES USING THE TYPE III MS FOR REPL*MIX AS AN ERROR TERM

SOURCE	DF	TYPE III SS	F VALUE	PR > F
MIX	2	145.71796296	133.38	0.0001

TESTS OF HYPOTHESES USING THE TYPE III MS FOR REPL*LAB*MIX AS AN ERROR TERM

SOURCE	DF	TYPE III SS	F VALUE	PR > F
LAB*MIX	4	0.33925926	0.35	0.8417

TESTS OF HYPOTHESES USING THE TYPE III MS FOR REPL*TEMP*MIX AS AN ERROR TERM

SOURCE	DF	TYPE III SS	F VALUE	PR > F
TEMP*MIX	4	43.68703704	58.75	0.0001

TESTS OF HYPOTHESES USING THE TYPE III MS FOR REPL*LAB*TEMP*MIX AS AN ERROR TERM

SOURCE	DF	TYPE III SS	F VALUE	PR > F
LAB*TEMP*MIX	8	1.07740741	0.65	0.7269

Problems

11.1 As part of the experiment discussed in Example 11.2, a chamber was set at 50 percent relative humidity and two boxes from each of three strips were inserted and the pull-off force determined. This was repeated for 70 percent and 90 percent, the order of humidities being randomized. Two replications of the whole experiment were made, and the results follow.

Replication	Strips	Humidity 50%		Humidity 70%		Humidity 90%	
		50%		70%		90%	
	1	1.12	1.75	3.50	0.50	1.00	0.75
I	2	1.13	3.50	0.75	1.00	0.50	0.50
	3	2.25	3.25	1.75	1.88	1.50	0.00
	1	1.75	1.88	1.75	0.75	1.50	0.75
II	2	5.25	5.25	0.75	2.25	1.50	1.50
	3	1.50	3.50	1.62	2.50	0.75	0.50

Set up a split-plot model for this experiment and outline the ANOVA table, including EMS.

11.2 Analyze Problem 11.1.

11.3 The following data layout was proposed to handle a factorial experiment designed to determine the effects of orifice size and flow rate on environmental chamber pressure in millimeters of mercury.

Flow Rate	Orifice Size (inches)		
	8	10	12
3.5			
4.0			
4.5			

a. If this experiment is run as a completely randomized design, how might data be collected to fill in the above table?
b. If this experiment were run as a split-plot design, how might data be collected?
c. Assuming three complete replications of the above run as a split-plot design, outline its ANOVA table. (You need *not* include the EMS column.)
d. Suggest how this experiment might be improved.

11.4 Interest centered on the yields of alfalfa in tons per acre from three varieties following four dates of final cutting. A large section of land was divided into three strips, and each strip was planted with one of the three varieties of alfalfa. Once a strip had been selected, one fourth of the plants were cut on the first cutting date, one fourth on the next, and so on, for the four dates studied. This whole experiment could be replicated several times as necessary by choosing sections at random.

a. Identify the type of design implied here.

b. Set up an appropriate ANOVA table for this situation with the minimum number of replications you feel might be adequate to detect differences, if they exist.

11.5 In a time study two types of stimuli are used—two- and three-dimensional film—and combinations of stimuli and job are repeated once to give eight strips of film. The order of these eight is completely randomized on one long film and presented to five analysts for rating at one sitting. Since this latter represents a restriction on the randomization, the whole experiment is repeated at three sittings (four in all) and the experiment is considered a split-plot design. Set up a model for this experiment and indicate the tests to be made. (Consider analysts as random.)

11.6 Determine an F' test for any effects in Problem 11.5 that cannot be run directly.

11.7 Three replications are made of an experiment to determine the effect of days of the week and operators on screen-color difference of a TV tube in K. On a given day, each of four operators measured the screen-color difference on a given tube. The order of measuring by the four operators is randomized each day for five days of the week. The results are

Replication	Operator	Monday	Tuesday	Wednesday	Thursday	Friday
I	A	800	950	900	740	880
	B	760	900	820	740	960
	C	920	920	840	900	1020
	D	860	940	820	940	980
II	A	780	810	880	960	920
	B	820	940	900	780	820
	C	740	900	880	840	880
	D	800	900	800	820	900
III	A	800	1030	800	840	800
	B	900	920	880	920	1000
	C	800	1000	920	760	920
	D	900	980	900	780	880

Set up the model and outline the ANOVA table for this problem.

11.8 Considering days as fixed, operators as random, and replications as random, analyze Problem 11.7.

11.9 Explain how a split-plot design differs from a nested design, since both have a factor within levels of another factor.

11.10 From the data of Table 11.9 on the rubber cure rate index experiment, verify the results of Table 11.11.

11.11 Because two qualitative factors in Problem 11.10 show significance, compare their means by a Newman–Keuls test.

11.12 Redo Table 11.10 assuming that the three laboratories are chosen at random from a large number of possible laboratories.

11.13 Analyze the results of Table 11.9, based on the EMS values of Problem 11.12.

11.14 In a study of the effect of baking temperature and recipe (or mix) on the quality of cake, three mixes were used: A, B, and C. One of these three was chosen at random and the mix made up and five cakes were poured, each placed in a different oven and baked at temperatures of 300°, 325°, 350°, 375°, and 400°. After these five cakes had been baked, removed, and their quality (Y) evaluated, another mix was chosen and again five cakes were poured and baked at the five temperatures. Then this was repeated on the third mix. This experiment was replicated r times.
   a. Identify this design. Show its data layout.
   b. Write the mathematical model for this situation and determine the ANOVA outline indicating tests to be made.
   c. Recommend how many replications r you would run to make this an acceptable experiment and explain why.

11.15 If in Problem 11.14 only one oven was available but cakes of all three recipes could be inserted in the oven when set at a given temperature, how would the problem differ from the setup in Problem 11.14? Explain which of the two designs you would prefer and why.

11.16 Running two random replications of a problem cited in Cochran and Cox [*Experimental Designs*, 2nd edition, Wiley, 1957] on the effect of temperature and mix on cake quality similar to Problem 11.14 resulted in the following data.

Replication	Mix	Temperatures					
		175°	185°	195°	205°	215°	225°
	1	47	29	35	47	57	45
I	2	35	46	47	39	52	61
	3	43	43	43	46	47	58
	1	26	28	32	25	37	33
II	2	21	21	28	26	27	20
	3	21	28	25	25	31	25

Outline the ANOVA table for this problem, show what tests can be run, and discuss these tests.

11.17 Analyze the data given in Problem 11.16 and state your conclusions.

11.18 What next steps would you recommend after your conclusions in Problem 11.17?

Factorial Experiment— Confounding in Blocks

12

12.1 Introduction

As seen in Chapter 11, there are many situations in which randomization is restricted. The split-plot design is an example of a restriction on randomization where a main effect is confounded with blocks. Sometimes these restrictions are necessary because the complete factorial cannot be run in one day, or in one environmental chamber. When such restrictions are imposed on the experiment, a decision must be made as to what information may be sacrificed and thus on what is to be confounded. A simple example may help to illustrate this point.

A chemist was studying the disintegration of a chemical in solution. He had a beaker of the mixture and wished to determine whether or not the depth in the beaker would affect the amount of disintegration and whether material from the center of the beaker would be different than material from near the edge. He decided on a 2^2 factorial experiment—one factor being depth, the other radius. A cross section of the beaker (Figure 12.1) shows the four treatment combinations to be considered.

In Figure 12.1, (1) represents radius zero, depth at lower level (near bottom of beaker); a represents the radius at near maximum (near the edge of the beaker) and a low depth level; b is radius zero, depth at a higher level; and ab is both radius and depth at higher levels. In order to get a reading on the amount of disintegration of material, samples of solution must be drawn from these four positions. This was accomplished by dipping into the beaker with an instrument designed to "capture" a small amount of solution. The only problem was that the experimenter had only two hands and could get only two samples at the same time. By

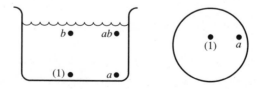

Figure 12.1 Beaker cross section.

the time he returned for the other two samples conditions might have changed and the amount of disintegration would probably be greater. Thus there is a physical restriction on this 2^2 factorial experiment; only two observations can be made at one time. The question is: Which two treatment combinations should be taken on the first dip? Considering the two dips as blocks, we are forced to use an incomplete block design, as the whole factorial cannot be performed in one dip.

Three possible block patterns are shown in Table 12.1. If plan I of Table 12.1 is used, the blocks (dips) are confounded with the radius effect, since all zero-radius readings are in dip 1 and all maximum radius readings are in dip 2. In plan II the depth effect is confounded with the dips, as low-level depth readings are both in dip 1 and high-level depth readings are in dip 2. In plan III neither main effect is confounded, but the interaction is confounded with the dips. To see that the interaction is confounded, recall the expression for the AB interaction in a 2^2 factorial found in Chapter 9.

$$AB = \frac{1}{2}[(1) - a - b + ab]$$

Note that the two treatment effects with the plus sign, (1) and ab, are both in dip 1, and the two with a minus sign, a and b, are in dip 2. Hence we cannot distinguish between a block effect (dips) and the interaction effect (AB). In most cases it is better to confound an interaction than to confound a main effect. The hope is, of course, that there is no interaction, and that information on the main effects can still be found from such an experiment.

Table 12.1 Three Possible Blockings of a 2^2 Factorial

	Plan					
Dip	I		II		III	
1	(1)	b	(1)	a	(1)	ab
2	a	ab	b	ab	a	b

12.2 Confounding Systems

To design an experiment in which the number of treatments that can be run in a block is less than the total number of treatment combinations, the experimenter must first decide on what effects he is willing to confound. If, as in the example above, he has only one interaction in the experiment and decides that he can confound this interaction with blocks, the problem is simply which treatment combinations to place in each block. One way to accomplish this is to place in one block those treatment combinations with a plus sign in the effect to be confounded, and those with a minus sign in the other block. A more general method is necessary when the number of blocks and the number of treatments increase.

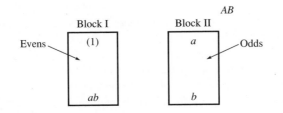

Figure 12.2 *AB* confounded in two blocks.

First a *defining contrast* is set up. This is merely an expression stating which effects are to be confounded with blocks. In the simple example above, to confound *AB*, write *AB* as the defining contrast. Once the defining contrast has been set up, several methods are available for determining which treatment combinations will be placed in each block. One method was shown above—place in one block the treatment combinations that have a plus sign in *AB* and those with a minus sign in the other block. Another method is to consider each treatment combination. Those that have an even number of letters in common with the effect letters in the defining contrast go in one block. Those that have an odd number of letters in common with the defining contrast go into the other block. Here (1) has no letters in common with *AB* or an even number. Type *a* has one letter in common with *AB* (*a* and *A*) or an odd number. *b* also has one letter in common with *AB*. *ab* has two letters in common with *AB*. Thus the block contents are those in Figure 12.2.

One disadvantage of the two methods given above is that they are only good on 2^f factorials. A more general method is attributed to Kempthorne [14]. Consider the linear expression

$$A_1 X_1 + A_2 X_2 + \cdots + A_n X_n = L$$

Here A_i is the exponent on the *i*th factor appearing in each independent defining contrast, and X_i is the level of the *i*th factor appearing in a given treatment combination. Every treatment combination with the same *L* value will be placed in the same block. For the simple example above where the defining contrast is *AB*, $A_1 = 1$, $A_2 = 1$, and all other A_i's $= 0$. Hence

$$L = 1 \cdot X_1 + 1 \cdot X_2$$

For each treatment combination the *L* values are

$$(1): \quad L = 1 \cdot 0 + 1 \cdot 0 = 0$$
$$a: \quad L = 1 \cdot 1 + 1 \cdot 0 = 1$$
$$b: \quad L = 1 \cdot 0 + 1 \cdot 1 = 1$$
$$ab: \quad L = 1 \cdot 1 + 1 \cdot 1 = 2 = 0 \text{ modulus } 2$$

Therefore, 0 and 2 are both 0, because a 2^f factorial is of modulus 2 (see Section 10.2). The assignment of treatment combinations to blocks is then

$L = 0$ block 1 | (1) *ab* |

$L = 1$ block 2 | *a* *b* |

For a more complex defining contrast such as ABC^2, the expression is

$$L = X_1 + X_2 + 2X_3$$

and this can be used to decide which treatment combinations in a 3^3 experiment go into a common block.

The block containing the treatment combination (1) is called the *principal block*. The treatment combinations in this block are elements of a group where the operation on the group is multiplication, modulus 2. The elements of the other block or blocks may be generated by multiplying one element in the new block by each element in the principal block. Multiplying elements together within the principal block will generate more elements within the principal block. This is best seen with a slightly more complex factorial, say 2^3. If the highest order interaction is to be confounded and the eight treatment combinations are to be placed into two blocks of four treatments each, then confounding ABC gives

$$L = X_1 + X_2 + X_3$$

Testing each of the eight treatment combinations gives

$$
\begin{aligned}
(1): \quad & L = 0 + 0 + 0 = 0 \\
a: \quad & L = 1 + 0 + 0 = 1 \\
b: \quad & L = 0 + 1 + 0 = 1 \\
ab: \quad & L = 1 + 1 + 0 = 2 = 0 \\
c: \quad & L = 0 + 0 + 1 = 1 \\
ac: \quad & L = 1 + 0 + 1 = 2 = 0 \\
bc: \quad & L = 0 + 1 + 1 = 2 = 0 \\
abc: \quad & L = 1 + 1 + 1 = 3 = 1
\end{aligned}
$$

The blocks are then

block 1	block 2
(1)	*a*
ab	*b*
ac	*c*
bc	*abc*
$L = 0$	$L = 1$

Group theory can simplify this procedure. Determine two elements in the principal block, for example, ab and bc. Multiply these two to get $ab^2c = ac$, the fourth element. Generate block 2 by finding one element, say a. Multiply this element by each element in the principal block, thus generating all the elements of the second block as follows:

$$a \cdot (1) = a$$
$$a \cdot ab = a^2b = b$$
$$a \cdot ac = a^2c = c$$
$$a \cdot bc = abc$$

These concepts may be extended to more complex confounding examples, as shown in later sections.

12.3 Block Confounding—No Replication

In many cases an experimenter cannot afford several replications of an experiment, and is further restricted in that the complete factorial cannot be run in one block or at one time. Again, the experiment may be blocked and information recovered on all but some high-order interactions, which may be confounded. Designs for such experiments will be considered for the special cases of 2^f and 3^f factorials because they appear often in practice.

Confounding in 2^f Factorials

The methods given earlier in this chapter can be used to determine the block composition for a specific confounding scheme. If only one replication is possible, one is run, and some of the higher order interaction terms must be used as experimental error unless some independent measure of error is available from previous data. This type of design is used mostly when there are several factors involved (say four or more), some high-order interactions may be confounded with blocks, and some others are yet available for an error estimate. For example, consider a 2^4 factorial where only eight treatment combinations can be run at one time. One possible confounding scheme would be

$$\text{confound} \quad ABCD \quad \text{and} \quad L = X_1 + X_2 + X_3 + X_4$$

block 1	(1)	ab	bc	ac	$abcd$	cd	ad	bd	$L = 0$
block 2	a	b	abc	c	bcd	acd	d	abd	$L = 1$

and the analysis would be as shown in Table 12.2. The three-way interactions may be pooled here to give 4 df for error, assuming that these interactions are actually nonexistent. All main effects and first-order interactions could then be tested with 1 and 4 df. After examining the results it might be possible to pool some of the two-way interactions that have small mean squares with the three-way interactions if more degrees of freedom are desired in the error term.

Table 12.2 2^4 Factorial in Two Blocks

Source	df
A	1
B	1
C	1
D	1
AB	1
AC	1
AD	1
BC	1
BD	1
CD	1
ABC	1 ⎫
ABD	1 ⎬ 4 df for error
ACD	1
BCD	1 ⎭
Blocks or $ABCD$	1
Total	15 df

If only four treatment combinations can be run in a block, one may confound the 2^4 factorial in four blocks of four treatment combinations each. For four blocks 3 df must be confounded with blocks. If two interactions are confounded, the product (modulus 2) of these two is also confounded, since the product of the signs in two effects gives the signs for the product. Thus, if AB and CD are confounded, so also will $ABCD$ be confounded. This scheme would confound two first-order interactions and one third order. It might be better to confound two second-order interactions and one first order:

$$ABC \quad BCD \quad AD$$

Note here that $(ABC)(BCD) = AB^2C^2D = AD$ (modulus 2). Hence 3 df are confounded with blocks. Before proceeding to the block compositions, consider a problem where four factors (A, B, C, and D) are involved, each at two levels, and only four treatment combinations can be run in one day. The measured variable is yield. The confounding scheme given above gives two independent equations

$$L_1 = X_1 + X_2 + X_3$$
$$L_2 = X_2 + X_3 + X_4$$

Each treatment combination will give one of the following four pairs when substituted into both L_1 and $L_2 = 00, 01, 10, 11$. For example,

$$
\begin{aligned}
(1): \quad & L_1 = 0 & L_2 = 0 \\
a: \quad & L_1 = 1 & L_2 = 0 \\
b: \quad & L_1 = 1 & L_2 = 1 \\
ab: \quad & L_1 = 2 = 0 & L_2 = 1
\end{aligned}
$$

and so on. All treatment combinations with the same pair of L values will be placed together in one block. Thus the 16 treatment combinations will be assigned to four blocks as in Table 12.3.

Table 12.3 Blocking a 2^4 Factorial in Four Blocks of Four Confounding ABC, BCD, and AD

block 1	block 2	block 3	block 4
(1) = 82	a = 76	b = 79	ab = 85
bc = 55	abc = 74	c = 71	ac = 84
acd = 81	cd = 72	abcd = 89	bcd = 84
abd = 88	bd = 73	ad = 79	d = 80
$L_1 = 0$	$L_1 = 1$	$L_1 = 1$	$L_1 = 0$
$L_2 = 0$	$L_2 = 0$	$L_2 = 1$	$L_2 = 1$

Here the order of experimentation within each block is randomized and the resulting responses are given for each treatment combination. The proper treatment combinations for each block can be generated from the principal block. If bc and acd are in the principal block, then their product

$$(bc)(acd) = abc^2d = abd$$

is also in the principal block. For another block, consider a and multiply a by all of the combinations in the principal block, giving

$$a(1) = a \qquad a(bc) = abc \qquad a(acd) = cd \qquad a(abd) = bd$$

the entries in the second block. In the same manner, the other two blocks may be generated. This means that only a few readings in the principal block need to be determined by the L_1, L_2 values, and all the rest of the design can be generated from these few treatment combinations. Tables are also available for designs when certain effects are confounded [10].

This experiment is then a 2^4 factorial experiment. The design is a randomized incomplete block design where blocks are confounded with the interactions ABC, BCD, and AD. An example will illustrate the method.

■ **EXAMPLE 12.1** A systematic test was to be made on the effects of four variables, each at two levels, on coil breakdown voltage. The variables were A—firing furnace 1 or 3; B—firing temperature 1650°C or 1700°C; C—gas humidification, no or yes; and D—coil outside diameter small (<0.0300 inch) or large (>0.0305 inch). Since only four coils could be tested in a given short time interval, it was necessary to run four blocks of four coils for such testing. Only one complete test of 2^4 in four blocks of four was run and the results (after dividing the voltages through by 10) are shown in Table 12.3, which confounds ABC, BCD, and AD with the four blocks. For the analysis one might use the Yates method given in Chapter 9 to give Table 12.4.

Table 12.4 Yates Method Analysis for 2^4 Example

Treatment Combination	Response	(1)	(2)	(3)	(4)	SS
(1)	82	158	322	606	1252	—
a	76	164	284	646	+60	225.00
b	79	155	320	+32	2	0.25
ab	85	129	326	28	30	56.25
c	71	159	0	−20	−32	64.00
ac	84	161	+32	22	+32	64.00
bc	55	153	14	+18	−14	12.25
abc	74	173	14	12	−26	42.25*
d	80	−6	6	−38	40	100.00
ad	79	+6	−26	6	−4	1.00*
bd	73	+13	2	+32	42	110.25
abd	88	+19	20	0	−6	2.25
cd	72	−1	12	−32	44	121.00
acd	81	15	+6	18	−32	64.00
bcd	84	9	16	−6	50	156.25*
abcd	89	5	−4	−20	−14	12.25
Total						1031.00

* Confounded with blocks.

In Table 12.4 the sums of squares in the last column correspond to the effect for the treatment combination on the left. That is,

$$SS_A = 225.00 \qquad SS_B = 0.25 \qquad \text{and so on}$$

Note that the effects ABC, BCD, and AD are confounded with blocks. Their total sum of squares is $42.25 + 156.25 + 1.00 = 199.50$. If the above block totals are computed, they yield

block 1 — 306   block 3 — 318
block 2 — 295   block 4 — 333

and the sum of squares between these blocks is

$$\frac{(306)^2 + (295)^2 + (318)^2 + (333)^2}{4} - \frac{(1252)^2}{16}$$

$$\text{SS}_{\text{block}} = 98{,}168.50 - 97{,}969.00 = 199.50$$

which is identical with the sum of the three interactions with which blocks are confounded. If all three-way and four-way interactions that are not confounded are pooled for an error term, the resulting analysis is that shown in Table 12.5.

Table 12.5 ANOVA for 2^4 Example in Four Blocks

Source	df	SS	MS
A	1	225.00	225.00
B	1	0.25	0.25
C	1	64.00	64.00
D	1	100.00	100.00
AB	1	56.25	56.25
AC	1	64.00	64.00
BC	1	12.25	12.25
BD	1	110.25	110.25
CD	1	121.00	121.00
Error or ABD, ACD, $ABCD$	3	78.50	26.17
Blocks or ABC, BCD, AD	3	199.50	66.50
Totals	15	1031.00	

With only 1 and 3 df, none of the four main effects or the five two-way interactions can be declared significant at the 5 percent significance level. Because the B effect and the BC interaction are not significant even at the 25 percent level, we might wish to pool these with the error to increase the error degrees of freedom. The resulting error sum of squares would then be

$$0.25 + 12.25 + 78.50 = 91.00$$

with 5 df. The error mean square then $= 91.00 / 5 = 18.20$. Using this error mean square, A and CD are now significant at the 5 percent level. We might conclude that the effect of factor A is important, that B is not present, and that there may be a CD interaction. As 2^f experiments are often run to get an overall picture of the important factors, each set at the extremes of its range, another experiment might now be planned without factor B, since its effect is negligible over the range considered here. The next experiment might include factors A, C, and D at, perhaps, three levels each, or a 3^3 experiment. This experiment might also require blocking. ■

12.4 Block Confounding with Replication

Whenever an experiment is restricted so that all treatments cannot appear in one block, some interaction is usually confounded with blocks. If several replications of the whole experiment (all blocks) are possible, as in the case of the split plot, the same interaction may be confounded in all replications. In this case, the design is said to be *completely confounded*. If, on the other hand, one interaction is confounded in the first replication, a different interaction is confounded in the second replication, and so on, the design is said to be *partially confounded*.

Complete Confounding

Considering a 2^3 factorial experiment in which only four treatment combinations can be finished in one day yields a 2^3 factorial run in two incomplete blocks of four treatment combinations each. We might confound the highest order interaction *ABC* as follows:

$$ABC$$

As seen in Section 12.2, the blocks would be

block 1	block 2
(1)	a
ab	b
ac	c
bc	abc

If this whole experiment (2^3 in two blocks of four each) can be replicated, say three times, the layout might be

replication I		replication II		replication III	
block 1	block 2	block 1	block 2	block 1	block 2
ac	a	c	(1)	ab	c
(1)	c	abc	ac	(1)	b
ab	abc	b	bc	ac	abc
bc	b	a	ab	bc	a

The confounding scheme in the above (*ABC*) is the same for all three replications, but the order of experimentation has been randomized within each block. Also, the decision as to which block is to be run first in each replication is made at random. An analysis layout for this experiment appears in Table 12.6.

Here the replication interaction with all three main effects and their interactions are usually taken as the error term for testing the important effects. The replication effect and the block (or *ABC*) effect could be tested against the replication $\times$ block interaction, but the degrees of freedom are low and the

Table 12.6 Analysis Layout for Completely Confounded 2^3 Factorial

Source	df	df
Replications	2	
Blocks or *ABC*	1	
Replications × block interactions	2	5 between plots
A	1	
B	1	
AB	1	
C	1	
AC	1	
BC	1	
Replications × all others	12	18 within plots
Totals	23	23

power of such a test is poor. Such a design is quite powerful, however, in testing the main effects *A*, *B*, and *C* and their first-order interactions. However, it will give no clear information on the *ABC* interaction.

Partial Confounding

In the example just considered, we might be concerned with some test on the *ABC* interaction. This could be found by confounding some interactions other than *ABC* in some of the replications. We might use four replications and confound *ABC* in the first one, *AB* in the second, *AC* in the third, and *BC* in the fourth. Thus the four replications will yield full information on *A*, *B*, and *C* but three-fourths information on *AB*, *AC*, *BC*, and *ABC*, since we can compute an unconfounded interaction such as *AB* in three out of four of the replications. The following layout shows the proper block entries for each replication.

replication I confound *ABC*		*replication II* *AB*		*replication III* *AC*		*replication IV* *BC*	
(1)	*a*	(1)	*a*	(1)	*a*	(1)	*b*
ab	*b*	*c*	*b*	*ac*	*c*	*bc*	*c*
ac	*c*	*ab*	*ac*	*abc*	*bc*	*a*	*ab*
bc	*abc*	*abc*	*bc*	*b*	*ab*	*abc*	*ac*

$$L = X_1 + X_2 + X_3 \qquad L = X_1 + X_2 \qquad L = X_1 + X_3 \qquad L = X_2 + X_3$$

The analysis layout would be that shown in Table 12.7.

Table 12.7 ANOVA for Partially Confounded 2^3 Factorial

Source	df		df
Replications	3		
Blocks within replications	4		7 between plots
ABC (Replication I)		1	
AB (Replication II)		1	
AC (Replication III)		1	
BC (Replication IV)		1	
A	1		
B	1		
C	1		
AB	1	only from	
AC	1	replications	
BC	1	where not	
ABC	1	confounded	
Replications × all effects	17		24 within plots
Totals	31		31

The residual term with its 17 df can be explained as follows. In all four replications the main effects *A*, *B*, and *C* can be isolated. This gives 3×1 df for each replication by main-effect interaction, or a total of 9 df. The *AB*, *AC*, *BC*, and *ABC* interactions can be isolated in only three out of the four replications. Hence the replications by *AC* have $2 \times 1 = 2$ df, or a total of 8 df for all such interactions. This gives $9 + 8 = 17$ df for effects by replications, which is usually taken as the error estimate. The blocks and replications are separated out in the top of Table 12.7 only to reduce the experimental error. The numerical analysis of problems such as this is carried out in the same manner as given in earlier chapters.

■ **EXAMPLE 12.2** Coils were wound from each of two winding machines using either of two wire stocks and their outside diameters were measured at two positions on an optical comparator. Since only four readings could be made in a given time period, the $2^3 = 8$ treatment combinations were divided into two blocks of four. However, four replications were possible. At first the experimenter decided to confound *ABC* in all four replicates. Here *A* represents machines, *B* stocks, and *C* positions. The results are shown in Table 12.8.

Analyzing by computer or by direct methods gives Table 12.9. Table 12.9 shows only positions on the coil as having a highly significant effect on coil outside diameter.

Table 12.8 Layout of Example 12.2 in Four Blocks of Four Confounding *ABC* in All Replicates

	replication I				*replication II*	
	block 1		*block 2*		*block 1*	*block 2*
	ab = 2173	b = 2300			a = 2228	ac = 3495
	(1) = 2249	c = 3538			abc = 3592	(1) = 2094
	ac = 3532	abc = 3524			c = 3116	bc = 2249
	bc = 2948	a = 2319			b = 2386	ab = 2373

	replication III				*replication IV*	
	block 1		*block 2*		*block 1*	*block 2*
	(1) = 2382	c = 3528			abc = 2996	ab = 2393
	ab = 2240	b = 2118			c = 1934	bc = 2814
	bc = 3495	abc = 3350			b = 2234	ac = 3400
	ac = 2995	a = 2272			a = 2215	(1) = 2297

All measurements multiplied by 10^5 so 2173 is an outside diameter of 0.02173 inch.

Table 12.9 ANOVA for Example 12.2

Source	df	SS	MS	F	Probability
Replications	3	409,719.59			
Blocks of *ABC*	1	8,482.53			
$R \times B$ interaction	3	515,407.85			
	7				
A	1	364,444.53	364,444.53	3.13	0.094
B	1	5,227.53	5,227.53	<1	
AB	1	18,963.78	18,963.78	<1	
C	1	6,330,571.53	6,330,571.53	54.44**	0.000
AC	1	302,058.78	302,058.78	2.60	0.124
BC	1	16,698.78	16,698.78	<1	
Replications $\times$ all	18	2,093,039.32	116,279.96		
	24				
Totals	31	10,064,614.22			

■ **EXAMPLE 12.3** When coils from two other winding machines were to be tested, it was decided to partially confound the interactions. Results gave Table 12.10.

Analysis gives Table 12.11. In this table, position on the coil is again highly

Table 12.10 Layout of Example 12.3 in Four Blocks of Four with Partial Confounding

replication I (confounding ABC)		replication II (confounding AB)	
block 1	block 2	block 3	block 4
(1) = 2208	a = 2196	(1) = 2004	a = 2179
ab = 2133	b = 2086	ab = 2112	b = 2073
ac = 2459	c = 3356	c = 3073	ac = 3474
bc = 3096	abc = 2776	abc = 2631	bc = 3360

replication III (confounding AC)		replication IV (confounding BC)	
block 5	block 6	block 7	block 8
c = 2839	(1) = 1916	a = 2056	b = 1878
a = 2189	ac = 2979	(1) = 2010	ab = 2156
bc = 3522	b = 2151	bc = 3209	c = 3423
ab = 2095	abc = 2500	abc = 3066	ac = 2524

Table 12.11 ANOVA for Example 12.3

Source	df	SS	MS	F	Probability
Replications	3	38,717.59			
Blocks in					
replications	4	401,060.13			
	7				
A	1	224,282.53	224,282.53	3.97	0.063
B	1	52.53	52.53	<1	
AB	1	737.04	737.04	<1	
C	1	6,886,688.28	6,886,688.28	121**	0.000
AC	1	416,066.67	416,066.67	7.36*	0.015
BC	1	2,667.04	2,667.04	<1	
ABC	1	70,742.04	70,742.04	1.25	0.279
Replications × all	17	961,283.32	56,546.08		
	24				
Totals	31	9,002,296.97			

significant, but an AC interaction is also significant at the 5 percent significance level.

In both problems the effect of A—machines—is approaching significance. It may be that if more replications are run, this machine effect will emerge as significant. ■

Confounding in 3^f Factorials

When a 3^f-factorial experiment cannot be completely randomized, it is usually blocked in blocks that are multiples of 3. The use of the I and J components of interaction introduced in Section 10.2 is helpful in confounding only part of an interaction with blocks. Kempthorne's rule also applies, using such interactions as AB, AB^2, ABC^2, ..., and treatment combinations in the form 00, 10, 01, 11 instead of (1), a, b, ab as in 2^f. Here treatment combinations can be multiplied that will actually add these exponents (modulus 3).

If a 3^2 factorial is restricted so that only three of the nine treatment combinations can be run in one block, we usually confound AB or AB^2, as each carries 2 df. The defining contrast might be AB^2, which gives $L = X_1 + 2X_2$.

Each of the nine treatment combinations then yields

$$00, L = 0 \qquad 01, L = 2 \qquad 02, L = 4 = 1$$
$$10, L = 1 \qquad 11, L = 3 = 0 \qquad 12, L = 5 = 2$$
$$20, L = 2 \qquad 21, L = 4 = 1 \qquad 22, L = 6 = 0$$

Placing treatment combinations with the same L value in common blocks gives

block 1	*block 2*	*block 3*
$L = 0$	$L = 1$	$L = 2$
00	10	20
11	21	01
22	02	12

Block 1 with 00 in it is the principal block. Block 2 is generated by *adding* one treatment combination in block 2 to all of those in block 1, giving

$$00 + 10 = 10 \qquad 11 + 10 = 21 \qquad 22 + 10 = 32 = 02$$

Similarly for block 3,

$$00 + 20 = 20 \qquad 11 + 20 = 31 = 01 \qquad 22 + 20 = 42 = 12$$

Since there are three blocks, 2 df must be confounded. Here AB^2 is confounded with blocks. If AB with its 2 df were confounded, $L = X_1 + X_2$, and the blocking would be

block 1	*block 2*	*block 3*
$L = 0$	$L = 1$	$L = 2$
00	10	20
12	22	02
21	01	11

If the data of Table 10.1 are used as an example, and AB^2 is confounded, the responses are

<table>
<tr><td>block 1</td><td>block 2</td><td>block 3</td></tr>
<tr>
<td>00 = 1
11 = 4
22 = 2</td>
<td>10 = −2
21 = 1
02 = 2</td>
<td>20 = 3
01 = 0
12 = −1</td>
</tr>
</table>

The block sums of squares are determined from the three block totals of 7, 1, and 2:

$$SS_{block} = \frac{7^2 + 1^2 + 2^2}{3} - \frac{(10)^2}{9} = 6.89$$

which is identical with the $I(AB)$ interaction component computed in Section 10.2 (Table 10.6), as it should be. The remainder of the analysis proceeds the same as in Section 10.2, and the resulting ANOVA table is that exhibited in Table 12.12.

Table 12.12 ANOVA for 3^2 Example

Source	df	SS	MS
A	2	4.22	2.11
B	2	1.56	0.78
AB	2	16.22	8.11
Blocks or AB^2	2	6.89	3.44
Totals	8	28.89	

If these were real data, the AB part of the interaction might be considered as error, but then neither main effect is significant. The purpose here is only to illustrate the blocking of a 3^2 experiment.

Consider now a 3^3 experiment in which the 27 treatment combinations (such as given in Table 10.7) cannot all be completely randomized. If nine can be randomized and run on one day, nine on another day, and so on, we might use a 3^3 factorial in three blocks of nine treatment combinations each. This requires 2 df to be confounded with blocks. Since the ABC interaction with 8 df can be partitioned into ABC, ABC^2, AB^2C, and AB^2C^2, each with 2 df, one of these four could be confounded with blocks. If AB^2C is confounded,

$$L = X_1 + 2X_2 + X_3$$

and the three blocks are

$L = 0$	$L = 1$	$L = 2$
000	100	200
011	111	211
110	210	010
121	221	021
102	202	002
212	012	112
220	020	120
022	122	222
201	011	101

The analysis of data for this design would yield Table 12.13.

Table 12.13 3^3 Factorial in Three Blocks

Source	df
A	2
B	2
AB	4
C	2
AC	4
BC	4
Error or ABC, ABC^2, AB^2C^2	6
Blocks or AB^2C	2
Total	26

This is a useful design, because we can retrieve all main effects (A, B, C) and all two-way interactions, if we are willing to pool the 6 df in ABC as error. The determination of sums of squares, and so forth, is the same as for a complete 3^3 given in Example 10.3. Confounding other parts of the ABC interaction would, of course, yield different blocking arrangements, although the outline of the analysis would be the same. In practice, different blocking arrangements might yield different responses.

More complex confounding schemes can be found in tables such as in Chapter 9 of Davies [10].

For practice, consider confounding a 3^3 in nine blocks of three treatment combinations each. Here 8 df must be confounded with blocks. One confounding scheme might be

$$AB^2C^2 \qquad AB \qquad BC^2 \qquad AC$$

Note that

$$(AB^2C^2)(AB) = A^2C^2 = AC$$

and

$$(AB)(BC^2) = AB^2C^2$$

These are not all independent. In fact, only two of the expressions are. For this reason we need only two expressions for the L values:

$$L_1 = X_1 + 2X_2 + 2X_3$$
$$L_2 = X_1 + X_2$$

as these two will yield nine pairs of numbers—one pair for each block. First determine the principal block, where both L_1 and L_2 are zero. One treatment combination is obviously 000. Another is 211, as

$$L_1 = 1(2) + 2(1) + 2(1) = 6 = 0$$
$$L_2 = 1(2) + 1(1) = 3 = 0$$

A third is 122, as

$$L_1 = 1(1) + 2(2) + 2(2) = 9 = 0$$
$$L_2 = 1(1) + 1(2) = 3 = 0$$

The other eight blocks can now be generated from this principal block, giving

block	*1*	*2*	*3*	*4*	*5*	*6*	*7*	*8*	*9*
	000	001	022	010	020	100	200	110	101
	211	212	210	221	201	011	111	021	012
	122	120	121	102	112	222	022	202	220

An analysis layout would be that shown in Table 12.14.

Table 12.14 3^3 Factorial in Nine Blocks

Source	df	
A	2	
B	2	
C	2	
AB^2	2	
AC^2	2	
BC	2	
ABC	2	
AB^2C	2	} 6 df for error
ABC^2	2	
Blocks or AB, BC^2, AC, AB^2C^2	8	
Total	26	

   This design might be reasonable if interest centered only in the main effects A, B, and C. Such a design might be necessary when three factors are involved, each at three levels, but only three treatment combinations can be run in one day. These three might involve an elaborate environmental test in which at best only three sets of environmental conditions—temperature, pressure, and humidity— can be simulated in one day.

12.5 Summary

The summary of Chapter 11 may now be extended.

Experiment	Design	Analysis

II. Two or
    more factors
  A. Factorial
    (crossed)		

	1. Completely randomized	
	2. Randomized block	2.
	a. Complete	a. Factorial
	$Y_{ijk} = \mu + R_k + A_i$	ANOVA
	$+ B_j + AB_{ij} + \varepsilon_{ijk}$	
	b. Incomplete, confounding:	b.
	i. Main effect—split plot	i. Split-plot
	$Y_{ijk} = \mu + \underbrace{R_i + A_j + RA_{ij}}_{\text{whole plot}}$	ANOVA
	$+ \underbrace{B_k + RB_{ik} + AB_{jk} + RAB_{ijk}}_{\text{split plot}}$	
	ii. Interactions in 2^f and 3^f	ii.
	(1) Several replications	(1) Factorial
	$Y_{ijkq} = \mu + R_i + \beta_j + R\beta_{ij}$	ANOVA

	$+ A_k + B_q + AB_{kq} + \varepsilon_{ikq}$	with replica-
		        tions R_i and
		        blocks β_j or
		        confounded
		        interaction

	(2) One replication only	(2) Factorial
	$Y_{ijk} = \mu + \beta_i + A_j$	ANOVA

	$+ B_k + AB_{jk}, \ldots$	with blocks
		        β_j or con-
		        founded
		        interaction

12.6 SAS Programs

The analyses of Examples 12.2 and 12.3 are now handled by SAS as shown:

Example 12.2

Two models are shown in the SAS program. In the first model, $Y = $ REP|BLK $A|B|C$, the analysis shows blocks with the same sum of squares as the ABC interaction (8,482.53). This is as expected because ABC was confounded with blocks in *all* four replications.

In the second model, $Y = $ REP|BLK A B $A*B$ C $A*C$ $B*C$, the ABC term is left out and we get the same results as in Table 12.9.

```
DATA A;
INPUT REP BLK A B C Y @@;
CARDS;
-(DATA)-
;
TITLE EXAMPLE 12.2;
PROC PRINT;
PROC ANOVA;
CLASS REP BLK A B C;
MODEL Y=REP|BLK A|B|C;
PROC ANOVA;
CLASS REP BLK A B C;
MODEL Y= REP|BLK A B A*B C A*C B*C;
```

The analyses of the two models are on page 298.

Example 12.3

Here the model, $Y = $ REP BLK(REP) $A|B|C$, gives some information on all interactions since it is only partially confounded. The results agree with Table 12.10.

```
DATA A;
INPUT REP BLK A B C Y @@;
CARDS;
-(DATA)-
;
TITLE EXAMPLE 12.3
PROC PRINT;        .
PROC GLM;
CLASS REP BLK A B C;
MODEL Y=REP BLK(REP)A|B|C;
```

This program uses GLM, but ANOVA could just as well have been used. The output for Example 12.3 is on page 299.

ANALYSIS OF VARIANCE PROCEDURE

DEPENDENT VARIABLE: Y

SOURCE	DF	SUM OF SQUARES	MEAN SQUARE	F VALUE	PR > F	R-SQUARE	C.V.
MODEL	14	7980057.43750000	570004.10267857	4.65	0.0018	0.792883	12.9127
ERROR	17	2084556.78125000	122620.98713235		ROOT MSE		Y MEAN
CORRECTED TOTAL	31	10064614.21875000			350.17279611		2711.84375000

SOURCE	DF	ANOVA SS	F VALUE	PR > F
REP	3	409719.59375000	1.11	0.3710
BLK	1	8482.53125000	0.07	0.7957
REP*BLK	3	515407.84375000	1.40	0.2768
A	1	364444.53125000	2.97	0.1028
B	1	5227.53125000	0.04	0.8389
A*B	1	18963.78125000	0.15	0.6990
C	1	6330571.53125000	51.63	0.0001
A*C	1	302058.78125000	2.46	0.1350
B*C	1	16698.78125000	0.14	0.7167
A*B*C	1	8482.53125000	0.07	0.7957

ANALYSIS OF VARIANCE PROCEDURE

DEPENDENT VARIABLE: Y

SOURCE	DF	SUM OF SQUARES	MEAN SQUARE	F VALUE	PR > F	R-SQUARE	C.V.
MODEL	13	7971574.90625000	613198.06971154	5.27	0.0008	0.792040	12.5744
ERROR	18	2093039.31250000	116279.96180556		ROOT MSE		Y MEAN
CORRECTED TOTAL	31	10064614.21875000			340.99847772		2711.84375000

SOURCE	DF	ANOVA SS	F VALUE	PR > F
REP	3	409719.59375000	1.17	0.3471
BLK	1	8482.53125000	0.07	0.7902
REP*BLK	3	515407.84375000	1.48	0.2542
A	1	364444.53125000	3.13	0.0936
B	1	5227.53125000	0.04	0.8345
A*B	1	18963.78125000	0.16	0.6911
C	1	6330571.53125000	54.44	0.0001
A*C	1	302058.78125000	2.60	0.1244
B*C	1	16698.78125000	0.14	0.7092

GENERAL LINEAR MODELS PROCEDURE
DEPENDENT VARIABLE: Y

SOURCE	DF	SUM OF SQUARES	MEAN SQUARE	F VALUE	PR > F	R-SQUARE	C.V.
MODEL	14	8041013.85416667	574358.13244048	10.16	0.0001	0.893218	9.3105
ERROR	17	961283.11458333	56546.06556373		ROOT MSE		Y MEAN
CORRECTED TOTAL	31	9002296.96875000			237.79416638		2554.03125000

SOURCE	DF	TYPE I SS	F VALUE	PR > F	DF	TYPE III SS	F VALUE	PR > F
REP	3	38717.59375000	0.23	0.8754	3	38717.59375000	0.23	0.8754
BLK(REP)	4	401060.12500000	1.77	0.1808	4	264696.54166667	1.17	0.3587
A	1	224282.53125000	3.97	0.0627	1	224282.53125000	3.97	0.0627
B	1	52.53125000	0.00	0.9760	1	52.53125000	0.00	0.9760
A*B	1	737.04166667	0.01	0.9104	1	737.04166667	0.01	0.9104
C	1	6886688.28125000	121.79	0.0001	1	6886688.28125000	121.79	0.0001
A*C	1	416066.66666667	7.36	0.0148	1	416066.66666667	7.36	0.0148
B*C	1	2667.04166667	0.05	0.8307	1	2667.04166667	0.05	0.8307
A*B*C	1	70742.04166667	1.25	0.2789	1	70742.04166667	1.25	0.2789

Problems

12.1 Assuming that a 2^3 factorial such as the problem of Chapters 1, 5, and 9 involving two tool types, two bevel angles, and two types of cut cannot all be run on one lathe in one time period but must be divided into four treatment combinations per lathe, set up a confounding scheme for confounding the AC interaction in these two blocks of four. Assuming the results of one replication give $(1) = 5$, $a = 0$, $b = 4$, $ab = 2$, $c = -3$, $ac = 0$, $bc = -1$, $abc = -2$, show that the scheme you have set up does indeed confound AC with blocks.

12.2 Assuming three replications of the 2^3 experiment in Problem 12.1 can be run but each one must be run in two blocks of four, set up a scheme for the complete confounding of ABC in all three replicates and show its ANOVA layout.

12.3 Repeat Problem 12.2 using partial confounding of AB, AC, and BC.

12.4 Data involving four replications of the 2^3 factorial described in the foregoing problems and in Section 12.3 gave the following numerical results.

	replication I			replication II	
confound	ABC			AB	

$(1) = 2$	$a = 0$	$(1) = -3$	$a = 1$
$ab = 3$	$b = 1$	$c = -6$	$b = 1$
$ac = -7$	$c = 0$	$ab = 8$	$ac = -6$
$bc = -2$	$abc = -1$	$abc = 0$	$bc = 2$

	replication III			replication IV	
confound	AC			BC	

$(1) = 5$	$a = 0$	$(1) = -2$	$b = 9$
$ac = 0$	$c = -3$	$bc = -1$	$c = -3$
$abc = -2$	$bc = -1$	$a = -6$	$ab = 0$
$b = 4$	$ab = 2$	$abc = -4$	$ac = -4$

Analyze these data and state your conclusions.

12.5 A 2^3 factorial is to be used to study three factors and only four treatments can be run on each of two days. On the first day the following data are collected:

$$(1) = 2 \qquad a = 1 \qquad bc = 3 \qquad abc = 4$$

and on the next day:

$$c = 1 \qquad ac = 4 \qquad b = 6 \qquad ab = 0$$

By examining the variation between the two days and your knowledge of a 2^f experiment, find out which term in the model (main effect or interaction) is confounded with days.

12.6 For the data of Problem 9.6 consider running the 16 treatment combinations in four blocks of four (two observations per treatment). Confound *ACD*, *BCD*, and *AB* and then determine the block compositions.

12.7 From the results of Problem 9.6 show that the blocks are confounded with the interactions of Problem 12.6.

12.8 For Problem 10.11 consider this experiment as run in three blocks of nine with two replications for each treatment combination. Confound *ABC* (2 df) where *A* is surface thickness, *B* is base thickness, and *C* is subbase thickness, and determine the design. Also use the numerical results of Problem 10.13 to show that *ABC* is indeed confounded in your design.

12.9 Set up a scheme for confounding a 2^4 factorial in two blocks of eight each, confounding *ABD* with blocks. L 1χ

12.10 Repeat Problem 12.9 and confound *BCD*.

12.11 Work out a confounding scheme for a 2^5 factorial, confounding in four blocks of eight each. Show the outline of an ANOVA table.   *See Answers*

12.12 Repeat Problem 12.11 in four replications confounding different interactions in each replication. Show the ANOVA table outline.

12.13 Work out the confounding of a 3^4 factorial in three blocks of 27 each.

12.14 Repeat Problem 12.13 in nine blocks of nine each.

12.15 In Example 12.2, since *ABC* is completely confounded, the analysis could be easily run on a computer or by using the Yates method on sections *A*, *B*, *C* and general methods for replications and blocks. Verify the results given in Table 12.9.

12.16 Since Example 12.3 has partial confounding, it may be difficult for the computer. However, a Yates method can be used if care is taken as to how it is applied. Try to verify Table 12.11 results using a modified Yates procedure.

12.17 A systematic test was made to determine the effects on coil breakdown voltage of the following six factors, each at two levels as indicated:
 a. Firing furnace   Number 1 or 3
 b. Firing temperature   1650°C or 1700°C
 c. Gas humidification   No or yes
 d. Coil outer diameter   Below 0.0300 inch or above 0.0305 inch
 e. Artificial chipping   No or yes
 f. Sleeve   Number 1 or 2

All 64 experiments could not be performed under the same conditions so the whole experiment was run in eight subgroups of eight experiments. Set up a reasonable confounding scheme for this problem and outline its ANOVA.

12.18 An experiment was run on the effects of several annealing variables on the magnetic characteristic of a metal. The following factors were to be considered.
 a. Temperature   1375°, 1450°, 1525°, 1600°
 b. Time in hours   2, 4
 c. Exogas ratio   6.5 to 1, 8.5 to 1

    d. Dew point   10°, 30°
    e. Cooling rate in hours   4, 8
    f. Core material   *A, B*
Since all factors are at two levels or multiples of two levels, outline a scheme for this experiment in two blocks of 64 observations.

12.19 Outline a Yates method for handling the data of Problem 12.18.

12.20 If X is the surface rating of a steel sample and material is tested from eight heats, two chemical treatments, and three positions on the ingot, the resulting factorial is an $8 \times 2 \times 3$ or $2^3 \times 2 \times 3$ or $2^4 \times 3$, which is called a mixed factorial in the form $2^m \cdot 3^n$. Set up an ANOVA table for such a problem treating the eight heats as 2^3 pseudo factors.

12.21 Assuming Problem 12.20 results must be fired in furnaces that can handle only 24 steel samples at a time, devise a confounding scheme for the experiment.

12.22 Complete the analysis of Problem 12.21 for the following data.

	Position					
	1		**2**		**3**	
	Chemical	Treatment	Chemical	Treatment	Chemical	Treatment
Heat	1	2	1	2	1	2
1	1	0	0	0	0	2
2	−3	2	−1	−2	−2	−2
3	1	2	0	3	0	4
4	0	0	−2	1	0	1
5	−3	1	0	2	−2	0
6	−2	1	−3	2	0	0
7	3	3	3	3	1	−1
8	1	1	0	5	1	2

12.23 Consider a 2^5 factorial experiment with all fixed levels to be run in four blocks of eight treatment combinations each. The whole experiment can be replicated in a total of three replications. The following interactions are to be confounded in each replication.

Replication I—*ABC, CDE, ABDE*

Replication II—*BCDE, ACD, ABE*

Replication III—*ABCD, BCE, ADE*

Determine the block compositions for each block in each replication and outline the ANOVA for this problem.

Fractional Replication 13

13.1 Introduction

As the number of factors to be considered in a factorial experiment increases, the number of treatment combinations increases very rapidly. This can be seen with a 2^f factorial where $f = 5$ requires 32 experiments for one replication, $f = 6$ requires 64, $f = 7$ requires 128, and so on. Along with this increase in the amount of experimentation comes an increase in the number of high-order interactions. Some of these high-order interactions may be used as error, as those above second order (three way) would be difficult to explain if found significant. Table 13.1 gives some idea of the number of main effects, first-order, second-order, . . . , interactions that can be recovered if a complete 2^f factorial can be run.

Considering $f = 7$, there will be 7 df for the seven main effects, 21 df for the 21 first-order interactions, 35 df for the 35 second-order interactions, leaving

$$35 + 21 + 7 + 1 = 64 \text{ df}$$

for an error estimate, assuming no blocking. Even if this experiment were confounded in blocks, there is still a large number of degrees of freedom for the error estimate. In such cases, it may not be economical to run a whole replicate of 128 observations. Nearly as much information can be gleaned from half as

Table 13.1 Buildup of 2^f-Factorial Effects

			Order of Interaction						
f	2^f	Main Effect	1st	2nd	3rd	4th	5th	6th	7th
5	32	5	10	10	5	1			
6	64	6	15	20	15	6	1		
7	128	7	21	35	35	21	7	1	
8	256	8	28	56	70	56	28	8	1

many observations. When only a fraction of a replicate is run, the design is called a *fractional replication* or a *fractional factorial*.

For running a fractional replication, the methods of Chapter 12 are used to determine a confounding scheme such that the number of treatment combinations in a block is within the economic range of the experimenter. If, for example, the experimenter can run 60 or 70 experiments and is interested in seven factors, each at two levels, a one-half replicate of a 2^7 factorial can be used. The complete $2^7 = 128$ experiment is laid out in two blocks of 64 by confounding some high-order interaction. Then only one of these two blocks is run; the decision is made as to which one by the toss of a coin.

13.2 Aliases

To see how a fractional replication is run, consider a simple case. Three factors are of interest, each at two levels. However, the experimenter cannot afford $2^3 = 8$ experiments, but will, however, settle for four. This suggests a one-half replicate of a 2^3 factorial. Suppose ABC is confounded with blocks. The two blocks are then

$$I = ABC$$

block 1	(1)	ab	bc	ac

block 2	a	b	c	abc

A coin is flipped and the decision is made to run only block 2. What information can be gleaned from block 2 and what information is lost when only half of the experiment is run?

Referring to Table 6.7, which shows the proper coefficients ($+1$ or -1) for the treatment combinations in a 2^3 factorial to give the effects desired, box only those treatment combinations in block 2 that are to be run (Table 13.2)

Note, in the boxed area, that ABC has all plus signs in block 2, which is a result of confounding blocks with ABC. Note also that the effect of A is

$$A = +a - b - c + abc$$

and

$$BC = +a - b - c + abc$$

so that we cannot distinguish between A and BC in block 2. Two or more effects that have the same numerical value are called *aliases*. We cannot tell them apart. B and AC likewise are aliases, as are C and AB. Because of this confounding when only a fraction of the experiment is run, we must check the aliases and be reasonably sure they are not both present if such a design is to be of value. An analysis of this one-half replication of a 2^3 would be as shown in Table 13.3.

Table 13.2 One-Half Replication of a 2^3 Factorial

Treatment Combination	A	B	AB	C	AC	BC	ABC
(1) *A0 B0*	−	−	+	−	+	+	−
a *A1 B0*	+	−	−	−	−	+	+
b *B0 A1*	−	+	−	−	+	−	+
ab *A1b1*	+	+	+	−	−	−	−
c	−	−	+	+	−	−	+
ac	+	−	−	+	+	−	−
bc	−	+	−	+	−	+	−
abc	+	+	+	+	+	+	+

Table 13.3 One-Half Replication of a 2^3 Factorial

Source	df
A (or BC)	1
B (or AC)	1
C (or AB)	1
Total	3

This would hardly be a practical experiment unless the experimenter is sure that no first-order interactions exist and has some external source of error to use in testing A, B, and C. The real advantage of such fractionating will be seen on designs with larger f values.

If block 1 were run for the experiment instead of block 2, A would have the value $-(1) + ab + ac - bc$, and BC the value $+(1) - ab - ac + bc$, which gives the same total except for sign. The sums of squares due to A and BC are therefore the same; again they are said to be aliases. Here,

$$A = -BC \qquad B = -AC \qquad C = -AB$$

The definition given above still holds when one effect is the alias of another if they have the same *numerical* value, or value regardless of sign.

A quick way to find the aliases of an effect in a fractional replication of a 2^f factorial experiment is to multiply the effect by the terms in the defining contrast, modulus 2. The results will be aliases of the original effect. In the example above,

$$I = ABC$$

The alias of A is

$$A(ABC) = A^2BC = BC$$

The alias of B is

$$B(ABC) = AB^2C = AC$$

The alias of C is

$$C(ABC) = ABC^2 = AB$$

This simple rule works for any fractional replication for a 2^f factorial. It works also with a slight modification for a 3^f factorial run as a fractional replication.

13.3 2^f Fractional Replications

As an example, consider the problem suggested in the introduction. An experimenter wishes to study the effect of seven factors, each at two levels, but cannot afford to run all 128 experiments and so will settle for 64, or a one-half replicate of a 2^7. Deciding to confound the highest order interaction with blocks, the experimenter has

$$I = ABCDEFG$$

The two blocks are found by placing (1) and all pairs, quadruples, and sextuples of the seven letters in one block and the single letters, triples, quintuples, and one septuple of the seven letters in the other block. One of the two blocks is chosen at random and run. Before carrying out this experiment, the experimenter should check on the aliases. From the defining contrast

$$I = ABCDEFG$$

the alias of A is $A(ABCDEFG) = BCDEFG$, and all main effects are likewise aliased with fifth-order interactions.

AB is aliased with $AB(ABCDEFG) = CDEFG$, a fourth-order interaction. So all first-order interactions are aliased with fourth-order interactions. A second-order interaction such as ABC is aliased with a third-order interaction as

$$ABC(ABCDEFG) = DEFG$$

If the second- and third-order interactions are taken as error, the analysis will be as in Table 13.4.

Table 13.4 One-Half Replication of a 2^7 Factorial

Source	df
Main effects $A, B, \ldots, G$ (or fifth order)	1 each for 7
First-order interaction AB, AC (or fourth order)	1 each for 21
Second-order interaction ABC, ABD (or third order)	1 each for 35 $\}$ use as error
Total	63

This is a very practical design, as there are good tests on all main effects and first-order interactions, assuming all higher order interactions are zero. The degrees of freedom for each test would be 1 and 35. If, for some reason known to the experimenter, some second-order interaction were suspected, the experimenter could leave it out of the error estimate and still have sufficient degrees of freedom for the error estimate. The analysis of such an experiment follows the methods given in Chapter 9.

If this same experimenter is further restricted and can only afford to run 32 experiments, a one-fourth replication of a 2^7 might be tried. Here 3 df must be confounded with blocks. If two fourth-order interactions are confounded with blocks, one third-order is automatically confounded also, as

$$I = ABCDE = CDEFG = ABFG$$

which confounds 3 df with the four blocks. In this design, if only one of the four blocks of 32 observations is run, each effect has three aliases. These are

$A = BCDE = ACDEFG = BFG$ $CD = ABE = EFG = ABCDFG$

$B = ACDE = BCDEFG = AFG$ $CE = ABD = DFG = ABCEFG$

$C = ABDE = DEFG = ABCFG$ $CF = ABDEF = DEG = ABCG$

$D = ABCE = CEFG = ABDFG$ $CG = ABDEG = DEF = ABCF$

$E = ABCD = CDFG = ABEFG$ $DE = ABC = CFG = ABDEFG$

$F = ABCDEF = CDEG = ABG$ $DF = ABCEF = CEG = ABDG$

$G = ABCDEG = CDEF = ABF$ $DG = ABCEG = CEF = ABDF$

$AB = CDE = ABCDEFG = FG$ $EF = ABCDF = CDG = ABEG$

$AC = BDE = ADEFG = BCFG$ $EG = ABCDG = CDF = ABEF$

$AD = BCE = ACEFG = BDFG$ $ACF = BDEF = ADEG = BCG$

$AE = BCD = ACDFG = BEFG$ $ACG = BDEG = ADEF = BCF$

$AF = BCDEF = ACDEG = BG$ $ADF = BCEF = ACEG = BDG$

$$AG = BCDEG = ACDEF = BF \quad ADG = BCEG = ACEF = BDF$$
$$BC = ADE = BDEFG = ACFG \quad AEG = BCDG = ACDF = BEF$$
$$BD = ACE = BCEFG = ADFG \quad BEG = ACDG = BCDF = AEF$$
$$BE = ACD = BCDFG = AEFG$$

This is quite a formidable list of aliases; but when only one block of 32 is run, there are 31 df within the block. The above list accounts for these 31 df. If, in this design, all second-order (three-way) and higher interactions can be considered negligible, the main effects are all clear of first-order interactions. Three first-order interactions (AB, AF, and AG) are each aliased with another first-order interaction (FG, BG, and BF). If the choice of factors A, B, F, and G can be made with the assurance that the above interactions are either negligible or not of sufficient interest to be tested, these 3 df can be either pooled with error or left out of the analysis. The remaining 15 first-order interactions are all clear of other interactions except second order or higher. There are also 6 df left over for error involving only second-order interactions or higher. An analysis might be that shown in Table 13.5.

Table 13.5 One-Fourth Replication of a 2^7 Factorial

Source	df
Main effects A, B, . . . , G	1 each for 7
First-order interaction AC, AD, . . .	1 each for 15
AB (or FG), AF (or BG), AG (or BF), . . .	1 each for 3
Second-order interaction or higher (ACF, . . .)	1 each for 6
Total	31

Here tests can be made with 1 and 6 df or 1 and 9 df if the last two lines can be pooled for error. This is a fairly good design when it is necessary to run only a one-fourth replication of a 2^7 factorial.

■ **EXAMPLE 13.1**   In an experiment on slicing yield from pork bellies, four factors were considered as possible sources of significant variation. These were A—press effect; B—molding temperature; C—slicing temperature; D—side of the hog. If each of these four factors were set at two levels, it would require $2^4 = 16$ experiments to run one complete replication. If only eight bellies could be handled at one time, it was decided to run a one-half replicate of this 2^4 factorial and confound $ABCD$ with blocks. If $1 = ABCD$, the blocks are

block 1	(1)	ab	cd	abcd	ac	bc	ad	bd

block 2	a	b	acd	bcd	c	abc	d	abd

Now if only block 1 is run, Table 13.6 shows the data where entries are for the treatment combinations from block I only.

Table 13.6 Slicing Yields in Percent from Pork Belly Experiment (Example 13.1)

		Press (*A*)			
		Normal		**High**	
		Molding Temperature (*B*)		**Molding Temperature**	
Slicing Temperature	**Side**	**23°**	**40°**	**23°**	**40°**
23°	L	(1) = 95.29	*b*	*a*	*ab* = 86.58
	R	*d*	*bd* = 89.38	*ad* = 96.45	*abd*
30°	L	*c*	*bc* = 86.79	*ac* = 88.70	*abc*
	R	*cd* = 90.35	*bcd*	*acd*	*abcd* = 89.57

The analysis could be handled by a Yates method using only the one-half replication data available as shown in Table 13.7. Note that the numerical alias of *A* is $A(ABCD) = BCD$, of $B = ACD$, etc. (shown by the arrows). The analysis follows in Table 13.8.

Table 13.7 Yates Solution of Example 13.1

Treatment Combination	Y	(1)	(2)	(3)	(4)	SS
(1)	95.29	95.29	181.87	357.36	723.11	
a		86.58	175.49	365.75	−0.51	0.03
b		88.70	185.83	−6.80	−18.47	42.64
ab	86.58	86.79	179.92	6.29	0.47	0.03
c		96.45	−8.71	−10.62	−12.29	18.88
ac	88.70	89.38	1.91	−7.85	2.77	0.96
bc	86.79	90.35	7.07	6.38	13.09	21.42
abc		89.57	−0.78	−5.91	8.39	8.78
d		−95.29	−8.71	−6.38	8.39	
ad	96.45	86.58	−1.91	−5.91	13.09	
bd	89.38	88.70	−7.07	10.62	2.77	
abd		−86.79	−0.78	−7.85	−12.29	
cd	90.35	96.45	181.87	6.80	0.47	
acd		−89.38	−175.49	6.29	−18.47	
bcd		−90.35	−185.83	−357.36	−0.51	
abcd	89.57	89.57	179.92	365.75	723.11	

Table 13.8 ANOVA for Example 13.1

Source	df	SS	MS	F	Prob.
A (or BCD)	1	0.03	0.03	<1	
B (or ACD)	1	42.64	42.64	5.71	0.097
C (or ABD)	1	18.88	18.88	2.53	0.210
D (or ABC)	1	8.78	8.78	1.18	0.357
AB or CD	1	0.03			
AC or BD	1 } 3	0.96 } 22.41	7.47		
AD or BC	1	21.42			
Total	7				

This experiment is not very good because the two-way interactions are hope-lessly confounded and the degrees of freedom on error are so small. Nothing shows significance but the analysis concept is seen by this example. ■

13.4 Design Resolution

The use of the concept of resolution in an experimental design is to determine what fractional factorial will result and what effects are aliased in a given 2^f fractional factorial. The letter R is used to designate the resolution for a given design and its value is usually shown as a Roman numeral. The most common resolutions are III, IV, and V.

Resolution III results in retrieving all the main effects in a 2^f fractional fac-torial, but each one will be aliased with some two-way interactions. An example would be a half replication of a 2^3 factorial. This is designated as $\frac{1}{2} \times 2^3 = 2^{3-1}_{III}$. As seen in Section 13.2, when taking a half replicate of a 2^3 factorial, one usually confounds the ABC interaction with blocks. Here $I = ABC$ and the aliases are $A = BC$, $B = AC$, $C = AB$ since all main effects are aliased, or confounded, with two-way interactions. This would only be a viable design if no interactions exist and we had some previous estimate of the error.

Resolution IV results in retrieving all main effects clear of two-way interac-tions, but some two-way interactions are aliased with other two-way interactions, 2^{4-1}_{IV} for example. Here the four factors A, B, C, and D can be found in a half replicate of a 2^4 factorial, and they will only be aliased with three-way interac-tions. However, some two-way interactions will be aliased with other two-way interactions. For a half replicate of a 2^4 factorial one usually confounds the four-way interaction with blocks: $I = ABCD$. Here $A = BCD$, and so on, so that all main effects, or single-factor effects, are aliased with three-way effects. But $AB = CD$, and so on, so that all two-way interactions are aliased with other two-way interactions. This is not a bad design if there is no interest in two-way or higher interactions. Two-way and three-way effects here are usually lumped into an error term with 3 degrees of freedom.

Resolution V results in retrieving all single-factor effects (main effects) and all two-way factor effects (interactions) free of each other. Only three-way factor effects and higher are confounded or aliased. For example 2_V^{5-1} is a half replicate of a 2^5 factorial of resolution V. Here $I = ABCDE$ and all main effects (single-factor) are aliased with four-way effects: $A = BCDE$, and so forth. All two-factor effects are aliased with three-factor effects: $AB = CDE$, and so forth. This design is quite good because one can study all main effects and two-way effects clear of each other. As seen, the value of f must increase in order to get a higher resolution value. It is common to assume that all interactions of three factors and higher are negligible so that the higher the resolution the better.

In general, the resolution is R if no p-factor effect (main or interaction) is aliased with another effect containing less than $R - p$ factors.

Thus, in 2_{III}^{3-1} no main effect ($p = 1$) is aliased with another effect with less than $3 - 1 = 2$ factors—that is, with other main effects.

In 2_V^{5-1} main effects, $p = 1$, are aliased with $5 - 1 = 4$ factor effects, and so on.

In our example in Section 13.3, $I = ABCDEFG$, a half replicate would be 2^{7-1} and $R = 7$ (or VII) because the main ($p = 1$) effects are aliased with six-factor effects: two-way with five-factor, three-way with four-way, and so on, so that $R = $ VII. In the case of half replications of a 2^f, the number of letters in the defining contrast equals the resolution. Therefore, we call this design 2_{VII}^{7-1}. The higher the resolution the better, but we often have to settle for designs of resolution III, IV, or V.

In the case of a half replicate of a 2^2 factorial, 2^{2-1}, we write $I = AB$, and $A = B$ are aliased and the resolution is II, written 2_{II}^{2-1}, is a poor and trivial design.

A good, detailed explanation of resolution is found in Box and Hunter [7].

When using a quarter replicate, as in Table 13.5, we write the design as 2_{IV}^{7-2} because we set $I = ABCDE = CDEFG = ABFG$, and as seen in the aliases, some two-way effects are aliased with other two-way effects: $AB = FG$, $AF = BG$, $AG = BF$, so that the resolution is IV. In the case of quarter replicates and smaller fractions like $1/8$, $1/16$, . . . , $1/2^k$, the resolution is equal to the "word" in the defining contrast with the smallest number of letters, in our example, $ABFG$ or IV.

Another use of the resolution concept is to determine what the minimum f would be for a "good" experiment. "Good" here usually means $R = $ IV, V, or higher. It can be shown that the minimum f is given by

$$f \geq \frac{R(2^k - 1)}{2^{k-1}}$$

where k is the power of 2 in the desired fraction. If $k = 1$, we have a half replicate and

$$f \geq \frac{R(2' - 1)}{2' - 1} = R$$

so the degree of the factorial equals R.

If $k = 2$, that is, a quarter replicate, $f \geq 3/2R$, and if $R = $ III, f must be $\geq \frac{9}{2}$ or $4\frac{1}{2}$, thus requiring $f = 5$ or higher, giving 2^{5-2}_{III}. If $R = $ IV, $f \geq \frac{3}{2}(4) = 6$, so the lowest number of factors to consider is 2^{6-2}_{IV}. If $R = $ V, $f \geq \frac{3}{2}(5) = 7\frac{1}{2}$ and we need 2^{8-2}_{V}.

In the case of a one-eighth replicate, $k = 3$ and $f \geq R((2^3 - 1)/2^{3-1}) = \frac{7}{4}R$. If $R = $ IV, f must be $\geq \frac{7}{4}(4) = 7$, giving 2^{7-2}_{IV}. If $R = $ V is desired, $f \geq \frac{7}{4}(5) = 8$, giving 2^{9-2}_{V}.

This procedure may be extended to any 2^f run as a $1/2^k$ fractional factorial.

Based on these ideas, the National Bureau of Standards has published extensive tables of 2^f fractional factorial designs [20]. In these tables, usually resolutions of IV or higher are required. These tables give the block compositions for the indicated fractionals and include a way to further block the resulting treatment combinations if the number of combinations in the fraction is quite large. Resolution may also be used to determine the block composition, but this has been achieved in Chapter 12 on confounding.

13.5 3^f Fractional Factorials

The concepts of Sections 13.2, 13.3, and 13.4 may easily be extended to fractions of a 3^f experiment. These will be $\frac{1}{3}$, $\frac{1}{9}$, $\frac{1}{27}$, . . . , to $\frac{1}{3^k}$ fractions of a 3^f experiment. The simplest 3^f factorial is a 3^2 requiring nine experiments for a complete factorial. If only a fraction of these can be run, we might consider a one-third replication of a 3^2 factorial. First we confound either AB or AB^2 with the three blocks. Confounding AB^2 gives

$$I = AB^2$$
$$L = X_1 + 2X_2$$

and the three blocks are

$L = 0$	$L = 1$	$L = 2$
00	10	20
11	21	01
22	02	12

If one of these three is run, the aliases are

$$A(AB^2) = A^2B^2 = A^4B^4 = AB$$

because the exponent on the first element is never left greater than one. A^2B^2 is squared to give $A^4B^4 = AB$ (modulus 3). However, the alias of B is

$$B(AB^2) = AB^3 = A$$

Therefore, all three effects A, B, and AB are aliases and mutually confounded. To get both aliases from one multiplication using the defining contrast, one multiplies the effect by the square of I as well as I. This gives

$$A[(AB^2)]^2 = A(A^2B^4) = A^3B = B$$

Thus, $A = AB = B$; this is a poor experiment, since, within the block that is run, both degrees of freedom confound A, B, and AB. In addition, no degrees of freedom are left for error. This example is cited only to show the slight modification necessary when determining the aliases in a 3^f factorial run as a fractional replication. The value of these fractional replications is seen when more factors are involved.

Consider a 3^3 factorial in three blocks of nine treatment combinations each. The effects can be broken down into thirteen 2-df effects: A, B, C, AB, AB^2, AC, AC^2, BC, BC^2, ABC, AB^2C, ABC^2, and AB^2C^2. Confounding ABC^2 with the three blocks gives

$$I = ABC^2 \qquad L = X_1 + X_2 + 2X_3$$

and the blocks are

$L = 0$	000	011	022	101	112	120	210	221	202
$L = 1$	100	111	122	201	212	220	010	021	002
$L = 2$	200	211	222	001	012	020	110	121	102

If only one of these blocks is now run, the aliases are

$$A = A(ABC^2) = A^2BC^2 = AB^2C$$

and

$$A = A(ABC^2)^2 = A^3B^2C^4 = B^2C = B^4C^2 = BC^2$$
$$B = B(ABC^2) = AB^2C^2$$

and

$$B = B(ABC^2)^2 = A^2B^3C^4 = A^4C^8 = AC^2$$
$$C = C(ABC^2) = ABC^3 = AB$$

and

$$C = C(ABC^2)^2 = A^2B^2C^5 = A^4B^4C^{10} = ABC$$
$$AB^2 = AB^2(ABC^2) = A^2B^3C^2 = A^4B^6C^4 = AC$$

and

$$AB^2 = AB^2(ABC^2)^2 = A^3B^4C^4 = BC$$

The analysis would be that in Table 13.9.

Table 13.9 One-Third Replication of a 3^3 Factorial

Source	df
A (or BC^2 or AB^2C)	2
B (or AB^2C^2 or AC^2)	2
C (or AB or ABC)	2
AC (or AB^2 or BC)	2
Total	8

This design would be somewhat practical if we could consider all interactions negligible and be content with 2 df for error.

These methods are easily extended to higher order 3^f experiments. Analysis proceeds as in Chapter 10.

It might be instructive to examine somewhat further this design of a one-third replication of a 3^3. A layout for 3^3 as a complete factorial was given in Table 10.7. If only one third of this were run, say block $L = 1$, suppose the other entries in Table 10.7 were crossed out, giving the results shown in Table 13.10.

Table 13.10 One-Third Replication of a 3^3 Factorial

		Factor A		
Factor B	Factor C	0	1	2
0	0	~~000~~	100	~~200~~
	1	~~001~~	~~101~~	201
	2	002	~~102~~	~~202~~
1	0	010	~~110~~	~~210~~
	1	~~011~~	111	~~211~~
	2	~~012~~	~~112~~	212
2	0	~~020~~	~~120~~	220
	1	021	~~121~~	~~221~~
	2	~~022~~	122	~~222~~

The remaining nine entries in Table 13.10 can now be summarized by levels of factor C only, to give the result shown in Table 13.11.

Table 13.11 One-Third Replication of a
3^3 in Terms of Factor C

Factor B	Factor A		
	0	1	2
0	2	0	1
1	0	1	2
2	1	2	0

A second glance at Table 13.11 reveals that this design is none other than a Latin square! In a sense our designs have now come full circle, from a Latin square as two restrictions on a single-factor experiment (Chapter 4) to a Latin square as a one-third replication of a 3^3 factorial. The designs come out the same, but they result from entirely different objectives. In a single-factor experiment the general model

$$Y_{ij} = \mu + \tau_j + \varepsilon_{ij}$$

is partitioned by refining the error term due to restrictions on the randomization to give $Y_{ijk} = \mu + \tau_j + \beta_i + \gamma_k + \varepsilon'_{ij}$ where the block and position effects are taken from the original error term ε_{ij}.

In the one-third replication of a 3^3, there are three factors of interest that comprise the treatments, giving

$$Y_{ijk} = \mu + A_i + B_j + C_k + \varepsilon_{ijk}$$

where A, B, and C are taken from the treatment effect. Both models look alike but derive from different types of experiments. In both cases the assumption is made that there is no interaction among the factors. This is more reasonable when the factors represent only randomization restrictions such as blocks and positions, rather than when each of the three factors is of vital concern to the experimenter.

By choosing each of the other two blocks in the confounding of ABC^2 ($L = 0$ or $L = 2$), two other Latin squares are obtained. If another confounding scheme is chosen, such as AB^2C, ABC, or AB^2C^2, more and different Latin squares can be generated. In fact, with four parts of the three-way interaction and three blocks per confounding scheme, 12 distinct Latin squares may be generated. This shows how it is possible to select a Latin square at random for a particular design.

It cannot be overstressed that we must be sure that no interaction exists before using these designs on a 3^3 factorial. It is not enough to say that there is no interest in the interactions, because, unfortunately, the interactions are badly confounded with main effects as aliases.

■ **EXAMPLE 13.2** Torque after preheat in pounds was to be measured on some rubber material with factors A—temperature at 145°C, 155°C, and 165°C; B—mix I, II, and III; and C—laboratory 1, 2, and 3. The $3^3 = 27$ experiments seemed too costly, so a one-third replicate was run. Confounding AB^2C gives $L = X_1 + 2X_2 + X_3$ and block composition as follows:

$L = 0$

| 000 | 110 | 011 | 121 | 212 | 022 | 220 | 201 | 102 |

$L = 1$

| 100 | 210 | 111 | 221 | 012 | 122 | 020 | 001 | 202 |

$L = 2$

| 200 | 010 | 211 | 021 | 112 | 222 | 120 | 101 | 002 |

If, now, block $L = 1$ only is run, the numerical results are as given in Table 13.12 and the analysis as in Table 13.13.

Table 13.12 Rubber Torque Data of Example 13.2

| | **Temperature (A)** | | | | | | | | |
| | 145°
Mix | | | 155°
Mix | | | 165°
Mix | | |
Laboratory	I	II	III	I	II	III	I	II	III
1			16.8	11.2				9.9	
2	15.8				14.4				17.8
3		17.1				20.5	15.7		

Table 13.13 ANOVA for Example 13.2

Source	df	SS	MS
A (or ABC^2 or BC^2)	2	6.66	3.33
B (or AC or ABC)	2	38.13	19.06
C (or AB^2 or AB^2C^2)	2	40.81	20.40
AB (or AC^2 or BC)	2	0.12	0.06
Totals	8	85.72	

Because of the tiny error term all main effects are highly significant even with only 2 and 2 degrees of freedom. ■

The concept of resolution here would label Table 13.9 and Example 13.2 as 3_{III}^{3-1}, where I $= ABC^2$ because some two-factor effects are aliased with some single-factor effects. That is, $A = BC^2$, $B = AC^2$, and so on.

Tables for 3^f fractional factorials are also available from the Bureau of Standards [20].

13.6 Plackett–Burman Designs

In many cases, one can only expect to retrieve main effects from a fractional factorial experiment. These are considered to be screening designs that may be used to extract the most important main effects if no interactions are present.

The Plackett–Burman designs have been devised for 2^f factorials to be able to study $f = N - 1$ main effects with only N runs, where N is a multiple of 4. A brief but excellent discussion of these designs is given by Nelson [21].

If N is a power of 2, these designs are like our 2^f fractional factorials. If N is a multiple of 4 and not a power of 2, Plackett–Burman designs indicate the levels of each factor to be used in each run for retrieving main effects. Often one may choose a design with a higher f than an $N - 1$ factorial and use the extra factors as "dummy" factors and combine them as an error term.

A case of $N = 8$ or 2^7 is illustrated in Nelson's paper. When $N = 8$, the first run is designated by Plackett–Burman as

$$N = 8 + + + - + - -$$

which indicates the upper or lower levels of $A, B, C, \ldots, G$ respectively. In this case, the first run is for the treatment combination *abce*. The next run is found by shifting the first run one place to its right and putting the last sign of run 1 in the first position of run 2, or $- + + + - + -$. This procedure continues for seven runs $(N - 1)$, and the eighth $(N\text{th})$ run always contains the lowest level of all factors. The order of these eight runs should, of course, be randomized.

We illustrate this procedure with an example similar to Example 13.1 and Problem 13.12. Here we have only four factors A, B, C, D, and we will add three dummy factors E, F, G as an error term. The percent yield in a pork belly study is recorded in Table 13.14 for the treatment combinations required in an

Table 13.14 Plackett–Burman Design Example

Run	Trt. Comb.	A	B	C	D	E	F	G	Yield
1	abce	+	+	+	−	+	−	−	83.06
2	bcdf	−	+	+	+	−	+	−	83.16
3	cdeg	−	−	+	+	+	−	+	92.21
4	adef	+	−	−	+	+	+	−	98.26
5	befg	−	+	−	−	+	+	+	84.77
6	acfg	+	−	+	−	−	+	+	88.32
7	abdg	+	+	−	+	−	−	+	93.19
8	(1)	−	−	−	−	−	−	−	97.18
	Sums	5.51	−31.78	−26.63	13.49	3.55	−11.13	−3.17	
	SS or MS	3.80	126.25	88.65	22.75	1.58	15.48	1.26	

$N = 8$ Plackett–Burman design. The results are analyzed very simply as shown in Nelson's example.

To analyze these results we note that the vertical columns under each factor give a factor contrast. The sum under each factor can then be squared and divided by N to give a sum of squares, or a mean square here because df $= 1$. If we have an error term, these mean squares may be tested for significance. Here we combine the mean squares under the dummy factors E, F, and G to give

$$\text{Error Mean Square} = \frac{1.58 + 15.48 + 1.26}{3} = 6.11$$

Testing A, B, C, and D with this error term using an F with 1 and 3 df shows that, at the 5 percent level, B (molding temperature) and C (slicing temperature) are significant factors. A run of the complete factorial showed that B, C, D, and AD were significant, so we see how a screening design will extract only the most important main factors.

Nelson's paper also gives the first run for designs where $N = 12, 16, 20, 24$, and 32.

13.7 Summary

Continuing the summary at the end of Chapter 12 under incomplete randomized blocks gives

Experiment	Design	Analysis
II. Two or more factors A. Factorial (crossed)		
	1. Completely randomized	
	2. Randomized block	
	a. Incomplete, confounding	
	i. Main effect—split plot	
	ii. Interactions in 2^f and 3^f	
	1. Several replications	1. Factorial ANOVA with replications and blocks
	$\begin{aligned} Y_{ijkq} = {}& \mu + R_i + \beta_j \\ & + R\beta_{ij} + A_k + B_q \\ & + AB_{kq} + \varepsilon_{ikq} \end{aligned}$	
	2. One replication only	2. Factorial ANOVA with blocks or confounded interaction
	$\begin{aligned} Y_{ijk} = {}& \mu + \beta_i + A_j \\ & + B_k + AB_{jk}, \ldots \end{aligned}$	

Experiment	Design	Analysis
	3. Fractional replication— aliases a. in 2^f b. in 3^f; $f = 3$: Latin square $$Y_{ijk} = \mu + A_i + B_j$$ $$+ AB_{ij}, \ldots$$	3. Factorial ANOVA with aliases or aliased inter- actions

13.8 SAS Programs

Example 13.1 can be analyzed by SAS as follows:

```
DATA A;
INPUT A B C Y;
CARDS;
— (DATA) —
TITLE EXAMPLE 13.1;
PROC PRINT;
PROC GLM;
CLASS A B C D;
MODEL Y=A B C D A*B A*C A*D
PROC GLM;
CLASS A B C D;
MODEL Y=A B C D;
```

The analyses of the two models are on page 320.
These results agree with those of Table 13.8.

GENERAL LINEAR MODELS PROCEDURE
DEPENDENT VARIABLE: Y

SOURCE	DF	SUM OF SQUARES	MEAN SQUARE	F VALUE	PR > F	R-SQUARE	C.V.
MODEL	7	92.75998750	13.25141250	99999.99	0.0000	1.000000	0.0000
ERROR	0	0.00000000	0.00000000		ROOT MSE		Y MEAN
CORRECTED TOTAL	7	92.75998750			0.00000000		90.3875000

SOURCE	DF	TYPE I SS	F VALUE	PR > F	TYPE III SS	F VALUE	PR > F
A	1	0.03251250	.	.	0.03251250	.	.
B	1	42.64261250	.	.	42.64261250	.	.
C	1	18.88051250	.	.	18.88051250	.	.
D	1	8.79901250	.	.	8.79901250	.	.
A*B	1	0.02761250	.	.	0.02761250	.	.
A*C	1	0.95911250	.	.	0.95911250	.	.
A*D	1	21.41851250	.	.	21.41851250	.	.

GENERAL LINEAR MODELS PROCEDURE
DEPENDENT VARIABLE: Y

SOURCE	DF	SUM OF SQUARES	MEAN SQUARE	F VALUE	PR > F	R-SQUARE	C.V.
MODEL	4	70.35465000	17.58866250	2.36	0.2538	0.758460	3.0234
ERROR	3	22.40523750	7.46841250		ROOT MSE		Y MEAN
CORRECTED TOTAL	7	92.75998750			2.73283964		90.3875000

SOURCE	DF	TYPE I SS	F VALUE	PR > F	TYPE III SS	F VALUE	PR > F
A	1	0.03251250	0.00	0.9515	0.03251250	0.00	0.9515
B	1	42.64261250	5.71	0.0968	42.64261250	5.71	0.0968
C	1	18.88051250	2.53	0.2101	18.88051250	2.53	0.2101
D	1	8.79901250	1.18	0.3572	8.79901250	1.18	0.3572

Problems

13.1 Assuming that not all of Problem 12.1 can be run, determine the aliases if a one-half replication is run. Analyze the data for the principal block only.   *Run only one Block*

13.2 Assuming only a one-fourth replication can be run in Problem 12.6 determine the aliases.

13.3 For Problem 12.8 use a one-third replication and determine the aliases.

13.4 Run an analysis on the block in Problem 13.3 that contains treatment combination 022.

13.5 What would be the aliases in Problem 12.9 if a one-half replication were run?

13.6 Determine the aliases in a one-fourth replication of Problem 12.11.

13.7 Determine the aliases in Problem 12.13 where a one-third replication is run.

13.8 If a one-eighth replication of problem 12.17 is run, what are the aliases? Comment on its ANOVA table.

13.9 What are the aliases and the ANOVA outline if only one block of Problem 12.18 is run?

13.10 If data are taken from one furnace only in Problem 12.21, what are the aliases and the ANOVA outline?

13.11 What are the numerical results and conclusions from just one furnace of Problem 12.22?

13.12 The experiment of Example 13.1 was later continued and the second block was now run with results as follows:

		Press (A)			
		Normal		High	
Slicing		Molding Temperature (B)		Molding Temperature	
Temperature	Side	23°	40°	23°	40°
23°	L		$b = 88.08$	$a = 89.26$	
	R	$d = 90.87$			$abd = 93.16$
30°	L	$c = 85.95$			$abc = 85.68$
	R		$bcd = 83.65$	$acd = 95.25$	

Analyze these results.

13.13 Take the results of Example 13.1 and Problem 13.12 and combine into the full factorial in two blocks of eight and analyze the results and comment.

13.14 a. A four-factor experiment is to be run with each factor at two levels only. However, only eight treatment combinations can be run in each block. If one wishes to confound the *ABD* interaction, what are the block compositions?

b. Outline the ANOVA for just one run of this experiment. Consider all factors fixed and note what you might use as an error term.

c. If only one block of this experiment can be run, explain what complications this makes in the experiment and how the ANOVA table must be altered.

13.15 Consider a 2^5 factorial experiment with all fixed levels to be run in four blocks of eight treatment combinations each. This whole experiment can be replicated in a total of three replications. The following interactions are to be confounded in each replication:

Replication I—*ABC, CDE, ABDE*

Replication II—*BCDE, ACD, ABE*

Replication III—*ABCD, BCE, ADE*

a. Determine the block composition for the principal block in replication I.

b. If only a one-fourth replication were run of replication II, what would be the aliases of *BC* in this block?

c. Determine the number of degrees of freedom for the replication by three-way interactions in this problem.

13.16 Consider an experiment designed to study the effect of six factors—*A, B, C, D, E,* and *F*—each at two levels.

a. Determine at least six entries in the principal block if this experiment is to be run in two blocks of 32 and the highest order interaction is to be confounded.

b. If only one of the two blocks can be run, what are the aliases of the effects *A, BC,* and *ABC*?

c. Outline the ANOVA table for part (b) and indicate what effects might be tested and how they would be tested.

13.17 In determining which of four factors *A, B, C,* or *D* might affect the underpuffing of cereal in a plant, the following one-fourth replicate was run.

		A_1		A_2	
		B_1	B_2	B_1	B_2
C_1	D_1			96.6	125.7
	D_2	43.5	22.4		
C_2	D_1	14.1	9.5		
	D_2			28.8	52.5

a. Determine which interaction was confounded in this design.

b. Analyze these data and comment on the results and on the design.

13.18 A company manufacturing engines was concerned about the emission characteristics of these engines. To try to minimize various undesirable emission variables, the company proposed to build and operate experimental engines varying five factors: throat diameter to be set at three levels, ignition system at three levels, temperature at three levels, velocity of the jet stream at three levels, and timing system at three levels. As this would require 243 experiments, it was decided that a one-third replicate would be run. Set up a confounding scheme for this problem and indicate a few terms in the principal block. Also show an ANOVA outline when running one of the three blocks, indicating some aliases. Comment on the design.

13.19 In studying the surface finish of steel, five factors were varied each at two levels; B—slab width; C—percent sulfur; D—tundish temperature; E—casting speed; and F—mold level. The following data were recorded on the number of longitudinal face cracks observed.

D Tundish temperature, °F	E Casting speed, 1 pm	F Mold level, inches	B < 60, C < 0.015	B < 60, C > 0.017	B > 62, C < 0.015	B > 62, C > 0.017
> 2845	> 59	> 10		2	3	
> 2845	> 59	5–8	0			0
> 2845	< 54	> 10	5			0
> 2845	< 54	5–8		6	0	
< 2840	> 59	> 10	6			3
< 2840	> 59	5–8		2	0	
< 2840	< 54	> 10		7	0	
< 2840	< 54	5–8	1			0

(Table column headers as printed: *B* Slab width, inches — < 60 / > 62; *C* Sulfur, percent — < 0.015, > 0.017.)

Determine what interaction is confounded here in order to run this one-half replicate. Outline its ANOVA and comment on the design.

13.20 Using the data of Problem 13.19, do the analysis and state your conclusions.

13.21 For Problem 13.6, determine the resolution, explain how you found it, and express the experiment in resolution form.

13.22 For Problem 13.7, determine the resolution, explain how you found it, and express the experiment in resolution form.

13.23 Determine the resolution in Problem 13.8 and comment on your results.

13.24 Consider Problem 13.19. Explain how you might work this problem as a Plackett–Burman design where $N = 8$.

13.25 Using the data of the complete factorial in Problem 9.6, try to check the results using the Plackett–Burman design where $N = 8$. (Use only the first reading in each cell and add some error as you see fit.)

Taguchi Approach to the Design of Experiments

14

14.1 Introduction

Genichi Taguchi [27] has made several significant contributions to industrial statistics problems. These include:

1. Loss functions that convert performance characteristics into societal loss measured in monetary units.
2. Signal-to-noise ratios (inverse of the coefficient of variation) that summarize the performance characteristic in decibels.
3. Orthogonal arrays for gathering experimental results and their associated linear graphs for designing experiments.
4. Analysis of variance for analyzing the results of the experiment and making decisions.

We will center our attention on his experimental designs. Taguchi methods cannot really be understood until one has studied experimental design through fractional factorial experiments as in Chapter 13. Examples to illustrate his designs will generally be taken from problems discussed earlier in this text.

We begin with a simpler case than Taguchi would consider in order to see how the methods work.

14.2 $L_4(2^3)$

The notation $L_4(2^3)$ means that we take four observations and at most three factors each at two levels. All cases use the orthogonal array design of Table 14.1. Here "1" represents the low level of a factor and "2" the high level. We're more familiar with this if we take low level as -1 and high level as $+1$ as shown in Table 14.2.

Table 14.1 Orthogonal Arrays for $L_4(2^3)$

Experiment	Column		
	1	2	3
1	1	1	1
2	1	2	2
3	2	1	2
4	2	2	1

Table 14.2 Orthogonal Arrays Using + and −

Experiment	Column		
	1	2	3
1	−1	−1	−1
2	−1	+1	+1
3	+1	−1	+1
4	+1	+1	−1

Note that column 3 is the negative of the products of columns 1 and 2. If A is assigned to column 1 and B to column 2, then column 3 is $-AB$ or the interaction. To find the third column using the Taguchi notation of Table 14.1, add the level number; if the sum is even, enter 1 in column 3, if odd enter 2.

Table 14.3 shows how each column is the result of multiplying any two columns. From the main diagonal, (1) × (3) is the intersection 2 and so on.

Table 14.3 Interaction of Columns in an $L_4(2^3)$

Column	1	2	3
	(1)	3	2
		(2)	1
			(3)

Linear graphs are made up of three parts: dots, lines, and numbers. Each dot represents a main effect, a line is the two-way interaction between two main effects, and the numbers are columns in Taguchi's orthogonal array. We shall consider three cases for this $L_4(2^3)$ design.

CASE 1.

All three factors A, B, and C are to be studied and no interaction is assumed present.

Here we have a half replication of $2^3 = 4$ observations $= 2^{3-1}$ of resolution III. If the defining contrast is taken as $I = ABC$, the aliases are $A = BC$ or $(1) = (2) \times (3)$, $B = AC$ or $(2) = (1) \times (3)$, and $C = AB$ or $(3) = (1) \times (2)$, which agrees with Table 14.3 if A is placed in column 1, B in column 2, and C in column 3. The linear graph is

$$
\begin{array}{ccc}
\text{o} & \rule{2cm}{0.4pt} & \text{o} \\
1 & 3 & 2 \\
(A) & (C) & (B)
\end{array}
$$

CASE 2.

2^2 full factorial including one interaction.

A	B	$-AB$
1	2	3
1	1	1
1	2	2
2	1	2
2	2	1

$$
\begin{array}{ccc}
\text{o} & \rule{2cm}{0.4pt} & \text{o} \\
1 & 3 & 2 \\
(A) & (AB) & (B)
\end{array}
$$

CASE 3.

Single factor—A with error terms as pseudo factors E and AE:

$$
\begin{array}{ccc}
\text{o} & - - - - - & \text{o} \\
1 & 3 & 2 \\
(A) & (AE) & (E)
\end{array}
$$

Examples will show the three cases in reverse order.

■ **EXAMPLE 14.1**   Consider factor A at just two levels: 0 and 1 with two observations per level. See Table 14.4.

Table 14.4 Data for Example 14.1

	A	
0		**1**
3		−4
7		−6
10		−10

By the methods of Chapter 3:

$$\text{Total SS} = 3^2 + 7^2 + (-4)^2 + (-6)^2 = 110$$
$$\text{SS}_A = [(10)^2 + (-10)^2]/2 = 100$$
$$\text{SS}_{\text{error}} = 110 - 100 = 10$$

By $L_4(2^3)$, where columns 2 and 3 are error, we get Table 14.5.

Table 14.5 Taguchi Array for Example 14.1

Column	A 1	E 2	AE 3	Y
1	1	1	1	3
2	1	2	2	7
3	2	1	2	-4
4	2	2	1	-6
1 or − total	10	-1	-3	
2 or + total	-10	1	3	
difference	-20	2	6	
SS or (diff)^2/4	100	1	9	110

In ANOVA terms:

Factor	df	SS	MS	F
A	1	100	100	20
Error	2	10	5	
Total	3	110		

■

■ **EXAMPLE 14.2** A, B, and $AB - A$ and B are both at two levels. Make four observations as in Table 9.4 using the first observation only. See Table 14.6.

Table 14.6 Data for Example 14.2

		A 0	A 1	
B	0	4	2	6
	1	3	-4	-1
		7	-2	5

By the methods of Chapter 5:

$$\text{Total SS} = 4^2 + 3^2 + 2^2 + (-4)^2 - 5^2/4 = 38.75$$
$$SS_A = [7^2 + (-2)^2]/2 - 6.25 = 20.25$$
$$SS_B = [6^2 + (-1)^2]/2 - 6.25 = 12.25$$
$$SS_{AB} \text{ or error} = 38.75 - 20.25 - 12.25 = 6.25$$

By Taguchi's $L_4(2^3)$, we get Table 14.7.

Table 14.7 Taguchi Array for Example 14.2

	Column			
	A	*B*	*AB*	
Experiment	**1**	**2**	**3**	*Y*
1	1	1	1	4
2	1	2	2	3
3	2	1	2	2
4	2	2	1	−4
− total	7	6	0	5
+ total	−2	−1	5	
diff	9	7	−5	
SS	20.25	12.25	6.25	38.75

Taguchi rarely makes any *F* tests, but he does report percentage of total variation (SS) attributable to each factor as

$$A: \quad 20.25/38.75 = 52\%$$
$$B: \quad 12.25/38.75 = 32\%$$
$$AB: \quad 6.25/38.75 = 16\%$$

Thus, *A* is the most important factor, followed by *B*. ■

■ **EXAMPLE 14.3** *A*, *B*, and *C* to be assessed with only four observations: $L_4(2^3)$.

This, of course, assumes that there is no *AB* interaction. From Example 9.2 on ceramic tools data and Table 9.8, we take the cell totals for our data. Table 9.9 gives the usual SS breakdown. So, by Taguchi, using only half of the data we can fill in Table 14.8. This table shows *C* and *B* to be the main factors of interest, which confirms the results of the full factorial (percentages from Table 9.9 are *C* = 29%, *B* = 19%, *A* = 3%, and all interactions and error = 50%). This agrees well with the classical methods because there were no interactions present.

Table 14.8 Taguchi Array for Example 9.2

	Column			
	A	**B**	**C**	
Experiment	**1**	**2**	**3**	**Y**
1 (1)	1	1	1	2
2 bc	1	2	2	−2
3 ac	2	1	2	−17
4 ab	2	2	1	13
−	0	−15	15	−4
+	−4	11	−19	
D	−4	26	−34	
SS	4	169	289	462
%	8%	34%	58%	100%

■

14.3 $L_8(2^7)$

The $L_8(2^7)$ design requires only eight observations to examine up to seven factors: A, B, C, D, E, F, and G. If seven factors are to be studied, one must assume *all* interactions are nonexistent. Table 14.9 shows how the factors should be assigned to the columns in an orthogonal array. This leads to one of the linear graphs as shown in Figure 14.1. These assignments can be found from Table 14.9. For example, if we choose columns (2) and (4) on the main diagonal, their intersection is 6 in Figure 14.1(a). In Figure 14.1(b), (1) × (6) = 7. This assignment, of course, comes from the aliases, although Taguchi never mentions confounding or aliases. Because we are examining seven main effects with only eight observations, it is a one-sixteenth replication of a $2^7 = 128 / 16 = 8$.

Table 14.9 Interaction of Columns in an $L_8(2^7)$

Column	1	2	3	4	5	6	7
	(1)	3	2	5	4	7	6
		(2)	1	6	7	4	5
			(3)	7	6	5	4
				(4)	1	2	3
					(5)	3	2
						(6)	1
							(7)

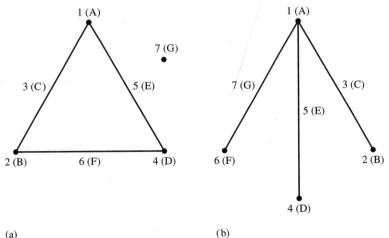

Figure 14.1 Linear graph of L_8.

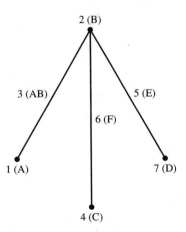

Figure 14.2 Six factors and one interaction.

If one wishes to include one interaction, say AB, the pattern would be as shown in Figure 14.2. Now other interactions are confounded with main effects but are assumed zero. Here we have a one-eighth replication of a 2^6 as $2^6/2^3 = 2^{6-3}$ of resolution III.

If two interactions are to be included and only five main effects, we have Figure 14.3. Letting column $3 = AB$, column $4 = C$, column $5 = D$, column $6 = E$, and column $7 = F$, we have a one-fourth replication of a $2^5 = 2^{5-2}$ also of resolution III, and as such it is not included in the Bureau of Standards tables.

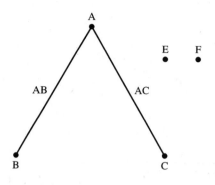

Figure 14.3 Five main effects and two interactions.

One could handle three interactions and four main effects as shown below:

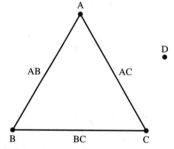

If $D = ABC$, then $I = ABCD$, so that main effects are aliased with three-way interactions and two-way interactions are aliased with each other. This design is of resolution IV, which is a bit better.

Fewer main effects could also be included, and the rest of the columns could be for error.

All of the above use Taguchi's orthogonal array $L_8(2^7)$, shown in Table 14.10. This table can be verified by the use of $1 = -$, $2 = +$, and then

Table 14.10 Taguchi Array for $L_8(2^7)$

Experiment	Column						
	1	2	3	4	5	6	7
1	1	1	1	1	1	1	1
2	1	1	1	2	2	2	2
3	1	2	2	1	1	2	2
4	1	2	2	2	2	1	1
5	2	1	2	1	2	1	2
6	2	1	2	2	1	2	1
7	2	2	1	1	2	2	1
8	2	2	1	2	1	1	2

(1)(2) = −(3), but the sum of squares is the same regardless of sign. Examples will be used to illustrate some of the possibilities.

■ **EXAMPLE 14.4** Consider the 2^2 factorial of Example 9.1 in Chapter 9 as an L_8 with the repeated measures as error. We are concerned about factors A and B and the AB interaction. Take the first observation in each cell as the low level of a pseudo-factor C', and then C', AC', BC', and ABC' are all part of the error term.

To handle this by Taguchi, we use the linear graph in Figure 14.4. We use the third column as the AB term, giving for the orthogonal array Table 14.11.

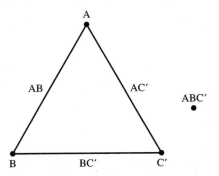

Figure 14.4 Two factors and error.

Table 14.11 Taguchi Array for Example 9.1

	Column								
	A	**B**	**AB**	**C'**	**AC'**	**BC'**	**ABC'**		
Experiment	**1**	**2**	**3**	**4**	**5**	**6**	**7**	**Y**	**Y²**
1 (1)	1	1	1	1	1	1	1	4	16
2 c'	1	1	1	2	2	2	2	6	36
3 b	1	2	2	1	1	2	2	3	9
4 bc'	1	2	2	2	2	1	1	7	49
5 a	2	1	2	1	2	1	2	2	4
6 ac'	2	1	2	2	1	2	1	−2	4
7 ab	2	2	1	1	2	2	1	−4	16
8 abc'	2	2	1	2	1	1	2	−6	36
−	20	10	0	5	−1	7	5	10	170
+	−10	0	10	5	11	3	5		
D	−30	−10	10	0	12	−4	0		
SS	112.5	12.5	12.5	0	18.0	2.0	0		

Summarizing from the array gives an ANOVA table:

Source	df	SS	MS	F	% of SS total
A	1	112.5	112.5	22.5	71
B	1	12.5	12.5	2.5	8
AB	1	12.5	12.5	2.5	8
Error	4	20.0	5.0		13
Total	7	157.5			100

Note:

$$SS_{total} = 170 - (10)^2 / 8 = 157.5$$

This agrees exactly, as we have the whole data here. It is only used to verify the Taguchi procedure. Taguchi is really used when you wish to examine more factors than possible with a 2^3 factorial. ■

■ **EXAMPLE 14.5** Consider a one-fourth replicate of a 2^5 factorial where we are concerned with two interactions and five main effects. This is a 2^{5-2} design. Let us try this on a problem like Problem 13.9, where we wish to include AD and BD along with A, B, C, D, and E in our study. Responses are similar to those in the problem in the text where response is the number of longitudinal cracks.

A linear graph might look like Figure 14.5 and the orthogonal array as in Table 14.12.

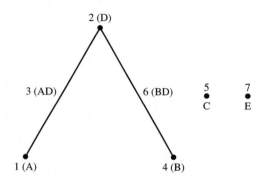

Figure 14.5 Graph of Example 14.5.

Table 14.12 Taguchi Array for Example 14.5

	Column							
Experiment	A 1	D 2	$-AD$ 4	B 4	C 5	$-BD$ 6	E 7	Y
1　(1)	1	1	1	1	1	1	1	1
2　bce	1	1	1	2	2	2	2	6
3　de	1	2	2	1	1	2	2	6
4　bcd	1	2	2	2	2	1	1	2
5　ace	2	1	2	1	2	1	2	0
6　ab	2	1	2	2	1	2	1	0
7　acd	2	2	1	1	2	2	1	3
8　abde	2	2	1	2	1	1	2	3
Totals	−9	7	−5	1	1	9	9	21
SS	10.125 ↓	↓	3.125 ↓	↓	0.125	↓	+10.125	
		6.125		0.125		10.125		

The results of Example 14.5 are summarized in Table 14.13.

Table 14.13 ANOVA for Example 14.5

Source or Factor	df	SS	% SS
A (slab width)	1	10.125	25%
D (casting speed)	1	6.125	15%
AD	1	3.125	8%
B (% sulfur)	1	0.125	<1%
C (tundish temp.)	1	0.125	<1%
BD	1	10.125	25%
E (mold level)	1	10.125	25%
Total	7	39.875	

$$\text{Total SS} = 1^2 + 6^2 + \cdots + 3^2 - 21^2/8 = 39.875$$

as above. Unfortunately, the factors in Problem 13.19 are labeled B through F instead of A through E. However, from the results of Problem 13.20, the following significant effects were found: B (slab width), F (mold level), and BE (slab width by casting speed). The Taguchi method agrees well on the two main effects using only 8 observations compared to 16 in Problem 15.20. The interaction effects fail to agree, but this may be expected when only half of the data are run. ∎

More examples could be used to show the many applications of an $L_8(2^7)$ design. Now we include a classic that has appeared in nearly all Taguchi papers, where he examines seven main effects with only eight observations.

■ **EXAMPLE 14.6** A Japanese company was concerned with excessive variability in the dimensions of tiles that it produced. Seven factors were identified as possible sources of serious variability. These were labeled A through G, where

A = amount of limestone (5 percent or 1 percent)

B = coarser (existing or finer)

C = amount of agalmatolite (43 percent or 53 percent)

D = type of agalmatolite (existing or new)

E = raw material charging quantity (1300 or 1200 kg)

F = amount of waste return (0 percent or 4 percent)

G = amount of feldspar (7 percent or 5 percent)

Here a full factorial would require $2^7 = 128$ experiments. Using Taguchi's approach and assuming *no* interactions present, one can use his orthogonal array of an $L_8(2^7)$ design. Taking the number of defective tiles per 1000 as the response (Y), results are shown in Table 14.14.

Table 14.14 Taguchi Arrays on Example 14.6

Experiment	Column							Y
	A 1	B 2	C 3	D 4	E 5	F 6	G 7	
1	1	1	1	1	1	1	1	16
2	1	1	1	2	2	2	2	17
3	1	2	2	1	1	2	2	12
4	1	2	2	2	2	1	1	6
5	2	1	2	1	2	1	2	6
6	2	1	2	2	1	2	1	68
7	2	2	1	1	2	2	1	42
8	2	2	1	2	1	1	2	26
Totals	91	−21	−9	41	−51	85	−71	193
SS	1035	55	10	210	325	903	630	3168
% SS	33	2	1	7	10	28	20	100

$$\text{Total SS} = 16^2 + \cdots + 26^2 - 193^2/8 = 3168$$

Clearly factors A, F, and G are the largest contributors to the variation in the dimensions of the tiles. We used only eight experiments to find this out. We really ran one-sixteenth of a 2^7 design or 2^{7-4} of resolution III. ■

14.4 $L_{16}(2^{15})$

Taguchi proposes an orthogonal array for examining 15 effects with only 16 experiments as seen in Table 14.15.

Table 14.15 Orthogonal Array for $L_{16}(2^{15})$

Experiment	Column														
	1	2	3	4	5	6	7	8	9	10	11	12	13	14	15
1	1	1	1	1	1	1	1	1	1	1	1	1	1	1	1
2	1	1	1	1	1	1	1	2	2	2	2	2	2	2	2
3	1	1	1	2	2	2	2	1	1	1	1	2	2	2	2
4	1	1	1	2	2	2	2	2	2	2	2	1	1	1	1
5	1	2	2	1	1	2	2	1	1	2	2	1	1	2	2
6	1	2	2	1	1	2	2	2	2	1	1	2	2	1	1
7	1	2	2	2	2	1	1	1	1	2	2	2	2	1	1
8	1	2	2	2	2	1	1	2	2	1	1	1	1	2	2
9	2	1	2	1	2	1	2	1	2	1	2	1	2	1	2
10	2	1	2	1	2	1	2	2	1	2	1	2	1	2	1
11	2	1	2	2	1	2	1	1	2	1	2	2	1	2	1
12	2	1	2	2	1	2	1	2	1	2	1	1	2	1	2
13	2	2	1	1	2	2	1	1	2	2	1	1	2	2	1
14	2	2	1	1	2	2	1	2	1	1	2	2	1	1	2
15	2	2	1	2	1	1	2	1	2	2	1	2	1	1	2
16	2	2	1	2	1	1	2	2	1	1	2	1	2	2	1

If the L_{16} is used with fewer than 15 main effects, Table 14.16 shows how the interactions should be selected.

Six types of linear graphs may be associated with an L_{16} design. These are shown in Figure 14.6. Here are some examples of when to use the six types of linear graphs in Figure 14.6:

Type 1   The interactions between main effects are equally important.

Type 2   One factor is very important and the interactions between this factor and other factors are required. Besides, two other interactions are also needed.

Type 3   Two groups of interactions in two stages of a process are required.

Type 4   Interactions of one and all other factors are important.

Type 5   Latin squares, Graeco–Latin squares, and hyper-Graeco–Latin squares may be prepared from this type.

Type 6   A convenient graph to use for pseudo factor design.

Table 14.16 Interactions Between Two Columns for $L_{16}(2^{15})$

Column	1	2	3	4	5	6	7	8	9	10	11	12	13	14	15
	(1)	3	2	5	4	7	6	9	8	11	10	13	12	15	14
		(2)	1	6	7	4	5	10	11	8	9	14	15	12	13
			(3)	7	6	5	4	11	10	9	8	15	14	13	12
				(4)	1	2	3	12	13	14	15	8	9	10	11
					(5)	3	2	13	12	15	14	9	8	11	10
						(6)	1	14	15	12	13	10	11	8	9
							(7)	15	14	13	12	11	10	9	8
								(8)	1	2	3	4	5	6	7
									(9)	3	2	5	4	7	6
										(10)	1	6	7	4	5
											(11)	7	6	5	4
												(12)	1	2	3
													(13)	3	2
														(14)	1

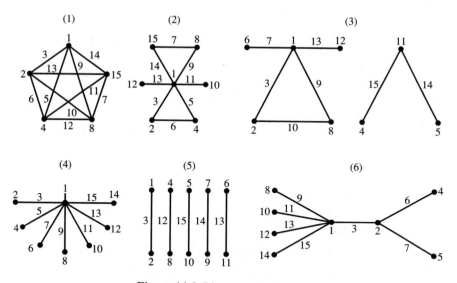

Figure 14.6 Linear graph of L_{16}.

We might note how Type 1 could be used. First it could be used for examining 15 factors if no interactions were believed present. At the other extreme it could be used for examining 5 factors (5 vertices of the pentagon) and all 10 two-way interactions ($C_2^5 = 10$, as sides and diagonals of the pentagon). This latter would be a 2^{5-1} of resolution V.

Examples of the L_{16} experiment are available but are not included, because one has to wonder about trying to handle 15 factors in one experiment. Are all 15 equally important? Probably not! One should then discuss with the team designing the experiment which are the five or six most important factors. If each member of the team listed their first four or five most important factors, the combined list would probably include only six or seven as most important. Then a smaller design could be used. After all, a $2^{15} = 32,768$ run in only $16 = 2^4$ experiments is a $2^{15-11} = 2^4$ or a 1/2048th replicate of a 2^{15} factorial.

14.5 $L_9(3^4)$

This design examines four factors with only nine experiments and less than four factors if some interactions are examined. Here each factor has 2 degrees of freedom, and the interactions are of the form AB and AB^2 as discussed in Chapter 10. We shall consider four applications:

1. Four factors, say A, B, C, and D, each at three levels. This would require $3^4 = 81$ experiments for a full factorial. To do this with nine experiments is a one-ninth replicate of a 3^4 experiment, or a 3^{4-2}.

2. Three factors and one interaction, say A, B, AB, and C, where AB is only part of the A by B interaction.

3. Two factors A and B with both parts of the interaction: AB and AB^2 or a complete 3^2 factorial.

4. One factor, A, with pseudo factors for error.

The orthogonal array for $L_9(3^4)$ is displayed in Table 14.17. Here 1 is the lowest level, 2 the middle level, and 3 the highest level. No linear graphs are available here because the effects can be quadratic, as seen in Chapter 10. In Chapter 10 the notation would be given as $-$, 0, $+$.

Table 14.17 Taguchi's $L_9(3^4)$ Array

Experiment	Column			
	1	2	3	4
1	1	1	1	1
2	1	2	2	2
3	1	3	3	3
4	2	1	2	3
5	2	2	3	1
6	2	3	1	2
7	3	1	3	2
8	3	2	1	3
9	3	3	2	1

■ **EXAMPLE 14.7**   Again, looking first at case 4, consider three levels of a factor to be vendors of capacitors and take three observations of leakage in milliamperes from each of the three vendors. Results appear in Table 14.18.

Table 14.18 Data for Example 14.7

	Vendor			
	1	2	3	
	7.3	10.7	10.5	
	8.0	10.2	10.1	
	8.1	10.2	10.8	
Totals	23.4	31.1	31.4	85.9

By the methods of Chapter 3:

$$\text{Total SS} = 7.3^2 + \cdots + 10.8^2 - 85.9^2/9 = 14.50$$
$$SS_V = (23.4^2 + 31.1^2 + 31.4^2)/3 - CT = 13.71$$
$$SS_{\text{error}} = 14.50 - 13.71 = 0.79$$

Using Taguchi's L_9 with error E as a pseudo factor where the first observation in each column is considered as the lowest level E and the second observation as the middle level, and so on, we obtain Table 14.19.

Table 14.19 Taguchi Array for Example 14.7

	Column				
	A	*E*	*AE*	*AE*^{2}	
Experiment	1	2	3	4	*Y*
1	1	1	1	1	7.3
2	1	2	2	2	8.0
3	1	3	3	3	8.1
4	2	1	2	3	10.7
5	2	2	3	1	10.2
6	2	3	1	2	10.2
7	3	1	3	2	10.5
8	3	2	1	3	10.1
9	3	3	2	1	10.8
					85.9

From column 1,

$$SS_A = (23.4)^2 + (31.1)^2 + (31.4)^2/3 - CT = 14.50$$

as before. From column 2,

$$SS_E = (28.5)^2 + (28.3)^2 + (29.1)^2 / 3 - CT = 0.11$$

From column 3,

$$SS_{AE} = (27.6^2 + 29.5^2 + 28.8^2) / 3 - CT = 0.61$$

From column 4,

$$SS_{AE2} = (28.3^2 + 28.7^2 + 28.9^2) / 3 - CT = 0.06$$

These give

$$SS_{error} = 0.11 + 0.61 + 0.06 = 0.78 \approx 0.79$$

as before. ■

■ **EXAMPLE 14.8**   If we want to examine A, B and its interactions AB and AB^2, it is a complete 3^2 factorial. As an example consider Example 10.1. Results are given in Table 10.2 and are given here by Taguchi's method in Table 14.20.

Table 14.20 Taguchi Array for Example 14.8

Exp. & Trt. Comb.	A 1	B 2	AB 3	AB^2 4	Y
1 00	1	1	1	1	1
2 01	1	2	2	2	0
3 02	1	3	3	3	2
4 10	2	1	2	3	−2
5 11	2	2	3	1	4
6 12	2	3	1	2	−1
7 20	3	1	3	2	3
8 21	3	2	1	3	1
9 22	3	3	2	1	2
					10

From Table 14.20, we find by our usual methods:

$$SS_A = (3^2 + 1^2 + 6^2) / 3 - 10^2 / 9 = 14.22$$
$$SS_B = (2^2 + 5^2 + 3^2) / 3 - 10^2 / 9 = 1.56$$
$$SS_{AB} = (1^2 + 0^2 + 9^2) / 3 - 10^2 / 9 = 16.22$$
$$SS_{AB2} = (7^2 + 2^2 + 1^2) / 3 - 10^2 / 9 = 6.89$$

which agree with the results of Table 10.2. So far there is little to be gained by using Taguchi's orthogonal arrays on Examples 14.7 and 14.8. When the number of factors exceeds 2, we see the merit of a Taguchi approach if we can assume few or no interactions. ∎

∎ **EXAMPLE 14.9**   For case 2 with three factors and part of an interaction, we have a one-third replicate of a 3^3 where $C = AB^2$. One could use $I = AB^2C^2$ as a defining contrast. Then the aliases are (by the methods of Chapter 10)

$$A = A^2B^2C^2 = ABC$$

and

$$A = A(AB^2C^2)^2 = BC$$
$$B = AC^2$$

and

$$B = B(AB^2C^2)^2 = A^2B^5C^4 = ABC^2$$
$$C = AB^2 \text{ as desired}$$

and

$$C = C(AB^2C^2)^2 = A^2B^4C^5 = AB^2C$$
$$AB = A^2B^3C^2 = AC$$

and

$$AB = AB(AB^2C^2)^2 = A^3B^5C^4 = BC^2$$

Consider part of Example 10.3 with day = factor A, operator = factor B, and concentration = factor C. As each main effect requires 2 degrees of freedom, there are 2 degrees of freedom left over in an L_9 design. We will assign AB interaction to column 3, noting that this is only part of the interaction A by B. We also assign C to column 4 as $C = AB^2$. The results are shown in Table 14.21.

Table 14.21 Taguchi Array for Example 14.9

	Column				
	A	B	C	D	
Experiment & Trt.	1	2	3	4	Y (Cell Totals)
1 000	1	1	1	1	3.9
2 011	1	2	2	2	10.4
3 022	1	3	3	3	20.4
4 102	2	1	2	3	18.7
5 110	2	2	3	1	1.0
6 121	2	3	1	2	11.9
7 201	3	1	3	2	14.2
8 212	3	2	1	3	19.5
9 220	3	3	2	1	3.2
					103.2

By the usual sum of squares:

$$SS_A = (34.7)^2 + (31.6)^2 + (36.9)^2 / 3 - 103.2^2 / 9 = 4.73$$
$$SS_B = (36.8)^2 + (30.9)^2 + (35.5)^2 / 3 - CT = 6.41$$
$$SS_{AB} = (35.3)^2 + (32.3)^2 + (35.6)^2 / 3 - CT = 2.22$$
$$SS_C = (8.1)^2 + (36.5)^2 + (58.6)^2 / 3 - CT = 427.25$$
$$\text{Total SS} = (3.9)^2 + (10.4)^2 + \cdots + (3.2)^2 - CT = 440.60$$

which checks with the addition of the SS terms within 0.01. Here we see that $427.25 / 440.60 = 97$ percent of the total sum of squares is accounted for by C, the concentration. This compares favorably with 95 percent for the complete factorial of Table 10.15 using only 9 totals instead of 27. ∎

■ **EXAMPLE 14.10**   As a final example for an L_9 experiment, consider the same data as in Table 10.14 where the order of readings in each cell will be considered as a fourth factor, D, since the whole table is a 3^3 with 3 observations per cell or a $3^4 = 81$ observations. An $L_9(3^4)$ would be a one-ninth replicate in order to examine these data with nine observations only. The results are seen in Table 14.22.

Following the usual SS procedures:

$$SS_A = 118.17 - 115.92 = 2.25$$
$$SS_B = = 0.80$$
$$SS_C = = 50.47$$
$$SS_D = = \underline{0.38}$$
$$\text{Total} \quad 53.90$$
$$\text{Total SS} = 1^2 + 3.7^2 + \cdots + 3.7^2 - CT = 53.89$$

Table 14.22 Taguchi Array for Example 14.10

| | Column | | | | |
| | A | B | C | D | |
Experiment	1	2	3	4	Y
1 0000	1	1	1	1	1.0
2 0111	1	2	2	2	3.7
3 0222	1	3	3	3	6.7
4 1012	2	1	2	3	3.5
5 1120	2	2	3	1	5.2
6 1201	2	3	1	2	0.0
7 2021	3	1	3	2	7.5
8 2102	3	2	1	3	1.0
9 2210	3	3	2	1	3.7
					32.3

Here again $C = 50.47 / 53.90 = 94$ percent of the variability accounted for is the big factor. Naturally factor D was only a repeat measure, so one doesn't expect to find any significance due to D. So we have gained some information on A, B, C, and D with only nine observations.

14.6 Some Other Taguchi Designs

So far, the designs have been for either a 2^f or a 3^f factorial experiment. Taguchi also has some orthogonal arrays for factorial experiments of the form $2^f 3^g$. One design is an $L_{12}(2^{11})$, where the factorial would be $2^2 3^1$ for 12 treatment combinations. This can be used to handle 11 factors with only 12 observations. See Table 14.23.

Table 14.23 Taguchi's $L_{12}(2^{11})$

Number	1	2	3	4	5	6	7	8	9	10	11
1	1	1	1	1	1	1	1	1	1	1	1
2	1	1	1	1	1	2	2	2	2	2	2
3	1	1	2	2	2	1	1	1	2	2	2
4	1	2	1	2	2	1	2	2	1	1	2
5	1	2	2	1	2	2	1	2	1	2	1
6	1	2	2	2	1	2	2	1	2	1	1
7	2	1	2	2	1	1	2	2	1	2	1
8	2	1	2	1	2	2	2	1	1	1	2
9	2	1	1	2	2	2	1	2	2	1	1
10	2	2	2	1	1	1	1	2	2	1	2
11	2	2	1	2	1	2	1	1	1	2	2
12	2	2	1	1	2	1	2	1	2	2	1
Group	1						2				

The $L_{12}(2^{11})$ is a specially designed array, in that interactions are distributed more or less uniformly to all columns. Note that there is no linear graph for this array. It should not be used to analyze interactions. The advantage of this design is its capability to investigate 11 main effects, making it a highly recommended array. (Taguchi Paper, No. 27.)

Another might be an $L_{18}(2^1 3^7)$ using $2 \times 3^2 = 18$ observations as shown in Table 14.24.

Table 14.24 Taguchi's $L_{18}(2^1 \times 3^7)$

Number	1	2	3	4	5	6	7	8
1	1	1	1	1	1	1	1	1
2	1	1	2	2	2	2	2	2
3	1	1	3	3	3	3	3	3
4	1	2	1	1	2	2	3	3
5	1	2	2	2	3	3	1	1
6	1	2	3	3	1	1	2	2
7	1	3	1	2	1	3	2	3
8	1	3	2	3	2	1	3	1
9	1	3	3	1	3	2	1	2
10	2	1	1	3	3	2	2	1
11	2	1	2	1	1	3	3	2
12	2	1	3	2	2	1	1	3
13	2	2	1	2	3	1	3	2
14	2	2	2	3	1	2	1	3
15	2	2	3	1	2	3	2	1
16	2	3	1	3	2	3	1	2
17	2	3	2	1	3	1	2	3
18	2	3	3	2	1	2	3	1
Group	1	2			3			

```
           (1)      1              2
                    O──────────────◎
```

Like the $L_{12}(2^{11})$, this is a specially designed array. An interaction is built in between the first two columns. This interaction information can be obtained without sacrificing any other column. Interactions between three-level columns are distributed more or less uniformly to all the other three-level columns, which permits investigation of main effects. Thus, it is a highly recommended array for experiments. (Taguchi Paper, No. 27.)

14.7 Summary

Several observations can be made concerning the use of Taguchi's experimental designs.

1. They present a procedure for examining many factors in an experiment using a very small number of observations.

2. In using such procedures one must usually assume that no interactions are present or, at least, be able to identify before the experiment some interactions and include them in the design.

3. Taguchi's orthogonal arrays are essentially fractional factorial experiments, but no mention is made either in Taguchi notes or in Taguchi courses of

confounding or aliases. No defining contrasts are shown, and yet aliases between main effects and two-way interactions are assumed to make up tables like 14.3, 14.9, and 14.16.

4. Using the Taguchi approach should require much more time to be spent in discussion before the experiment to ensure the experimenter that no interactions exist or that some that may exist can be worked into the design.

5. Taguchi's arrays when used on the maximum number of factors are all of resolution III, which means that all main effects are independent of other main effects but are confounded with two-way or higher interactions. The Bureau of Standards tables referred to in Chapter 13 never include any such designs because of their low resolution number.

6. Although it is true that any fractional factorial must confound some interactions, the usual designs attempt to confound only three-way or higher interactions and allow one to safely assess main effects and two-way interactions free of each other. Again, an investigating team needs to carefully consider what aliases (or confounding) do exist *before* running a fractional factorial experiment.

7. Taguchi-type designs should be considered only as a screening type of experiment. They will usually find the most important main effects. The Plackett–Burman design is, of course, an orthogonal array as shown in Chapter 13.

8. To really find out what factors and interactions are important, one should run either a complete factorial or a fractional factorial of resolution V or higher.

9. If using Taguchi methods does nothing more than make experimenters aware of the need for a good experimental design, it has accomplished an important objective.

Problems

14.1 Consider the first observation only in each cell of Table 1.2 and try a Taguchi $L_4(2^3)$ on these four observations. Compare the results with Example 14.3.

14.2 Use all the data of Problem 9.17 and solve the problem with a Taguchi $L_8(2^7)$; compare results with the text answers.

14.3 Use the cell totals in Example 10.2 to solve this problem with an L_9 and compare results.

14.4 Example 13.1 is handled as a half replicate of a 2^4 experiment. Set this up as an L_8 and include three interactions that you designate. Work this out and compare with the results of Table 13.7.

14.5 Do Example 13.2 as an L_9 experiment.

14.6 Try Problem 13.17 as a four factor with three interactions using Taguchi's approach.

Regression 15

15.1 Introduction

In previous chapters most of the analyses of the experiments have used analysis of variance (ANOVA) as the basic procedure. Our SAS programs have usually been handled with a PROC ANOVA format. Occasionally we have used PROC GLM (General Linear Model) for some procedures after ANOVA. In Figure 3.1, we noted several procedures that might follow an analysis of variance when the effects of factors were significant. We did not treat cases in which the factor levels were quantitative. Figure 3.1 suggests two procedures to follow when significant quantitative effects are present: (1) the method of orthogonal polynomials when the factor levels are equispaced and (2) general methods of curve fitting or regression.

Some experimenters prefer to treat all analyses as regression models because even ANOVA can be cast into a regression model. This is evidenced by the term GLM. If one or more independent variables (X's) is quantitative and its effect is statistically significant, then one should probably use regression analysis (either PROC GLM or PROC REG) in an attempt to determine a mathematical model that may be used to predict the magnitude of the response variable or criterion from values of the independent variables. This regression procedure will be examined for Y as a function of one X where the relationship may be linear, quadratic, or a higher order polynomial. Also considered will be the case of multiple regression where Y is a function of several X's.

SAS computer programs will be used in our search for an adequate prediction equation, because most real problems are too difficult to handle without a computer.

15.2 Linear Regression

To review linear regression and to become familiar with the notation used, consider an example where an experiment was designed to determine the effect of the amount of drug dosage on a person's reaction time to a given stimulus.

■ **EXAMPLE 15.1**   To study the effect of drug dosage in milligrams on a person's reaction time in milliseconds, 15 subjects were randomly assigned one of five dosages of a drug—0.5, 1.0, 1.5, 2.0, or 2.5 mg—with three subjects assigned to each level of drug dosage. The reaction times were recorded as shown in Table 15.1. An analysis of variance of these data yields Table 15.2.

Table 15.1   Reaction Time Data

| | Dosage (mg) | | | | | |
	0.5	1.0	1.5	2.0	2.5	
Reaction	26	28	28	32	38	
Time (ms)	28	26	30	33	39	
Y_{ij}	29	30	31	31	38	
n	3	3	3	3	3	$N = 15$
$T_{.j}$	83	84	89	96	115	$T_{..} = 467$
$\sum_i Y_{ij}^2$	2301	2360	2645	3074	4409	$\sum_i \sum_j Y_{ij}^2 = 14{,}789$

Table 15.2   Reaction Time ANOVA

Source	df	SS	MS	F	Significance
Between dosages	4	229.73	57.42	28.72	0.000
Error	10	20.00	2.00		
Totals	14	249.73			

Obviously dosage has a highly significant effect on reaction time. But since reaction time is a quantitative factor, the next question might well be: How does reaction time vary with drug dosage? Can one find a functional relationship between these two variables that might enable one to predict reaction time from drug dosage?

As a first step one might plot a scattergram of all 15 X, Y pairs of points. This is shown in Figure 15.1. In Figure 15.1 the small x's denote the average Y for each X_j or the $\overline{Y}_{.j}$'s. The plot shows that Y increases with increases in X and one may try a straight-line "fit" as a first approximation for predicting Y from X. Such a line is shown in Figure 15.1. One can see how close the line comes to the average Y values for each X. Note that for a given X_j (say 2.5 mg) the observed Y_{ij} can be partitioned into three parts: Y_X' is the predicted value of Y_{ij} for a given X_j as found on the straight line, $\overline{Y}_{.j} - Y_X'$ is the amount by which the mean Y for a given X_j departs from its predicted value Y_X' (referred to as the

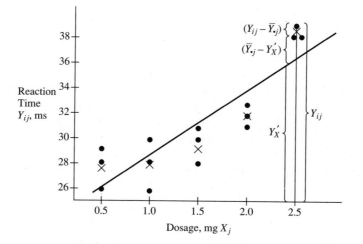

Figure 15.1 Reaction time versus drug dosage.

departure from linear regression in this model), and $Y_{ij} - \bar{Y}_{\cdot j}$ is the amount by which an observed Y_{ij} varies from its treatment mean.

In the analysis of variance model

$$Y_{ij} = \mu + (\mu_j - \mu) + (Y_{ij} - \mu_j) \tag{15.1}$$

and now one tries to predict μ_j from some function based on X_j. If this predicted mean is labeled $\mu_{Y/X}$, the model can be expended to read

$$Y_{ij} = \mu + (\mu_{Y/X} - \mu) + (\mu_j - \mu_{Y/X}) + (Y_{ij} - \mu_j) \tag{15.2}$$

where $\mu_{Y/X}$ is the true predicted mean based on X and $\mu_j - \mu_{Y/X}$ is the amount by which the true mean μ_j departs from its predicted value.

For the sample model Equation (15.2) becomes

$$Y_{ij} = \bar{Y}_{\cdot\cdot} + (Y'_X - \bar{Y}_{\cdot\cdot}) + (\bar{Y}_{\cdot j} - Y'_X) + (Y_{ij} - \bar{Y}_{\cdot j}) \tag{15.3}$$

If the second and third terms on the right of this equation are combined, we have the ANOVA sample model. Thus, we attempt here to partition the mean of a given treatment into two parts: that which can be predicted by regression on X and the amount by which the mean shows a departures from such a regression model. The purpose, of course, is to find a model for predicting the response means such that the departures from these means are very small. By choosing a polynomial of high enough degree one can find a model that actually goes through all the Y means. However, it is hoped that an adequate lower degree model can be found whose departures are small and insignificant when compared with the error of individual Y's around their means.

In Equation 15.3 Y'_X represents the predicted Y for a given X for any assumed model. The straight-line model is only one possible attempt to "fit" a curve to the data. The straight-line model is given by

$$Y'_X = b_0 + b_1 X_j \tag{15.4}$$

where b_0 is the Y intercept and b_1 is its slope. The usual method for determining the values of b_0 and b_1 is the *method of least squares*. The values of b_0 and b_1 are determined from the data in such a way as to minimize the sum of squares of the deviations of each Y_{ij} from its predicted value Y'_X. [One could also find b_0 and b_1 such that the sum of squares of the departures from regression $(\bar{Y}_{.j} - Y'_X)$ is minimized, but this leads to the same results.] Here

$$Y_{ij} - Y'_X = Y_{ij} - b_0 - b_1 X_j$$

and

$$SS_{\text{deviations}} = \sum_i \sum_j (Y_{ij} - Y'_X)^2 = \sum_i \sum_j (Y_{ij} - b_0 - b_1 X_j)^2$$

Because the summations of X's and Y's can be found from a given set of data, $SS_{\text{deviations}}$ is a function of b_0 and b_1. To minimize $SS_{\text{deviations}}$ then, one differentiates $SS_{\text{deviations}}$ with respect to b_0 and b_1

$$\frac{\partial(SS_{\text{deviations}})}{\partial b_0} = 2 \sum_i \sum_j (Y_{ij} - b_0 - b_1 X_j)(-1) = 0$$

$$\frac{\partial(SS_{\text{deviations}})}{\partial b_1} = 2 \sum_i \sum_j (Y_{ij} - b_0 - b_1 X_j)(-X_j) = 0$$

Dividing through by (-2) and simplifying

$$\left.\begin{aligned}
\sum_i \sum_j Y_{ij} &= b_0 N + b_1 n \sum_j X_j \\
\sum_i \sum_j X_j Y_{ij} &= b_0 n \sum_j X_j + b_1 n \sum_i X_j^2
\end{aligned}\right\} \tag{15.5}$$

These equations are often called the least squares normal equations, which can now be solved for b_0 and b_1.

From Table 15.1, $\sum_i \sum_j Y_{ij} = T_{..} = 467$, $N = 15$, $n = 3$.

$$\sum_j X_j = 0.5 + 1.0 + 1.5 + 2.0 + 2.5 = 7.5$$

$$\sum_j X_j^2 = (0.5)^2 + (1.0)^2 + (1.5)^2 + (2.0)^2 + (2.5)^2 = 13.75$$

To find the cross product $\sum_i \sum_j X_j Y_{ij}$ note that $\sum_i Y_{ij} = T_{.j}$, so $\sum_j X_j T_{.j}$ gives

$$(0.5)(8.3) + (1.0)(84) + (1.5)(89) + (2.0)(96) + (2.5)(115) = 738.5$$

Substituting in Equation (15.5)

$$467 = 15b_0 + 22.5b_1$$
$$738.5 = 22.5b_0 + 41.25b_1$$

To solve one might multiply the first equation by 1.5, giving

$$700.5 = 22.5b_0 + 33.75b_1$$
$$738.5 = 22.5b_0 + 41.25b_1$$

Subtracting and solving for b_1,

$$b_1 = \frac{700.5 - 738.5}{33.75 - 41.25} = \frac{-38.0}{-7.5} = 5.07$$

and from the original first equation

$$b_0 = \frac{467 - (22.5)(5.07)}{15} = 23.53$$

and the linear model for this set of sample data is

$$Y'_X = 23.53 + 5.07X_j$$

Some previous work on a straight-line fit may have used the formula for slope as

$$b_1 = \frac{SP_{XY}}{SS_X}$$

of the ratio of the sum of cross products of X and Y to the sum of the squares of X. And b_0 as

$$b_0 = \bar{Y} - b_1\bar{X}$$

Here

$$b_1 = \frac{SP_{XY}}{SS_X} = \frac{\sum_i \sum_j X_j Y_{ij} - \dfrac{\sum_i \sum_j X_j \sum_i \sum_j Y_{ij}}{N}}{\sum_i \sum_j X_j^2 - \dfrac{\left(\sum_i \sum_j X_j\right)^2}{N}}$$

In terms of Table 15.1 notation,

$$b_1 = \frac{\sum_j X_j T_{.j} - n\sum_j X_j T_{..}/N}{n\sum_j X_j^2 - (n\sum_j X_j)^2/N} = \frac{738.5 - 3(7.5)(467)/15}{3(13.75) - 3(7.5)^2/15} = 5.07$$

as before and

$$b_0 = \bar{Y}_{..} - b_1\bar{X}_{..} = \frac{467}{15} - (5.07)\frac{3(7.5)}{15} = 23.53$$

as before.

To see how well this linear model predicts the Y's from the X's, we construct Table 15.3. With the predicted values taken from the model

$$Y'_X = 23.53 + 5.07X_j$$

Table 15.3 Departures from Linear Regression

X_j	$\bar{Y}_{.j}$	Y'_X	$\bar{Y}_{.j} - Y'_X$	$(\bar{Y}_{.j} - Y'_X)^2$
0.5	27.667	26.067	1.600	2.560
1.0	28.000	28.600	−0.600	0.360
1.5	29.667	31.134	−1.467	2.152
2.0	32.000	33.667	−1.667	2.779
2.5	38.333	36.200	2.133	4.550
Totals			0	12.401

as, for example, when $X_j = 1.0$.

$$Y'_X = 23.53 + 5.07(1.0) = 28.600$$

The table shows the departures from linear regression, for example,

$$\bar{Y}_{.j} - Y'_X = 28.000 - 28.600 = -0.600$$

Note that the sum of the departures adds to zero and the sum of squares of departures from linear regression is

$$\text{SS}_{\text{departures}} = n \sum_j (\bar{Y}_{.j} - Y'_X)^2 = 3(12.40) = 37.20$$

as each departure of a mean must be weighted with three observed values for each X_j. Knowing the sum of squares of departures from linear regression, the ANOVA Table 15.2 can be expanded to give Table 15.4. Table 15.4 shows a highly significant linear effect but also a significant departure from the linear model at the 5 percent significance level. With such a departure it would be appropriate to try a second-degree equation or a quadratic of the form

$$Y'_X = b_0 + b_1 X_j + b_2 X_j^2$$

Because there is a strong linear effect in this example and, in some problems, the linear model is adequate to explain most of the variation in the Y variable, some useful statistics can be determined from Table 15.4.

Table 15.4 ANOVA on Reaction Time with Linear Regression

Source	df	SS	MS	F	Significance
Between dosages	4	229.73			
Linear	1		192.53	192.53	96.3***	0.000
Departure from linear	3		37.20	12.40	6.2*	0.012
Error	10	20.00		2.00	
Totals	14	249.73			

The proportion of the total sum of squares that can be accounted for by linear regression is sometimes called the *coefficient of determination* r^2, and its positive square root r the *Pearson product-moment correlation coefficient*. Here

$$r^2 = 192.53 / 249.73 = 0.7712$$

and

$$r = 0.88$$

Thus linear regression will account for about 77 percent of the variation seen in the reaction time Y.

By taking the ratio of the sum of squares between means to the total sum of squares, one finds the maximum amount of the total variation that could be accounted for by a curve or model that passes through all the mean Y's for each X. This is called *eta squared* (η^2). Here

$$\eta^2 = 229.73 / 249.73 = 0.9199$$

or approximately 92 percent of the variation in reaction time could be accounted for by a model through all the means for each dosage X_j.

Another statistic used with a linear model is the standard error of estimate ($s_{Y.X}$). It is the standard deviation of the deviations of the Y_{ij}'s from their predicted values

$$Y_{ij} - Y'_X = (Y_{ij} - \bar{Y}_{.j}) + (\bar{Y}_{.j} - Y'_X)$$

This expression adds the error about the mean and the departure of the mean from the predicted curve, which is appropriate only if the departure is nonsignificant. Then

$$s_{Y.X} = \sqrt{\frac{SS_{\text{departure}} + SS_{\text{error}}}{N - 2}}$$

$$s_{Y.X} = \sqrt{\frac{37.20 + 20.00}{13}} = 2.10$$

This statistic is often used to set confidence limits around a line of best fit when linear regression is appropriate.

In this discussion of Example 15.1 no use has been made of the fact that the dosages (X_j's) are equispaced. When this is true, one can code the X'_j's by considering their mean $\bar{X}_{..}$ and the width of the interval between them c. Let

$$u_j = \frac{X_j - \bar{X}_{..}}{c} = \frac{X_j - 1.5}{0.5}$$

and our model becomes

$$Y'_u = b'_0 + b'_1 u \tag{15.6}$$

whose least squares normal equations are given by

$$\left.\begin{array}{l} \sum_i \sum_j Y_{ij} = b_0' N + b_1' N \sum_j u_j \\[2mm] \sum_i \sum_j u_j Y_{ij} = b_0' n \sum_j u_j + b_1' n \sum_j u_j^2 \end{array}\right\} \qquad (15.7)$$

By equispacing the X_j''s, the u_j''s are $-2, -1, 0, 1, 2$, and $\sum_j u_j = 0$. Because this is always the case, Equations (15.7) become

$$\left.\begin{array}{l} b_0' = \dfrac{\sum_i \sum_j Y_{ij}}{N} = \bar{Y}_{..} \\[4mm] b_1' = \dfrac{\sum_i \sum_j u_j Y_{ij}}{n \sum_j u_j^2} = \dfrac{\sum_j u_j T_{.j}}{n \sum_j u_j^2} \end{array}\right\} \qquad (15.8)$$

and they are relatively easy to solve. Note, too, that because $\sum_j u_j = 0$, the expression $\sum_j u_j T_{.j}$ is a contrast and its sum of squares is $(\sum_j u_j T_{.j})^2 / n \sum u_j^2$. For our data

$$b_0' = \bar{Y}_{..} = 467 / 15 = 31.133$$

$$b_1' = \frac{-2(83) - 1(84) + 0(89) + 1(96) + 2(115)}{3(10)} = \frac{76}{30} = 2.533$$

and

$$Y_u' = 31.13 + 2.53 u_j$$

That this is the same model, substitute for u_j and

$$\begin{aligned} Y_X' &= 31.133 + 2.533(X_j - 1.5) / 0.5 \\ &= 23.53 + 5.07 X_j \end{aligned}$$

as before. Note also that the sum of squares due to linear regression can be found from the linear contrast

$$SS_{\text{due to linear regression}} = \frac{(76)^2}{3(10)} = 192.53$$

as shown in Table 15.4. ∎

15.3 Curvilinear Regression

When a linear regression model is not sufficient to explain all the significant variation in the means, the next logical step is to consider a second-degree (or quadratic) model:

$$Y_X' = b_0 + b_1 X_j + b_2 X_j^2 \qquad (15.9)$$

By the methods of least squares, the normal equations are

$$\sum_i \sum_j Y_{ij} = b_0 N + b_1 n \sum_j X_j + b_2 n \sum_j X_j^2$$

$$\sum_i \sum_j X_j Y_{ij} = b_0 n \sum_j X_j + b_1 n \sum_j X_j^2 + b_2 n \sum_j X_j^3 \qquad (15.10)$$

$$\sum_i \sum_j X_j^2 Y_{ij} = b_0 n \sum_j X_j^2 + b_1 n \sum_j X_j^3 + b_2 n \sum_j X_j^4$$

These equations can be solved for b_0, b_1, and b_2, but it is difficult. Again coded by u_j's,

$$Y_u' = b_0' + b_1' u_j + b_2' u_j^2 \qquad (15.11)$$

$$\sum_i \sum_j Y_{ij} = b_0' N + b_1' n \sum_j u_j + b_2' n \sum_j u_j^2$$

$$\sum_i \sum_j u_j Y_{ij} = b_0' n \sum_j u_j + b_1' n \sum_j u_j^2 + b_2' n \sum_j u_j^3 \qquad (15.12)$$

$$\sum_i \sum_j u_j^2 Y_{ij} = b_0' n \sum_j u_j^2 + b_1' n \sum_j u_j^3 + b_2' n \sum_j u_j^4$$

Because of the choice of the u_j's, the sums of all odd powers of the u_j's are zero ($\sum u_j = \sum u_j^3 = 0$). The equations become

$$\sum_i \sum_j Y_{ij} = b_0' N + b_2' n \sum_j u_j^2$$

$$\sum_i \sum_j u_j Y_{ij} = b_1' n \sum_j u_j^2 \qquad (15.13)$$

$$\sum_i \sum_j u_j^2 Y_{ij} = b_0' n \sum_j u_j^2 + b_2' n \sum_j u_j^4$$

For the data of Example 15.1, the middle equation of (15.13) gives

$$b_1' = \frac{\sum_i \sum_j u_j Y_{ij}}{n \sum_j u_j^2} = \frac{76}{30} = 2.53$$

The other two equations of (15.13) give

$$467 = 15 b_0' + 30 b_2'$$
$$972 = 30 b_0' + 102 b_2'$$

since

$$\sum_i \sum_j u_j^2 Y_{ij} = \sum_j u_j^2 T_{.j} = 4(83) + 1(84) + 0(89) + 1(96) + 4(115) = 972$$

and

$$\sum_j u_j^4 = (-2)^4 + (-1)^4 + (0)^4 + (1)^4 + (2)^4 = 34$$

$$n \sum_j u_j^4 = 102$$

Multiplying the first equation by 2 and subtracting it from the second:

$$972 = 30b_0' + 102b_2'$$
$$934 = 30b_0' + 60b_2'$$
$$38 = 42b_2'$$

$$b_2' = \frac{38}{42} = 0.90$$

and

$$b_0' = \frac{467 - (30)(0.90)}{15} = 29.33$$

and the quadratic model in terms of u_j is

$$Y_u' = 29.33 + 2.53u_j + 0.90u_j^2$$

One need not write this in terms of X_j to see how well it predicts the Y's from the X's. Table 15.5 shows the predicted Y's and the departure of the means from the quadratic model.

Table 15.5 Departures from Quadratic

X_j	u_j	$\bar{Y}_{.j}$	Y_X'	$\bar{Y}_{.j} - Y_X'$	$(\bar{Y}_{.j} - Y_X')^2$
0.5	-2	27.67	27.87	-0.20	0.040
1.0	-1	28.00	27.70	0.30	0.090
1.5	0	29.67	29.33	0.34	0.116
2.0	1	32.00	32.76	-0.76	0.578
2.5	2	38.33	37.99	0.34	0.116
Totals				$0.02 \approx 0$	0.940

From Table 15.5 the sum of the squares of the departure from the quadratic is

$$SS_{departure} = 3(0.94) = 2.82$$

and this step can be added to refine Table 15.4 further and give Table 15.6. Now the departure from the quadratic is not significant at the 5 percent level so it is appropriate to stop with the quadratic for predicting reaction time from dosage. One may also note from Table 15.6 that linear and quadratic terms together account for $192.53 + 34.38 = 226.91$ of the total sum of squares in Y of 249.73.

This is $226.91 / 249.73 = 0.9086$ or approximately 91 percent of the variability accounted for. The square root of this statistic is sometimes referred to as a *correlation ratio R* and $R^2 = 0.91$. The error and nonsignificant departure term could be "pooled" to give a standard error of estimate for the quadratic

$$s_{Y.X,X^2} = \sqrt{\frac{20.00 + 2.82}{10 + 2}} = 1.38$$

which might be used to set confidence limits on the Y's around the quadratic curve.

Table 15.6 ANOVA with Quadratic Regression

Source	df	SS	MS	F	Significance
Between dosages	4	229.73			
Linear	1	192.53	192.53	96.3***	0.000
Quadratic	1	34.38	34.38	17.2**	0.002
Departure	2	2.82	1.41	<1	0.517
Error	10	20.00	2.00		
Totals	14	249.73			

15.4 Orthogonal Polynomials

In Section 10.1, it was shown that one could extract a linear and quadratic effect by the proper use of coefficients in orthogonal contrasts if the X's were equispaced. This concept can be extended to cubic, quartic, and so on, contrasts as the number of levels (k) of the X's increase. Statistical Table F gives the proper coefficients for such contrasts for $k = 3$ through 10. The use of these coefficients will provide orthogonal contrasts that can be treated independent of each other and from which a polynomial of the highest order whose coefficients are significant will provide an adequate equation to use for predicting Y and X.

Once the numerical value of the contrast in treatment totals is found, its sum of squares can be tested against the error mean square to determine the highest order polynomial to consider as an adequate prediction equation. Then one can run an SAS program to find this equation.

We illustrate with the data of Example 15.1. Because we have five levels of dosage, $k = 5$, Table F gives the coefficients of Table 15.7 for our orthogonal contrasts. In this table, the term $\sum (\xi_j')^2$ simply gives the sum of squares of the coefficients in a given contrast. For example, for a cubic contrast with $k = 5$, coefficients squared are $(-1)^2 + (2)^2 + (0)^2 + (-2)^2 + (1)^2 = 10$. The λ term is only used as a scaling factor when finding the adequate equation by hand. SAS will allow us to find this equation by using the given X's.

Table 15.7 Orthogonal Coefficients for Example 15.1

$T_{.j}$	83	84	89	96	115	$\sum_j (\xi'_j)^2$	λ_j	Contrast	SS
Linear	-2	-1	0	1	2	10	1	76	192.53
Quadratic	2	-1	-2	-1	2	14	1	38	34.38
Cubic	-1	2	0	-2	1	10	5/6	8	2.13
Quartic	1	-4	6	-4	1	70	35/12	12	0.69
								Total	229.73

From Table 15.7, each contrast may be squared and divided by $n \sum (\xi'_j)^2$ to give the sum of squares for this contrast. For example, for a cubic the cubic contrast is $-1(83) + 2(84) + 0(89) + (-2)(96) + 1(115) = 8$, and its sum of squares is

$$SS_{cubic} = 8^2 / 3(10) = 2.13 \qquad \text{as shown}$$

From Table 15.7, one can then generate Table 15.8. Here only linear and quadratic effects are significant at the 5 percent level, so one would use a quadratic as an adequate equation.

Table 15.8 ANOVA for Polynomial Model

Source	df	SS	MS	F	Significance
Between dosages	4	229.72			
Linear	1	192.53	192.53	96.3	0.000
Quadratic	1	34.38	34.38	17.2	0.002
Cubic	1	2.13	2.13	1.1	0.325
Quartic	1	0.69	0.69	<1	0.570
Error	10	20.00	2.00		
Totals	14	249.73			

15.5 SAS Programs

To illustrate the use of SAS PROC GLM programs on polynomial regression, we create three new variables for X^2, X^3, and X^4. Here DOSE 2=DOSE*DOSE, a quadratic variable. DOSE 3=DOSE2*DOSE or DOSE*DOSE*DOSE, a cubic variable, and DOSE 4=DOSE3*DOSE, which is the same as DOSE*DOSE* DOSE*DOSE, a quartic variable. The **PROC PRINT** gives the square, cube, and fourth power of the X's, but one need only to insert data on X and Y. The first model considered is linear, Y versus X. The equation given is $Y' = 5.07X + 23.53$, the same as the formula in Section 15.2. We note that both terms in this model are highly significant, and the straight line accounts for about 77 percent of the variation in reaction time.

```
Example 15.1 by SAS
DATA A;
INPUT DOSE TIME @@;
CARDS;
     0.5 26   0.5 28   0.5 29
     1.0 28
               etc.
;
DOSE2=DOSE*DOSE;
DOSE3=DOSE2*DOSE;
DOSE4=DOSE3*DOSE;
TITLE'EXAMPLE15.1';
PROC PRINT;
PROC GLM;
MODEL TIME=DOSE;
PROC GLM;
MODEL TIME=DOSE DOSE2 DOSE3 DOSE4;
PROC GLM;
MODEL TIME=DOSE DOSE2;
```

The output continues on page 360.
The next model considered is the full model or quartic, $Y' = f(X, X^2, X^3, X^4)$. Here $Y' = 33.33 - 22.44X + 28.67X^2 - 14.22X^3 + 2.67X^4$, which accounts for about 92 percent of the variation in reaction time. However, a glance at the PR column shows only the intercept significant at the 5 percent level. One should then try a cubic, and (as in this case) some of these coefficients would prove nonsignificant. Then we go to a quadratic, which our use of orthogonal polynomials showed to be an adequate prediction equation.

Our third model gives the equation as $Y' = 29.87 - 5.79X + 3.62X^2$ with all coefficients significant at the 5 percent level and accounting for about 91 percent of the variation in reaction time. That this is the same prediction equation as the one found in Section 15.3 can be verified by decoding the u_j's, where $u_j = X_j - 1.5/0.5$. The results from SAS also agree with the results in Table 15.6 if the departure is added to the error term to give an error mean square of

$$\text{MS}_{\text{error}} = \frac{20.00 + 2.82}{10 + 2} = \frac{22.02}{12} = 1.90$$

Two more models of Example 15.1 are on page 361.

OBS	DOSE	TIME	DOSE2	DOSE3	DOSE4
1	0.5	26	0.25	0.125	0.0625
2	0.5	28	0.25	0.125	0.0625
3	0.5	29	0.25	0.125	0.0625
4	1.0	28	1.00	1.000	1.0000
5	1.0	26	1.00	1.000	1.0000
6	1.0	30	1.00	1.000	1.0000
7	1.5	28	2.25	3.375	5.0625
8	1.5	30	2.25	3.375	5.0625
9	1.5	31	2.25	3.375	5.0625
10	2.0	32	4.00	8.000	16.0000
11	2.0	33	4.00	8.000	16.0000
12	2.0	31	4.00	8.000	16.0000
13	2.5	38	6.25	15.625	39.0625
14	2.5	39	6.25	15.625	39.0625
15	2.5	38	6.25	15.625	39.0625

GENERAL LINEAR MODELS PROCEDURE

DEPENDENT VARIABLE: TIME

SOURCE	DF	SUM OF SQUARES	MEAN SQUARE	F VALUE	PR > F	R-SQUARE	C.V.
MODEL	1	192.53333333	192.53333333	43.76	0.0001	0.770956	6.7375
ERROR	14	57.20000000	4.40000000		ROOT MSE		TIME MEAN
CORRECTED TOTAL	14	249.73333333			2.09761770		31.13333333

SOURCE	DF	TYPE I SS	F VALUE	PR > F	TYPE III SS	F VALUE	PR > F
DOSE	1	192.53333333	43.76	0.0001	192.53333333	43.76	0.0001

PARAMETER	ESTIMATE	T FOR HO: PARAMETER=0	PR > |T|	STD ERROR OF ESTIMATE
INTERCEPT	23.53333333	18.53	0.0001	1.27017059
DOSE	5.06666667	6.61	0.0001	0.76594169

GENERAL LINEAR MODELS PROCEDURE
DEPENDENT VARIABLE: TIME

SOURCE	DF	SUM OF SQUARES	MEAN SQUARE	F VALUE	PR > F	R-SQUARE	C.V.
MODEL	4	229.73333333	57.43333333	28.72	0.0001	0.919915	4.5424
ERROR	10	20.00000000	2.00000000		ROOT MSE		TIME MEAN
CORRECTED TOTAL	14	249.73333333			1.41421356		31.13333333

SOURCE	DF	TYPE I SS	F VALUE	PR > F	TYPE III SS	F VALUE	PR > F
DOS	1	192.53333333	96.27	0.0001	0.44598191	0.22	0.6469
DOS*DOS	1	34.38095238	17.19	0.0020	0.50630370	0.25	0.6258
DOS*DOS*DOS	1	2.13333333	1.07	0.3260	0.53965744	0.27	0.6147
DOS*DOS*DOS*DOS	1	0.68571429	0.34	0.5712	0.68571429	0.34	0.5712

PARAMETER	ESTIMATE	T FOR H0: PARAMETER=0	PR > |T|	STD ERROR OF ESTIMATE
INTERCEPT	33.33333333	2.58	0.0276	12.93573861
DOS	-22.44444444	-0.47	0.6469	47.52971780
DOS*DOS	28.66666667	0.50	0.6258	56.97530329
DOS*DOS*DOS	-14.22222222	-0.52	0.6147	27.37936563
DOS*DOS*DOS*DOS	2.66666667	0.59	0.5712	4.55420034

GENERAL LINEAR MODELS PROCEDURE
DEPENDENT VARIABLE: TIME

SOURCE	DF	SUM OF SQUARES	MEAN SQUARE	F VALUE	PR > F	R-SQUARE	C.V.
MODEL	2	226.91428571	113.45714286	59.66	0.0001	0.908626	4.4293
ERROR	12	22.81904762	1.90158730		ROOT MSE		TIME MEAN
CORRECTED TOTAL	14	249.73333333			1.37898053		31.13333333

SOURCE	DF	TYPE I SS	F VALUE	PR > F	TYPE III SS	F VALUE	PR > F
DOSE	1	192.53333333	101.25	0.0001	9.41339445	4.95	0.0460
DOSE2	1	34.38095238	18.08	0.0011	34.38095238	18.08	0.0011

PARAMETER	ESTIMATE	T FOR H0: PARAMETER=0	PR > |T|	STD ERROR OF ESTIMATE
INTERCEPT	29.86666667	17.49	0.0001	1.70756177
DOSE	-5.79047619	-2.22	0.0460	2.60255133
DOSE2	3.61904762	4.25	0.0011	0.85112526

■ **EXAMPLE 15.2** Two factors are considered: one qualitative and one quantitative. We wish to determine the effect of both depth and position in a tank (Fig. 15.2) on the concentration of a cleaning solution in ounces per gallon. Concentrations are measured at three depths from the surface of the tank, 0 inch, 15 inches, and 30 inches. At each depth measurements are taken at five different lateral positions in the tank. These are considered as five qualitative positions, although probably some orientation measure might be made on them. At each depth and position, two observations are taken. This is then a 5×3 factorial with two replications per cell (total of 30 observations). The design is a completely randomized design and the model is

$$Y_{ijk} = \mu + D_i + P_j + DP_{ij} + \varepsilon_{k(ij)}$$

with

$$i = 1, 2, 3 \qquad j = 1, 2, \ldots, 5 \qquad k = 1, 2$$

where D_i represents the depth and P_j the position. The data collected are shown in Table 15.9. The first step is to run a PROC ANOVA on a model that breaks

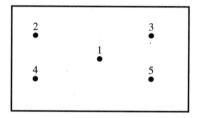

Figure 15.2 Positions in tank at each depth.

Table 15.9 Cleaning-Solution Concentration Data

Position P_j	Depth from Top of Tank D_i		
	0 in.	15 in.	30 in.
1	5.90	5.90	5.94
	5.91	5.89	5.80
2	5.90	5.89	5.75
	5.91	5.89	5.83
3	5.94	5.91	5.86
	5.90	5.91	5.83
4	5.93	5.94	5.83
	5.91	5.90	5.89
5	5.90	5.94	5.83
	5.87	5.90	5.86

depth into linear and quadratic parts and the interaction (8 df) into depth linear by position (4 df) and depth quadratic by position (4 df), as shown in the first SAS model.

```
Example 15.2
DATA A;
INPUT DEPTH POS CONC @@;
CARDS;
     0 1 5.90 0 1 5.91 15 1 5.90 15 1 5.89
         etc. to   30 5 5.86
;
PROC GLM (OR ANOVA)
CLASS POS;
MODEL CONC=DEPTH DEPTH*DEPTH POS DEPTH*POS DEPTH*DEPTH*POS;
PROC GLM;
MODEL CONC=DEPTH DEPTH*DEPTH;
```

From the analysis only the depth factor (linear and quadratic) needs to be considered, since the other factors are not significant. A second model is then run with concentration as a quadratic function of depth. This last model gives an adequate equation of

$$Y'(\text{concentration}) = 0.7000 + 0.2167D - 0.0144D^2$$

This equation accounts for about 49 percent of the variation in concentration, whereas the full model would account for 67 percent of the variation. Table 15.10 shows how well this model predicts average concentration from depth.

Table 15.10 Curve Fit for Example 15.2

D	Y'	Y
0	0.70	0.70
15	0.69	0.70
30	−5.80	−5.80

Had there been a significant interaction, we would have had to find a quadratic equation in depth for each position. Two analyses are on page 364. Had position been significant with no interaction, one could use the quadratic equation above with addition of a constant for each position.

GENERAL LINEAR MODELS PROCEDURE
DEPENDENT VARIABLE: CONC

SOURCE	DF	SUM OF SQUARES	MEAN SQUARE	F VALUE	PR > F	R-SQUARE	C.V.
MODEL	14	390.46666667	27.89047619	2.21	0.0694	0.673838	242.0214
ERROR	15	189.00000000	12.60000000		ROOT MSE		CONC MEAN
CORRECTED TOTAL	29	579.46666667			3.54964787		-1.46666667

SOURCE	DF	TYPE I SS	F VALUE	PR > F	DF	TYPE III SS	F VALUE	PR > F
DEPTH	1	211.25000000	16.77	0.0010	1	16.25000000	1.29	0.2739
DEPTH*DEPTH	1	70.41666667	5.59	0.0320	1	70.41666667	5.59	0.0320
POS	4	50.46666667	1.00	0.4374	4	16.60000000	0.33	0.8539
DEPTH*POS	4	41.50000000	0.82	0.5303	4	14.73076923	0.29	0.8794
DEPTH*DEPTH*POS	4	16.83333333	0.33	0.8508	4	16.83333333	0.33	0.8508

GENERAL LINEAR MODELS PROCEDURE
DEPENDENT VARIABLE: CONC

SOURCE	DF	SUM OF SQUARES	MEAN SQUARE	F VALUE	PR > F	R-SQUARE	C.V.
MODEL	2	281.66666667	140.83333333	12.77	0.0001	0.486079	226.4379
ERROR	27	297.80000000	11.02962963		ROOT MSE		CONC MEAN
CORRECTED TOTAL	29	579.46666667			3.32108862		-1.46666667

SOURCE	DF	TYPE I SS	F VALUE	PR > F	DF	TYPE III SS	F VALUE	PR > F
DEPTH	1	211.25000000	19.15	0.0002	1	16.25000000	1.47	0.2353
DEPTH*DEPTH	1	70.41666667	6.38	0.0177	1	70.41666667	6.38	0.0177

PARAMETER	ESTIMATE	T FOR H0: PARAMETER=0	PR > |T|	STD ERROR OF ESTIMATE
INTERCEPT	0.70000000	0.67	0.5107	1.05022044
DEPTH	0.21666667	1.21	0.2353	0.17850315
DEPTH*DEPTH	-0.01444444	-2.53	0.0177	0.00571668

15.6 Multiple Regression

We now consider a more general situation in which Y may be a function of several independent variables $X_1, X_2, \ldots, X_k$ with no restrictions on the settings of these k independent variables. In fact, in most such multiple regression situations the X variables have already acted and we simply record their values along with the dependent variable Y. This is, of course, *ex-post-facto* research as opposed to experimental research in which one manipulates the X's and observes the effect on Y.

In practice there are many studies of this type. For example, one may wish to predict the surface finish of steel from dropout temperature and back-zone temperature. Here Y is a function of two recorded temperatures X_1 and X_2. It may be of interest to predict college grade-point average (GPA) in the freshman year for students whose input data include rank in high school, high school regents' average, SAT (Scholastic Aptitude Test) verbal score, and SAT mathematics score. Here Y, the freshman year GPA, is to be predicted from four independent variables: X_1—high school rank, X_2—high school regents' average, X_3—SAT verbal, and X_4—SAT mathematical.

As in the previous sections a mathematical model is written and the coefficients in the model are determined from the observed sample data by the method of least squares, making the sum of squares of deviations from this model a minimum.

To predict the value of a dependent variable (Y) from its regression on several independent variables ($X_1, X_2, \ldots, X_k$) the linear population model is given as

$$Y' = \beta_0 + \beta_1 X_1 + \beta_2 X_2 + \cdots + \beta_k X_k$$

where the β's are the true coefficients to be used to weight the observed X's. In practice a sample of N is chosen and all variables ($Y, X_1, X_2, \ldots, X_k$) are recorded for each of the N items in the sample and the corresponding sample model is

$$Y' = B_0 + B_1 X_1 + B_2 X_2 + \cdots + B_k X_k \tag{15.14}$$

where the B's are estimators of the β's. (Capital B's are used here because computer programs do not usually provide for lowercase letters). Note that if $k = 1$, Equation (15.14) gives

$$Y' = B_0 + B_1 X_1$$

or the straight-line model of Section 15.2.

Because each observed Y can be expressed as

$$Y_i = Y_i' + e_i \qquad i = 1, 2, \ldots, N$$

by making the sum of the squares of the errors of estimate ($\sum e_i^2$) a minimum, one can derive the following least squares equations.

$$\sum Y = B_0 N + B_1 \sum X_1 + B_2 \sum X_2 + \cdots + B_k \sum X_k$$

$$\sum X_1 Y = B_0 \sum X_1 + B_1 \sum X_1^2 + B_2 \sum X_1 X_2 + \cdots + B_k \sum X_1 X_k$$

$$\sum X_2 Y = B_0 \sum X_2 + B_1 \sum X_2 X_1 + B_2 \sum X_2^2 + \cdots + B_k \sum X_2 X_k \quad (15.15)$$

$$\vdots \qquad \vdots \qquad \vdots \qquad \vdots \qquad \vdots$$

$$\sum X_k Y = B_0 \sum X_k + B_1 \sum X_k X_1 + B_2 \sum X_k X_2 + \cdots + B_k \sum X_k^2$$

This set of $k+1$ equations in $k+1$ unknowns $(B_0, B_1, \ldots, B_k)$ can be solved for the B's. The solution is tedious if more than two or three independent variables are involved and so these equations are usually solved by a computer program that can handle a large number of independent variables.

To see how such problems are handled consider a problem from Burr.

■ **EXAMPLE 15.3** Twenty-four samples of a steel alloy are checked as to their percent elongation under stress and for each sample a chemical analysis is made and the percentages of five specific chemical elements are recorded. These are identified only as X_1, X_2, X_3, X_4, and X_5. The data are given in Table 15.11. To enter Equations (15.15) the following sums are computed from Table 15.11.

$$N = 24$$

$\sum X_1 = 11.68$	$\sum X_1^2 = 5.9038$	$\sum X_1 X_2 = 35.553$
$\sum X_2 = 73.7$	$\sum X_2^2 = 240.85$	$\sum X_1 X_3 = 10.101$
$\sum X_3 = 22.1$	$\sum X_3^2 = 31.17$	$\sum X_1 X_4 = 47.644$
$\sum X_4 = 96.1$	$\sum X_4^2 = 408.09$	$\sum X_1 X_5 = 0.18405$
$\sum X_5 = 0.397$	$\sum X_5^2 = 0.0145$	$\sum X_1 Y = 67.318$
$\sum Y = 136.2$	$\sum Y^2 = 943.78$	$\sum X_2 X_3 = 71.17$
$\sum X_2 X_4 = 297.57$	$\sum X_3 X_4 = 74.58$	$\sum X_4 X_5 = 1.2881$
$\sum X_2 X_5 = 1.2886$	$\sum X_3 X_5 = 0.5945$	$\sum X_4 Y = 563.35$
$\sum X_2 Y = 390.32$	$\sum X_3 Y = 102.53$	$\sum X_5 Y = 1.6384$

and the equations to be solved would be

$$136.2 = 24B_0 + 11.68B_1 + 73.7B_2 + 22.1B_3 + 96.1B_4 + 0.397B_5$$
$$67.318 = 11.68B_0 + 5.9038B_1 + 35.553B_2 + 10.101B_3 + 47.644B_4 + 0.18405B_5$$
$$390.32 = 73.7B_0 + 35.553B_1 + 240.85B_2 + 71.17B_3 + 297.57B_4 + 1.2886B_5$$
$$102.53 = 22.1B_0 + 10.101B_1 + 71.17B_2 + 31.17B_3 + 74.58B_4 + 0.5945B_5$$
$$1.2881 = 96.1B_0 + 47.644B_1 + 297.57B_2 + 74.58B_3 + 408.09B_4 + 1.2881B_5$$
$$1.6384 = 0.397B_0 + 0.18405B_1 + 1.2886B_2 + 0.5945B_3 + 1.2881B_4 + 0.0145B_5$$

Table 15.11 Percent Elongation Data

Item No.	Y	X_1	X_2	X_3	X_4	X_5
1	11.3	0.50	1.3	0.4	3.4	0.010
2	10.0	0.47	1.2	0.3	3.6	0.012
3	9.8	0.48	3.1	0.7	4.3	0.000
4	8.8	0.54	2.6	0.7	4.0	0.022
5	7.8	0.45	2.8	0.7	4.2	0.000
6	7.4	0.41	3.2	0.7	4.7	0.000
7	6.7	0.62	3.0	0.6	4.7	0.026
8	6.3	0.53	4.1	0.9	4.6	0.035
9	6.3	0.57	3.7	0.8	4.6	0.000
10	6.3	0.67	2.7	0.6	4.8	0.013
11	6.0	0.54	3.1	0.7	4.2	0.000
12	6.0	0.42	3.1	0.7	4.4	0.000
13	5.8	0.33	2.6	0.6	4.7	0.008
14	5.5	0.51	3.9	0.9	4.4	0.000
15	5.5	0.54	3.1	0.7	4.2	0.000
16	4.7	0.48	4.0	1.1	3.7	0.024
17	4.1	0.38	3.3	0.8	4.1	0.000
18	4.1	0.39	3.2	0.7	4.6	0.016
19	3.9	0.60	2.9	0.7	4.3	0.025
20	3.5	0.54	3.2	0.7	4.9	0.022
21	3.1	0.33	2.9	2.9	1.0	0.063
22	1.6	0.40	3.2	3.2	1.0	0.059
23	1.1	0.64	2.5	0.7	3.8	0.018
24	0.6	0.34	5.0	1.3	3.9	0.044

15.7 SAS Programs

Many computer programs are available to solve a set of simultaneous equations. We will try four such programs: **FORWARD (F)**, **BACKWARD (B)**, **STEPWISE**, and **MAXR**.

In the **FORWARD (F)** solution, a correlation coefficient is run between Y and each independent variable, and then the one with the largest r is run to give an equation for predicting Y from this most promising variable. In this example, X_2 contributes the most to the variation in Y (about 31 percent), and the equation is $Y' = 11.58 - 1.92X_2$. After this, partial r's are determined, and the variable which contributes the most to the variation in Y after the first one is then added. Here, X_5 is the next such variable and the equation is $Y' = 11.69 - 1.62X_2 - 63.18X_5$, accounting for about 49 percent of the variation in Y. This procedure continues until all variables are entered whose probabilities of being greater than F are 0.50 or less (a default value unless some other Pr > F is chosen). If the 5 percent significance level is desired, we examine the PROB

$> F$ column and label as an "adequate" equation the last one where the B's are significant at the 5 percent (0.05) or lower probability level. Here, we call the equation of Y as a function of X_2 and X_5 the adequate equation.

In the **BACKWARD (B)** solution, all X's are entered in the first step, and then those contributing the least to the variation in Y are dropped out until an equation is found where all B's are significant. Here, this procedure gives the same equation in step 3. Note that with all five X's in the equation (step 1) only 51 percent of the variation in Y would be accounted for, whereas the adequate equation would account for 49 percent of such variation, a nonsignificant difference.

```
Example 15.3 SAS programs
DATA A;
INPUT Y X1 X2 X3 X4 X5;
CARDS;
      11.3 0.50 1.3 0.4 3.4 0.010
      10.0 0.47 1.2 0.3 3.6 0.012
         etc
;
TITLE 'EXAMPLE 15.3';
PROC PRINT;
PROC STEPWISE;
MODEL Y=X1 X2 X3 X4 X5/ F B STEPWISE MAXR;
```

The program continues on pages 369–371.
The **STEPWISE** procedure follows the same steps as the **FORWARD** solution, except that in some rare cases it may drop out a variable after inserting a new one. In our example, the adequate equation is the same as found before. Note that it has a 0.15 default probability.

The **MAXR** procedure is a forward solution that at each step gives the "best" equation for the number of X's at that step, or the one with the most variation in Y accounted for as determined by the higher multiple regression coefficient squared (R^2). Here at step 2 the results are the same as found before.

From the printout of any one of these procedures, the square root of the mean square for error, $\sqrt{4.13} = 2.03$, gives the standard error of estimate when predicting Y from the adequate equation.

To see that this is a reasonable prediction we might insert X_2 and X_5 values into it and examine the errors of estimate, for example, if $X_2 = 1.3$ and $X_5 = 0.010$,

$$Y' = 11.69 - 1.62(1.3) - 63.18(.010)$$
$$= 8.957 \text{ versus an observed } Y = 11.3$$

FORWARD SELECTION PROCEDURE FOR DEPENDENT VARIABLE Y

STEP 1 VARIABLE X2 ENTERED   R SQUARE = 0.31420181   C(P) = 5.35523227

	DF	SUM OF SQUARES	MEAN SQUARE	F	PROB>F
REGRESSION	1	53.67980887	53.67980887	10.08	0.0044
ERROR	22	117.16519111	5.32569051		
TOTAL	23	170.84500000			

	B VALUE	STD ERROR	TYPE II SS	F	PROB>F
INTERCEPT	11.57748846				
X2	-1.92211293	0.60542644	53.67980887	10.08	0.0044

STEP 2 VARIABLE X5 ENTERED   R SQUARE = 0.49208798   C(P) = 0.77845055

	DF	SUM OF SQUARES	MEAN SQUARE	F	PROB>F
REGRESSION	2	84.07077035	42.03538517	10.17	0.0008
ERROR	21	86.77422965	4.13210617		
TOTAL	23	170.84500000			

	B VALUE	STD ERROR	TYPE II SS	F	PROB>F
INTERCEPT	11.69482586				
X2	-1.61999580	0.54479609	36.53686509	8.84	0.0073
X5	-63.17916865	23.29632535	30.39396148	7.35	0.0181

STEP 3 VARIABLE X3 ENTERED   R SQUARE = 0.5055419   C(P) = 2.28106031

	DF	SUM OF SQUARES	MEAN SQUARE	F	PROB>F
REGRESSION	3	86.36918442	28.78972814	6.82	0.0024
ERROR	20	84.47581558	4.22379078		
TOTAL	23	170.84500000			

	B VALUE	STD ERROR	TYPE II SS	F	PROB>F
INTERCEPT	11.82239196				
X2	-1.54994707	0.55893256	32.48009675	7.69	0.0117
X3	-0.74853877	1.01473201	2.29841407	0.54	0.4693
X5	-42.22569494	36.89984445	5.53104177	1.31	0.2660

NO OTHER VARIABLES MET THE 0.5000 SIGNIFICANCE LEVEL FOR ENTRY INTO THE MODEL.

BACKWARD ELIMINATION PROCEDURE FOR DEPENDENT VARIABLE Y

STEP 0   ALL VARIABLES ENTERED   R SQUARE = 0.51314320   C(P) = 6.000000000

	DF	SUM OF SQUARES	MEAN SQUARE	F	PROB>F
REGRESSION	5	87.66794928	17.53358986	3.79	0.0160
ERROR	18	83.17705072	4.62094726		
TOTAL	23	170.84500000			

	B VALUE	STD ERROR	TYPE II SS	F	PROB>F
INTERCEPT	14.91722299				
X1	-1.57654684	5.26157760	0.41487075	0.09	0.7679
X2	-1.27616946	0.96242424	8.12483590	1.76	0.2014
X3	-1.69396517	2.34231361	2.41685102	0.52	0.4788
X4	-0.57829883	1.52248084	0.66670232	0.14	0.7085
X5	-43.14470468	39.67043862	4.97076416	1.08	0.3134

STEP 1   VARIABLE X1 REMOVED   R SQUARE = 0.51071485   C(P) = 4.08978045

	DF	SUM OF SQUARES	MEAN SQUARE	F	PROB>F
REGRESSION	4	87.25307854	21.81326963	4.96	0.0066
ERROR	19	83.59192146	4.39957481		
TOTAL	23	170.84500000			

	B VALUE	STD ERROR	TYPE II SS	F	PROB>F
INTERCEPT	14.31358314				
X2	-1.22427578	0.92375797	7.72774578	1.76	0.2008
X3	-1.66044527	2.28291111	2.32745790	0.53	0.4759
X4	-0.65610814	1.46379635	0.88389412	0.20	0.6591
X5	-43.70034653	37.80329379	5.87923996	1.34	0.2620

STEP 2   VARIABLE X4 REMOVED   R SQUARE = 0.50554119   C(P) = 2.28106031

	DF	SUM OF SQUARES	MEAN SQUARE	F	PROB>F
REGRESSION	3	86.3691842	28.7897814	6.82	0.0024
ERROR	20	84.47581558	4.22379078		
TOTAL	23	170.84500000			

	B VALUE	STD ERROR	TYPE II SS	F	PROB>F
INTERCEPT	11.82239196				
X2	-1.54994707	0.55893256	32.48009675	7.69	0.0117
X3	-0.74853877	1.01473201	2.29841407	0.54	0.4693
X5	-42.22569494	36.89984445	5.53104177	1.31	0.2660

BACKWARD ELIMINATION PROCEDURE FOR DEPENDENT VARIABLE Y

STEP 3   VARIABLE X3 REMOVED   R SQUARE = 0.49208798   C(P) = 0.77845055

	DF	SUM OF SQUARES	MEAN SQUARE	F	PROB>F
REGRESSION	2	84.07077035	42.03538517	10.17	0.0008
ERROR	21	86.77422965	4.13210617		
TOTAL	23	170.84500000			

	B VALUE	STD ERROR	TYPE II SS	F	PROB>F
INTERCEPT	11.69482586				
X2	-1.61999580	0.54479609	36.53686509	8.84	0.0073
X5	-63.17916865	23.29632535	30.39096148	7.35	0.0131

ALL VARIABLES IN THE MODEL ARE SIGNIFICANT AT THE 0.1000 LEVEL.

Example 15.3 *(Continued)*

```
STEPWISE REGRESSION PROCEDURE FOR DEPENDENT VARIABLE Y
STEP 1    VARIABLE X2 ENTERED        R SQUARE = 0.31420181      C(P) = 5.35523227
```

	DF	SUM OF SQUARES	MEAN SQUARE	F	PROB>F
REGRESSION	1	53.67980887	53.67980887	10.08	0.0044
ERROR	22	117.16519113	5.32569051		
TOTAL	23	170.84500000			

	B VALUE	STD ERROR	TYPE II SS	F	PROB>F
INTERCEPT	11.57748846				
X2	-1.92211293	0.60542644	53.67980887	10.08	0.0044

```
STEP 2    VARIABLE X5 ENTERED        R SQUARE = 0.49208798      C(P) = 0.77845055
```

	DF	SUM OF SQUARES	MEAN SQUARE	F	PROB>F
REGRESSION	2	84.07077035	42.03538517	10.17	0.0008
ERROR	21	86.77422965	4.13210617		
TOTAL	23	170.84500000			

	B VALUE	STD ERROR	TYPE II SS	F	PROB>F
INTERCEPT	11.69482586				
X2	-1.61999580	0.54479609	36.53686509	8.84	0.0073
X5	-63.17916865	23.29632535	30.39096148	7.35	0.0131

```
NO OTHER VARIABLES MET THE 0.1500 SIGNIFICANCE LEVEL FOR ENTRY INTO THE MODEL.
```

(Continued)

Example 15.3 (*Continued*)

MAXIMUM R-SQUARE IMPROVEMENT FOR DEPENDENT VARIABLE Y

STEP 1   VARIABLE X2 ENTERED   R SQUARE = 0.31420181   C(P) = 5.35523227

	DF	SUM OF SQUARES	MEAN SQUARE	F	PROB>F
REGRESSION	1	53.67980887	53.67980887	10.08	0.0044
ERROR	22	117.16519113	5.32569051		
TOTAL	23	170.84500000			

	B VALUE	STD ERROR	TYPE II SS	F	PROB>F
INTERCEPT	11.57748846				
X2	-1.92211293	0.60542644	53.67980887	10.08	0.0044

THE ABOVE MODEL IS THE BEST 1 VARIABLE MODEL FOUND.

STEP 2   VARIABLE X5 ENTERED   R SQUARE = 0.49208798   C(P) = 0.77845055

	DF	SUM OF SQUARES	MEAN SQUARE	F	PROB>F
REGRESSION	2	84.07077035	42.03538517	10.17	0.0008
ERROR	21	86.77422965	4.13210617		
TOTAL	23	170.84500000			

	B VALUE	STD ERROR	TYPE II SS	F	PROB>F
INTERCEPT	11.69482586				
X2	-1.61999580	0.54479609	36.53686509	8.84	0.0073
X5	-63.17916865	23.29632535	30.39096148	7.35	0.0131

THE ABOVE MODEL IS THE BEST 2 VARIABLE MODEL FOUND.

STEP 3   VARIABLE X3 ENTERED   R SQUARE = 0.50554119   C(P) = 2.28106031

	DF	SUM OF SQUARES	MEAN SQUARE	F	PROB>F
REGRESSION	3	86.36918442	28.7897814	6.82	0.0024
ERROR	20	84.47581558	4.22379078		
TOTAL	23	170.84500000			

	B VALUE	STD ERROR	TYPE II SS	F	PROB>F
INTERCEPT	11.82239196				
X2	-1.54994707	0.55893256	32.48009675	7.69	0.0117
X3	-0.74853877	1.01473201	2.29841407	0.54	0.4693
X5	-42.22569494	36.89984445	5.53104177	1.31	0.2660

THE ABOVE MODEL IS THE BEST 3 VARIABLE MODEL FOUND.

(*Continued*)

Example 15.3 *(Continued)*

MAXIMUM R-SQUARE IMPROVEMENT FOR DEPENDENT VARIABLE Y

STEP 4   VARIABLE X4 ENTERED   R SQUARE = 0.51071485   C(P) = 4.08978045

	DF	SUM OF SQUARES	MEAN SQUARE	F	PROB>F
REGRESSION	4	87.25307854	21.81326963	4.96	0.0066
ERROR	19	83.59192146	4.39957481		
TOTAL	23	170.84500000			

	B VALUE	STD ERROR	TYPE II SS	F	PROB>F
INTERCEPT	14.31358314				
X2	-1.22427578	0.92375797	7.72774578	1.76	0.2008
X3	-1.66044527	2.28291111	2.32745790	0.53	0.4759
X4	-0.65610814	1.46379635	0.88389412	0.20	0.6591
X5	-41.70034653	37.80329379	5.87923996	1.34	0.2620

THE ABOVE MODEL IS THE BEST 4 VARIABLE MODEL FOUND.

STEP 5   VARIABLE X1 ENTERED   R SQUARE = 0.51314320   C(P) = 6.0000000

	DF	SUM OF SQUARES	MEAN SQUARE	F	PROB>F
REGRESSION	5	87.66794928	17.53358986	3.79	0.0160
ERROR	18	83.17705072	4.62094726		
TOTAL	23	170.84500000			

	B VALUE	STD ERROR	TYPE II SS	F	PROB>F
INTERCEPT	14.91722299				
X1	-1.57654684	5.26157760	0.41487075	0.09	0.7679
X2	-1.27616946	0.96242424	8.12483590	1.76	0.2014
X3	-1.69396517	2.34231361	2.41685102	0.52	0.4788
X4	-0.57829883	1.52248084	0.66670232	0.14	0.7085
X5	-43.14470468	39.67043862	4.97078416	1.08	0.3134

THE ABOVE MODEL IS THE BEST 5 VARIABLE MODEL FOUND.

374

giving an error of estimate $Y - Y' = 11.3 - 8.95 = 2.343$. For all 24 points we find as in Table 15.12. Here $\Sigma\, e = -0.003$ (close to zero) and $\Sigma\, e^2 = 86.77$ as given in the ANOVA table after X_2 and X_5 are accounted for.

Table 15.12 Deviations from Regression on Elongation Data

Y	Y'	e	Y	Y'	e
11.3	8.957	2.343	5.8	6.977	−1.177
10.0	8.993	1.007	5.5	5.377	+0.123
9.8	6.673	3.127	5.5	6.673	−1.173
8.8	6.093	2.707	4.7	3.699	1.001
7.8	7.159	0.641	4.1	6.349	−2.149
7.4	6.511	0.889	4.1	5.500	−1.400
6.7	5.192	1.508	3.9	5.417	−1.517
6.3	2.842	3.458	3.5	5.121	−1.621
6.3	5.701	0.599	3.1	3.017	0.083
6.3	6.500	−0.200	1.6	2.783	−1.183
6.0	6.673	−0.673	1.1	6.508	−5.408
6.0	6.673	−0.673	0.6	0.815	−0.215

One might also note in this table that the residuals or errors of estimate seem to be nonrandom as in general they are positive for large Y's and negative for small Y's. This may indicate a need to consider some higher order terms such as X_2^2, X_5^2, or $X_2 X_5$. These could be entered into the equation by proper instructions at the start of a program. Some computer programs will give a plot of these residuals when requested.

If one wishes only to check the significance of a given R^2, the proper test is

$$F_{k, N-k-1} = \frac{R^2_{y.12\ldots k}/k}{\left(1 - R^2_{y.12\ldots k}\right)/N - k - 1} \tag{15.16}$$

To test whether any more variables are needed after a subset r has been found significant the test is

$$F_{k-r, N-k-1} = \frac{\left(R_k^2 - R_r^2\right)/k - r}{\left(1 - R_k^2\right)/N - k - 1} \tag{15.17}$$

where R_k represents the correlation of all k values with Y and R_r the correlation of a subset r of the k variables with Y.

In our example $R_k^2 = 0.51314$ and $R_r^2 = 0.49209$. Testing for significance of all five variables:

$$F_{5, 18} = \frac{0.51314/5}{0.48686/18} = 3.79$$

as in ANOVA. Testing whether one needs to go beyond X_2 and X_5,

$$F_{3,\,18} = \frac{0.51314 - 0.49203 / 3}{0.48686 / 18} < 1$$

hence nonsignificant. ∎

15.8 Summary

The examples of this chapter give the basic adequate equations. The methods may be augmented to include terms in multiple variables that are more complex. For example, with three variables (X_1, X_2, X_3) one might create new variables such as $X_4 = X_1^2, X_5 = X_1 X_2^2,$ and $X_6 = X_3^3$, allowing for prediction surfaces that go beyond the rectilinear model of Equation (15.14). So there is much more that can be done when the X's are quantitative.

To summarize some of the procedures shown above, the following practical steps may be found helpful in such analyses.

CASE I.   Polynomial regression when X is set at equispaced levels.

1. Plot a rough scattergram of Y versus X to get some idea of the underlying polynomial.

2. Determine the Y treatment totals for each X.

3. Use the table of orthogonal polynomials to test for the highest degree polynomial for an adequate prediction equation.

4. Run a computer program such as SAS to find this adequate equation.

5. Note the size of R^2 to see how much of the variation in Y can now be predicted and whether this is worthwhile (even though statistically significant) from a practical point of view. Note also the standard error of estimate (square root of the MS_{error}) if desirable to place confidence limits around one's predictions.

CASE II.   Polynomial regression when the X's are not equispaced—usually an *ex-post-facto* situation.

1. Plot a rough scattergram to help in estimating what the highest degree polynomial is that might work.

2. Run a computer program to find this highest degree equation and note if all B's are statistically significant at some desired level.

3. If some B's are not significant, try a polynomial of one less degree and check the B's.

4. Continue as in step 3 until an adequate prediction is found.

5. Note R^2 and the standard error of estimate as in Case I.

Caution: When writing the model for a given polynomial, be sure to include as variables all lower powers of X as well.

CASE III. Multiple Regression

1. For each experimental unit, record the Y value and each X to be studied $(X_1$ to $X_k)$.

2. Use some STEPWISE procedure to find an adequate equation for predicting Y from all X's deemed to be significant.

3. Note R^2 and the standard error of estimate.

A word of caution seems necessary also when one plans a regression study. Often experimenters seem inclined to include every X value that they can think of as computers can deal with so many X's. However, it is difficult to believe that all X's would be deemed important in their effect on Y by those knowledgeable in the field. Thus the use of multiple regression should not substitute for careful "soaking" in a problem and searching out of past studies and theoretical consid- erations before including a given X in a multiple regression study. Sometimes simple scattergrams of Y versus a given X will give some idea as to whether they may be related and the scattergram may also indicate an association that is nonlinear. This latter may indicate a need for a term such as X^2. The literature in the area may also suggest that some X is logarithmically related to Y and one should use $\log X$ as the independent variable instead of X. Careful consideration of all these notions before the experiment is essential to good experimentation and increases the prospect of meaningful prediction.

Problems

15.1 For the following data on X and Y plot a scattergram and determine the least squares straight line of best fit.

		X			
	3	**4**	**5**	**6**	**7**
	7	8	10	11	10
Y	8	8	9	9	10
	9	9	9	10	9

15.2   For Problem 15.1 present your results in an ANOVA table and comment on these results.

15.3   From your table in Problem 15.2 find r^2, η^2, and the standard error of estimate.

15.4   Use the method of orthogonal polynomials on the data of Problem 15.1 and find the best fitting polynomial.

15.5   An experiment to determine the effect of planting rate in thousands of plants per acre on yield in bushels per acre of corn gave the following results.

Planting Rate	Yield
X	Y
12	130.5, 129.6, 129.9
16	142.5, 140.3, 143.4
20	145.1, 144.8, 144.1
24	147.8, 146.6, 148.4
28	134.8, 135.1, 136.7

Plot a scattergram of these data and comment on what degree polynomial might prove appropriate for predicting yield from planting rate.

15.6   From the data of Problem 15.5, use orthogonal polynomials and determine what degree polynomial is appropriate here.

15.7   Find the equation of the polynomial in Problem 15.6.

15.8   From an ANOVA table of the results of Problem 15.6, find r^2, R^2, and η^2, and comment on what these statistics tell you.

15.9   Seasonal indexes of hog prices taken in two different years for five months gave the following results.

			Month		
Year	1	2	3	4	5
1	93.9	94.3	101.2	96.7	107.5
2	95.4	96.5	94.7	97.1	107.2

a. Sketch a scatterplot of these data and comment on what type of curve might be used to predict hog prices when the month is known.

b. Do a complete analysis of these data and find the equation of the best-fitting curve for predicting hog price from month.

15.10   A bakery is interested in the effect of six baking temperatures on the quality of its cakes. The bakers agree on a measure of cake quality, Y. A large batch of cake mix is prepared and 36 cakes poured. The baking temperature is assigned at random to the cakes such that six cakes are baked at each temperature. Results are as follows:

Temperature					
175°	185°	195°	205°	215°	225°
22	4	10	22	32	20
10	21	22	14	27	36
18	18	18	21	22	33
26	28	32	25	37	33
21	21	28	26	27	20
21	28	25	25	31	25

Do a complete analysis of these data and write the "best" fitting model for predicting cake quality from temperature.

15.11 Compute statistics that will indicate how good your fit is in Problem 15.10 and comment.

15.12 Data on the effect of age on retention of information gave

Ages:	6–14	15–23	24–32	33–41
Total score of 10 people, $T_{.j}$:	80	92	103	90

Source	df	SS	MS
Between ages	3	26.675	8.892
Linear	1	8.405	8.405
Quadratic	1	15.625	15.625
Cubic	1	2.645	2.645
Error	36	72.000	2.000
Total	39	98.675	

a. Determine the significance of each prediction equation and state which degree equation (linear, quadratic, or cubic) will adequately predict retention score from age.
b. Write the equation for part (a).
c. Find the departure from regression of the mean retention score of a 37-year-old based on your answer to part (b).

15.13 A study of the effect of chronological age on history achievement scores gave the following results.

History Achievement	Chronological Age				
Score Coded by $\dfrac{Y - 18}{15}$	8	9	10	11	12
5			1	6	5
4		1	7	4	5
3		6	2		
2	3	2			
1	6	1			
0	1				

Do a complete analysis justifying your curve of best fit.

15.14 In studying the effect of the distance a road sign stands from the edge of the road on the amount by which a driver swerves from the edge of the road, five distances were chosen. Then the five distances were set at random, observing four cars at each set distance. The results gave

$$\eta^2 = 0.70$$
$$r^2 = 0.61$$

a. Test whether or not a linear function will "fit" these data. (You may assume a total SS of 10,000 if you wish, but this is not necessary.)

b. In fitting the proper curve, what are you assuming about the four readings at each specified distance?

15.15 Thickness of a film is studied to determine the effect of two types of resin and three gate settings on the thickness. Results showed:

	Gate Setting (mm)		
Resin Type	2	4	6
1	1.5	2.5	3.6
	1.3	2.5	3.8
2	1.4	2.6	3.7
	1.3	2.4	3.6

Do an analysis of variance of these data following the methods of Chapter 5.

15.16 For Problem 15.15, extract a linear and quadratic effect of gate setting and test for significance. Also partition the interaction and test its components.

15.17 Write any appropriate prediction equation or equations from the data of Problem 15.15.

15.18 If the experiment of Problem 15.15 is extended to include a third factor, weight fraction at three levels, the data are as on page 381. Compile an ANOVA table for this three-factor experiment as in Chapter 5.

Gate Setting (mm)	Resin Type					
	1			**2**		
	Weight Fraction			Weight Fraction		
	0.20	0.25	0.30	0.20	0.25	0.30
2	1.6	1.5	1.5	1.5	1.4	1.6
	1.5	1.3	1.3	1.4	1.3	1.4
4	2.7	2.5	2.4	2.4	2.6	2.2
	2.7	2.5	2.3	2.3	2.4	2.1
6	3.9	3.6	3.5	4.0	3.7	3.4
	4.0	3.8	3.4	4.0	3.6	3.3

15.19 Outline a further breakdown of the ANOVA table in Problem 15.18 based on how the independent variables are set. Show the proper degrees of freedom.

15.20 For the significant ($\alpha = 0.05$) effects in Problem 15.18 complete the ANOVA breakdown suggested in Problem 15.19.

15.21 Graph any significant effects found in Problem 15.20 and discuss.

15.22 Data on the effect of knife-edge radius R in inches and feedroll force F in pounds per inch on the energy necessary to cut 1-inch lengths of alfalfa are given as follows:

Feedroll Force F_j (lb/in)	Knife-Edge Radius, R_i (inches)				$T_{.j.}$
	0.000	0.005	0.010	0.015	
5	29	98	44	84	
	30	128	81	100	
	20	67	77	63	
	79	293	202	247	821
10	22	35	53	103	
	26	80	93	90	
	16	29	59	98	
	64	144	205	291	704
15	18	49	58	80	
	17	68	103	91	
	11	61	128	77	
	46	178	289	248	761
20	38	68	87	86	
	31	74	116	113	
	21	47	90	81	
	90	189	293	280	852
$T_{i..}$	279	804	989	1066	$T_{...} = 3138$

Write the mathematical model for this experiment and do an ANOVA on the two factors and their interaction.

15.23 Outline a further ANOVA based on Problem 15.22 with single-degree-of-freedom terms.

15.24 Do an analysis for each of the terms in Problem 15.23.

15.25 Plot some results of Problem 15.22 and try to argue that they confirm some results found in Problem 15.24.

15.26 The sales volume Y of a product in thousands of dollars, price X_1 per unit in dollars, and advertising expense X_2 in hundreds of dollars were recorded for $n = 8$ cases. The results are as follows:

	Case							
	1	2	3	4	5	6	7	8
Y	10.1	6.5	5.1	11.0	9.9	14.7	4.8	12.2
X_1	1.3	1.9	1.7	1.5	1.6	1.2	1.6	1.4
X_2	8.8	7.1	5.5	13.8	18.5	9.8	6.4	10.2

a. Find the means and standard deviations of each variable and the intercorrelation matrix.

b. On the basis of the discussion in Section 15.2 find the regression equation Y from X_1 alone.

15.27 In studying the percent conversion (Y) in a chemical process as a function of time in hours (X_1) and average fusion temperature in degrees Celsius (X_2) the following data were collected.

Y	X_1	X_2
62.7	3	297.5
76.2	3	322.5
80.8	3	347.5
80.8	6	297.5
89.2	6	322.5
78.6	6	347.5
90.1	9	297.5
88.0	9	322.5
76.1	9	347.5

Analyze these data using multiple linear regression. Re-examine these data and comment on any peculiarities noted.

15.28 A management consulting firm attempted to predict annual salary of executives from the executives' years of experience (X_1), years of education (X_2), sex (X_3), and number of employees supervised (X_4). A sample of 25 executives gave an average salary of \$29,700 with a standard deviation of \$1300. From a computer program the following statistics were recorded.

$$r_{Y4}^2 = 0.42$$
$$R_{Y.24}^2 = 0.78$$
$$R_{Y.124}^2 = 0.90$$
$$R_{Y.1234}^2 = 0.95$$

a. Explain how you would interpret these statistics.
b. Test whether one could stop after variables 2 and 4 have been entered into the equation.
c. If variables 1, 2, and 4 were used in the prediction equation, what would the limits on the salaries expected for a predicted salary have to be in order to be correct about 95 percent of the time?

15.29 Consider a prediction situation in which some dependent variable Y is to be predicted from four independent variables X_1, X_2, X_3, and X_4 and assume you have a stepwise regression printout on this problem involving 25 observations. Explain briefly each of the following with respect to this problem and its printout.

a. How does the computer decide which independent variable to enter first into a regression equation?
b. How does it decide which variable to enter next in the equation?
c. How do you decide when to stop adding variables based on the printout?
d. Do you need to add any more variables if $R_{Y.1234}^2 = 0.7$ and $R_{Y.24}^2 = 0.6$?

15.30 A study was made to determine the effect of four variables on the grade-point average of 55 high school students who attended a junior college after graduation. These variables were high school rank (X_1), SAT score (X_2), IQ (X_3), and age in months (X_4). From the computer printout the following statistics were recorded.

$$R_{Y.1234}^2 = 0.45 \qquad R_{Y.12}^2 = 0.33$$
$$R_{Y.123}^2 = 0.42 \qquad r_{Y1}^2 = 0.19$$

Use this information to test for the significance of each added variable and determine whether or not an ordered subset of these four will make a satisfactory prediction. If you wish, you may assume a total of sum of squares in this study of 10,000 units. Also explain how you would predict with approximate 2 to 1 odds the maximum GPA expected for an individual student based on his or her scores of the variables in the appropriate equation.

15.31 In a study of several variables that might affect freshman GPA (Y) a sample of 55 students reported their college entrance arithmetic test (X_1), their analogies test score on an Ohio battery (X_2), their high school average (X_3), and their interest score on an interest inventory (X_4). Results of this study are given in the following (oversimplified) ANOVA table.

Source*	df	SS	MS
Due to X_2	1	1360	1360
Due to X_3 / X_2	1	480	480
Due to $X_1 / X_2, X_3$	1	80	80
Due to $X_4 / X_1, X_2, X_3$	1	80	80
Error	50	2000	40
Total	54	4000	

*/means "given."

Assuming a stepwise procedure was used so that variables were entered in their order of importance, answer the following.

a. Determine $R_{Y.1234}$ and explain its meaning in this problem.

b. Determine which of the four variables are sufficient to predict GPA (Y) and show that your choice is sufficient.

c. If a regression equation based on your answer to (b) is used to predict student A's GPA based on test scores, how close do you think this prediction will be to the actual GPA? Justify your answer and note any assumptions you are making.

15.32 In an attempt to predict the average value per acre of farm land in Iowa in 1920 the following data were collected.

Y = average value in dollars per acre

X_1 = average corn yield in bushels per acre for ten preceding years

X_2 = percentage of farm land in small grain

X_3 = percentage of farm land in corn

Y	X_1	X_2	X_3	Y	X_1	X_2	X_3
87	40	11	14	193	41	13	28
133	36	13	30	203	38	24	31
174	34	19	30	279	38	31	35
385	41	33	39	179	24	16	26
363	39	25	33	244	45	19	34
274	42	23	34	165	34	20	30
235	40	22	37	257	40	30	38
104	31	9	20	252	41	22	35
141	36	13	27	280	42	21	41
208	34	17	40	167	35	16	23
115	30	18	19	168	33	18	24
271	40	23	31	115	36	18	21
163	37	14	25				

Analyze these data using multiple linear regression and comment on the results.

15.33 In a study involving handicapped students an attempt was made to predict GPA from two demographic variables, sex and ethnicity, and two independent measured variables, interview score and contact hours used in counseling and/or tutoring. A dummy variable for sex was taken as 1 = male, 2 = female. For ethnicity: 1 = African-American, 2 = Hispanic, and 3 = white, non-Hispanic. Do a complete analysis of the data below and justify the regression equation that will adequately "fit" the data.

Students	Ethnic	Sex	Interview	Hours	GPA
1	1	2	11.0	4.0	5.50
2	1	2	10.0	5.0	4.10
3	1	2	12.0	73.0	5.00
4	1	2	11.5	68.0	4.22
5	1	2	10.8	82.0	5.00
6	1	1	12.5	72.5	5.00
7	1	1	9.5	64.0	4.60
8	1	1	9.5	78.0	4.25
9	1	1	8.0	64.0	4.00
10	1	1	7.5	13.0	2.00
11	2	2	9.0	37.0	4.25
12	2	2	8.2	4.0	4.00
13	2	2	10.7	38.5	4.61
14	2	2	8.5	3.0	2.93
15	2	2	12.5	10.5	5.50
16	2	1	12.0	80.0	4.77
17	2	1	12.2	6.0	5.00
18	2	1	7.0	6.5	3.25
19	2	1	8.6	22.0	2.66
20	2	1	8.3	28.5	3.37
21	3	2	10.9	12.0	5.00
22	3	2	9.0	9.0	4.00
23	3	2	10.0	5.0	5.00
24	3	2	7.2	12.0	3.87
25	3	2	8.5	4.0	3.00
26	3	1	10.0	8.0	4.77
27	3	1	8.5	8.0	5.00
28	3	1	10.0	22.0	5.08
29	3	1	11.4	61.5	5.57
30	3	1	11.9	37.0	6.00

15.34 In a study on the effect of several variables on posttest achievement scores in biology in a rural high school involving 48 students the following table was presented. It included variables that had a significant effect on achievement at the 10 percent level of significance.

Step	Variable Entered	Multiple R	R^2	R^2 Change
1	Figural elaboration pretest	0.46587	0.21704	0.21704
2	Achievement pretest	0.83022	0.68927	0.47223
3	Piagetian pretest	0.86536	0.74885	0.05958
4	Attitudes pretest	0.90606	0.82094	0.07209

Discuss these results and test whether or not the last two variables were necessary.

15.35 To study the effect of planting rate on yield in an agricultural problem, points were taken with the following results.

$$X \quad 12 \quad 12 \quad 16 \quad 16 \quad 16 \quad 20 \quad 20$$
$$Y \quad 130.5 \quad 129.6 \quad 142.5 \quad 140.3 \quad 143.4 \quad 144.8 \quad 144.1$$

$$X \quad 23 \quad 23 \quad 23 \quad 23 \quad 28 \quad 28$$
$$Y \quad 145.1 \quad 147.8 \quad 146.6 \quad 148.4 \quad 134.8 \quad 135.1$$

a. Plot a scattergram and note what type of curve you might expect of Y versus X.
b. Taking $X^2 = X_2$ and $X = X_1$, use multiple regression to find the equation of the best fitting quadratic.
c. Compare this quadratic with the best linear fit.

15.36 The printout of a computer program yielded the following statistics.

$$B_0 = 140.27 \qquad r_{y1}^2 = 0.0907$$
$$B_1 = 1.24 \qquad R_{y.12}^2 = 0.3285$$
$$B_2 = 2.65 \qquad R_{y.123}^2 = 0.4211$$
$$B_3 = -0.30 \qquad R_{y.1234}^2 = 0.4548$$
$$B_4 = -1.27 \qquad S_y = 16.0$$
$$N = 45$$

a. Explain the meaning of the R squares given above.
b. Test and determine which regression model is adequate for predicting Y.
c. Compute the standard error of estimate for your model in (b).
d. Explain whether or not the program was a stepwise (step-up) program.

Miscellaneous Topics $\qquad$ 16

16.1 Introduction

Several techniques have been developed, based on the general design principles presented in the first 15 chapters of this book and on other well-known statistical methods. No attempt will be made to discuss these in detail, as they are presented very well in the references. It is hoped that a background in experimental design as given in the first 15 chapters will make it possible for the experimenter to read and understand the references.

The methods to be discussed are covariance analysis, response surface experimentation, evolutionary operation, analysis of attribute data, incomplete block design, and Youden squares.

16.2 Covariance Analysis

Philosophy

Occasionally, when a study is being made of the effect of one or more factors on some response variable, say Y, there is another variable, or variables, that vary along with Y. It is often not possible to control this other variable (or variables) at some constant level throughout the experiment, but the variable can be measured along with the response variable. This variable is referred to as a *concomitant variable* X as it "runs along with" the response variable Y. In order, then, to assess the effect of the treatments on Y, one should first attempt to remove the effects of this concomitant variable X. This technique of removing the effect of X (or several X's) on Y and then analyzing the residuals for the effect of the treatments on Y is called *covariance analysis*.

Several examples of the use of this technique might be cited. If one wishes to study the effect on student achievement of several teaching methods, it is customary to use the difference between a pretest and a posttest score as the response variable. It may be, however, that the gain Y is also affected by the student's pretest score X as gain, and pretest scores may be correlated. Covariance analysis will provide for an adjustment in the gains due to differing pretest scores assuming some type of regression of gain on pretest scores. The advantage of using this technique is that one is not limited to running an experiment on only those pupils who have approximately the same pretest scores or matching pupils with the same pretest scores and randomly assigning them to control and experimental groups. Covariance analysis has the effect of providing "handicaps" as if each student had the same pretest scores while actually letting the pretest scores vary along with the gains.

This technique has often been used to adjust weight gains in animals by their original weights in order to assess the effect of certain feed treatments on these gains. Many industrial examples might be cited such as the one given below in which original weight of a bracket may affect the weight of plating applied to the bracket by different methods.

Snedecor [26] gives several examples of the use of covariance analysis. Ostle [22] is an excellent reference on the extension of covariance analysis to many different experimental designs. The *Biometrics* article of 1957 [5] devotes most of its pages to a discussion of many of the fine points of covariance analysis. The problem discussed below will illustrate the technique for the case of a single-factor experiment run in a completely randomized design. Because the covariance technique represents a marriage between regression and the analysis of variance, the methods of Chapter 3 and Chapter 15 will be used.

■ **EXAMPLE 16.1**   Several steel brackets were sent to three different vendors to be zinc plated. The chief concern in this process is the thickness of the zinc plating and whether or not there was any difference in this thickness between the three vendors. Data on this thickness in hundred thousandths of an inch (10^{-5}) for four brackets plated by the three vendors are given in Table 16.1.

Table 16.1 Plating Thickness in 10^{-5} Inches from Three Vendors

| | Vendor | |
A	B	C
40	25	27
38	32	24
30	13	20
47	35	13

Assuming a single-factor experiment and a completely randomized design, an analysis of variance model for this experiment would be

$$Y_{ij} = \mu + \tau_j + \varepsilon_{ij} \tag{16.1}$$

where Y_{ij} is the observed thickness of the ith bracket from the jth vendor, μ is a common effect, τ_j represents the vendor effect, and ε_{ij} represents the random error. Considering vendors as a fixed factor and random errors as normally and independently distributed with mean zero and common variance σ_ε^2, an ANOVA table was compiled (Table 16.2).

Table 16.2 ANOVA on Plating Thickness

Source	df	SS	MS
Between vendors	2	665.2	332.6
Within vendors	9	543.5	60.4
Totals	11	1208.7	

The F statistic equals 5.51 and is significant at the 5 percent significance level with 2 and 9 df. One might conclude, therefore, that there is a real difference in average plating thickness among these three vendors, and steps should be taken to select the most desirable vendor.

During a discussion of these results, it was pointed out that some of this difference among vendors might be due to unequal thickness in the brackets before plating. In fact, there might be a correlation between the thickness of the brackets before plating and the thickness of the plating. To see whether or not such an idea is at all reasonable, a scatterplot was made on the thickness of the bracket before plating X and the thickness of the zinc plating Y. Results are shown in Figure 16.1.

A glance at this scattergram would lead one to suspect that there is a positive correlation between the thickness of the bracket and the plating thickness.

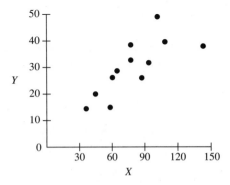

Figure 16.1 Scattergram of bracket thickness X versus plating thickness Y in 10^{-5} inches.

Because another variable, X, can be measured on each piece and may be related to the variable of interest Y, a covariance analysis may be run in order to remove the linear effect of X on Y when comparing vendors. A covariance analysis is used to remove the linear or higher order relationship of one or more independent variables from a dependent variable in order to assess the effect of a factor on the dependent variable.

Covariance

Since a positive correlation is suspected between X and Y above, a regression model might be written as

$$Y_{ij} = \mu + \beta(X_{ij} - \bar{X}) + \varepsilon_{ij} \tag{16.2}$$

where β is the true linear regression coefficient (or slope) between Y and X over all the data; $\bar{X}$ is the mean of the X values. In such a model it is assumed that a measure of variable X can be made on each unit along with a corresponding measure of Y. For a covariance analysis, models (16.1) and (16.2) can be combined to give the covariance model

$$Y_{ij} = \mu + \beta(X_{ij} - \bar{X}) + \tau_j + \varepsilon_{ij} \tag{16.3}$$

In this model the error term ε_{ij} should be smaller than in model (16.1) because of the removal of the effect of the covariate X_{ij}.

To determine how to adjust the analysis of variance to provide for the removal of this covariate, consider the regression model (16.2) in deviation form

$$y = bx + e \tag{16.4}$$

where x and y are now deviations from their respective means ($x = X_{ij} - \bar{X}$, $y = Y_{ij} - \bar{Y}$) and b is the sample slope. By the method of least squares it will be recalled that

$$b = \frac{\sum xy}{\sum x^2}$$

where summations are over all points. The sum of the squares of the errors of estimate, which have been minimized, is now, from Equation (16.4),

$$e = y - bx$$

$$\sum e^2 = \sum (y - bx)^2$$

$$= \sum y^2 - 2b \sum xy + b^2 \sum x^2$$

Substituting for b

$$b = \frac{\sum xy}{\sum x^2}$$

and

$$\sum e^2 = \sum y^2 - 2\frac{\sum xy}{\sum x^2} \cdot \sum xy + \left(\frac{\sum xy}{\sum x^2}\right)^2 \sum x^2$$

$$= \sum y^2 - \frac{(\sum xy)^2}{\sum x^2} \tag{16.5}$$

In this expression the term $(\sum xy)^2/\sum x^2$ is the amount of reduction in the sum of squares of the Y variable due to its linear regression on x. This term is then the sum of squares due to linear regression. It involves the sum of the cross products of the two variables $(\sum xy)$ and the sum of squares of the x variable alone $(\sum x^2)$. If a term of this type is subtracted from the sum of squares for the dependent variable y, the result will be a corrected or adjusted sum of squares for y.

The sums of squares and cross products can be computed on all the data (total), within each vendor or between vendors. In each case it will be helpful to recall that the sums of squares and cross products in terms of the original data are

$$\sum x^2 = \sum X^2 - \frac{(\sum X)^2}{N}$$

$$\sum y^2 = \sum Y^2 - \frac{(\sum Y)^2}{N}$$

$$\sum xy = \sum XY - \frac{(\sum X)(\sum Y)}{N}$$

Data on both variables appear with the appropriate totals in Table 16.3.

Table 16.3 Bracket Thickness X and Plating Thickness Y in 10^{-5} Inches from Three Vendors

	Vendor							
	A		**B**		**C**		**Total**	
	X	Y	X	Y	X	Y	X	Y
	110	40	60	25	62	27		
	75	38	75	32	90	24		
	93	30	38	13	45	20		
	97	47	140	35	59	13		
Totals	375	155	313	105	256	84	944	344

From Table 16.3 the sums of squares and cross products for the total data are[1]

$$T_{xx} = (110)^2 + (75)^2 + \cdots + (59)^2 - \frac{(944)^2}{12} = 9240.7$$

$$T_{yy} = (40)^2 + (38)^2 + \cdots + (13)^2 - \frac{(344)^2}{12} = 1208.7$$

$$T_{xy} = (110)(40) + (75)(38) + \cdots + (59)(13) - \frac{(944)(344)}{12} = 2332.7$$

The between-vendors sums of squares and cross products are computed by vendor totals as in an ANOVA:

$$V_{xx} = \frac{(375)^2 + (313)^2 + (256)^2}{4} - \frac{(944)^2}{12} = 1771.2$$

$$V_{yy} = \frac{(155)^2 + (105)^2 + (84)^2}{4} - \frac{(344)^2}{12} = 665.2$$

$$V_{xy} = \frac{(375)(155) + (313)(105) + (256)(84)}{4} - \frac{(944)(344)}{12} = 1062.2$$

By subtraction the error sums of squares are

$$E_{xx} = T_{xx} - V_{xx} = 9240.7 - 1771.2 = 7469.5$$
$$E_{yy} = T_{yy} - V_{yy} = 1208.7 - 665.2 = 543.5$$
$$E_{xy} = T_{xy} - V_{xy} = 2332.7 - 1062.2 = 1270.5$$

Using this information, the sums of squares of the dependent variable Y may be adjusted for regression on X. On the totals, the adjusted sum of squares is then

$$\text{adjusted } \sum y^2 = T_{yy} - \frac{T_{xy}^2}{T_{xx}} \quad \text{as in Equation (16.5)}$$

$$= 1208.7 - \frac{(2332.7)^2}{9240.7} = 619.8$$

and the adjusted sum of squares within vendors is

$$\text{adjusted } \sum y^2 = E_{yy} - \frac{(E_{xy})^2}{E_{xx}}$$

$$= 543.5 - \frac{(1270.5)^2}{7469.5} = 327.4$$

Then, by subtraction, the adjusted sum of squares between vendors is

$$619.8 - 327.4 = 292.4$$

These results are usually displayed as in Table 16.4.

---

[1] Here T, V, and E are used to denote sum of squares and cross products for totals, vendors, and error, respectively.

Table 16.4 Analysis of Covariance for Bracket Data

| | | SS and Products | | | Adjusted | | |
Source	df	$\sum y^2$	$\sum xy$	$\sum x^2$	$\sum y^2$	df	MS
Between vendors	2	665.2	1062.2	1771.2	—	—	—
Within vendors	9	543.5	1270.5	7469.5	327.4	8	40.9
Totals	11	1208.7	2332.7	9240.7	619.8	10	
Between vendors					292.4	2	146.2

With the adjusted sums of squares, the new F statistic is

$$F = \frac{146.2}{40.9} = 3.57$$

with 2 and 8 df. This is now not significant at the 5 percent significance level. This means that after removing the effect of bracket thickness, the vendors no longer differ in the average plating thickness on the brackets.

Table 16.4 should have a word or two of explanation. The degrees of freedom on the adjusted sums of squares are reduced by 2 instead of 1 as estimates of both the mean and the slope are necessary in their computation. Adjustments are made here on the totals and the within sums of squares rather than on the between sum of squares as we are interested in making the adjustment based on an overall slope of Y on X and a within-group average slope of Y on X as explained in more detail later.

In this analysis three different types of regression can be identified: the "overall" regression of all the Y's on all the X's, the within-vendors regression, and the regression of the three Y means on the three X means. This last regression could be quite different from the other two and is of little interest since we are attempting to adjust the Y's and the X's within the vendors. An estimate of this "average within vendor" slope is

$$b = \frac{\sum xy}{\sum x^2} = \frac{E_{xy}}{E_{xx}} = \frac{1270.5}{7469.5} = 0.17$$

If this slope or regression coefficient is used to adjust the observed Y means for the effect of X on Y, one finds

$$\text{adjusted } \overline{Y}_{.j} = \overline{Y}_{.j} - b(\overline{X}_{.j} - \overline{X}_{..})$$

For vendor A,

$$\text{adjusted } \overline{Y}_{.1} = \frac{155}{4} - (0.17)\left(\frac{375}{4} - \frac{944}{12}\right)$$
$$= 38.75 - (0.17)(93.75 - 78.67) = 36.19$$

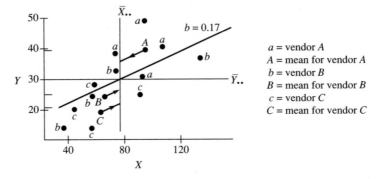

Figure 16.2 Plating thickness Y versus bracket thickness X plotted by vendors.

For vendor B,

$$\text{adjusted } \bar{Y}_{.2} = 26.25 - (0.17)(78.25 - 78.67) = 26.31$$

For vendor C,

$$\text{adjusted } \bar{Y}_{.3} = 21.00 - (0.17)(64.00 - 78.67) = 23.49$$

A comparison of the last column above with the column of unadjusted means just to the right of the equal signs shows that these adjusted means are closer together than the unadjusted means, which is confirmed in the preceding analysis. This can be seen graphically by plotting each vendor's data in a different symbol and "sliding" the means along lines parallel to this regression slope (Figure 16.2).

At the left-hand ordinate in Figure 16.2, the three unadjusted Y means are marked off. By moving these three means along lines parallel to the general slope ($b = 0.17$), a comparison is seen between the three means (on the $\bar{X}_{..}$ ordinate) after adjustment. This shows graphically the results of the covariance analysis. By examining each group of observations for a given vendor, it is seen that all three slopes within vendors are in the same direction and of similar magnitude. By considering all 12 points, the overall slope can be visualized. On the other hand, if the three means (indicated by the capital letters) are considered, the slope of these $\bar{Y}$'s on their corresponding $\bar{X}$'s is much steeper. Hence adjustments are made on the basis of the overall slope and the "average" or "pooled" within-groups slope.

One of the problems in covariance analysis is that there are many assumptions made in its application. These should be examined in some detail.

First, there are the usual assumptions of analysis of variance on the dependent variable Y: an additive model, normally and independently distributed error, and homogeneity of variance within the groups. This last assumption is often tested with Barlett's or a similar test. In this problem, since the ranges of the three vendor readings are 17, 22, and 14, the assumption seems tenable.

In covariance analysis it is further assumed that the regression is linear, that the slope is not zero (covariance was necessary), that the regression coefficients within each group are homogeneous so that the "average" or "pooled" within-groups regression can be used on all groups, and finally that the independent variable X is not affected by the treatments given to the groups.

The linearity assumption may be tested if the experiment is planned in such a way that there is more than one Y observation for each X. Since this was not done here, a look at the scattergram will have to suffice for this linearity assumption.

To test the hypothesis that the true slope β in Equation (16.3) is zero, consider whether or not the reduction in the error sum of squares is significant when compared with the error sum of squares. By an F test, the ratio would be

$$F_{1,\,N-k-1} = \frac{E_{xy}^2 / E_{xx}}{\text{adjusted } E_{yy} / (N - k - 1)}$$

$$F_{1,\,8} = \frac{216.1}{327.4 / 8} = \frac{216.1}{40.9} = 5.28$$

As the 5 percent F is 5.32, this result is nearly significant and certainly casts considerable doubt on the hypothesis that $\beta = 0$. Since the results after co-variance adjustment were not significant as compared with the results before adjustment, it seems reasonable to conclude that covariance was necessary.

To test the hypothesis that all three regression coefficients are equal, let us compute each sample coefficient and adjust the sum of squares within each group by its own regression coefficient.

For vendor A,

$$b_A = \frac{\sum xy}{\sum x^2} = \frac{\sum XY - (\sum X)(\sum Y)/n}{\sum X^2 - (\sum X)^2/n}$$

Within group A,

$$b_A = \frac{14,599 - 58,125/4}{35,783 - 140,625/4} = 0.108$$

$$\text{adjusted } \sum y^2 = \sum y^2 - \frac{(\sum xy)^2}{\sum x^2} = 146.75 - 7.32 = 139.43$$

For vendor B,

$$b_B = \frac{9294 - 32,865/4}{30,269 - 24,492.25/4} = 0.187$$

$$\text{adjusted } \sum y^2 = 286.75 - 201.72 = 85.03$$

For vendor C,

$$b_C = \frac{5501 - 21,504/4}{17,450 - 65,536/4} = 0.117$$

adjusted $\sum y^2 = 110.00 - 14.66 = 95.34$

Summarizing, we obtain Table 16.5.

Table 16.5 Adjusted Sums of Squares Within Each Vendor

Within Vendor	df	$\sum y^2$	b	Adjusted $\sum y^2$	df
A	3	146.75	0.108	139.43	2
B	3	286.75	0.187	85.03	2
C	3	110.00	0.117	95.34	2
Totals	9	543.50		319.80	6

Adjusting each within-vendor sum of squares by its own regression co-efficient reduces the degree of freedom to 2 per vendor. Adding the new adjusted sums of squares, Table 16.5 shows a within-vendor sum of squares of 319.80 based on 6 df. When the within-vendor sum of squares was adjusted by a "pooled" within-vendor regression, Table 16.4 showed an adjusted sum of squares within vendors of 327.4 based on 8 df. If there were significant differences in regression within the three vendors, this would show as a difference in these two figures: $327.4 - 319.8 = 7.6$. An F test for this hypothesis would then be

$$F_{k-1,\,N-2k} =$$
$$\frac{\left[\begin{array}{l} \text{adjusted } \sum y^2 \text{ (based on pooled within-groups regression)} \\ -\text{adjusted } \sum y^2 \text{ (based on regressions within each group)} \end{array}\right] \Big/ (k-1)}{\left[\text{adjusted } \sum y^2 \text{ (within each group)}\right] / (N - 2k)}$$

and

$$F_{2,\,6} = \frac{(327.4 - 319.8)/2}{319.8/6} = \frac{3.8}{53.3} = < 1$$

hence nonsignificant, and we conclude that the three regression coefficients are homogeneous.

The final assumption that the vendors do not affect the covariate X is tenable here on practical grounds as the vendor has nothing to do with the bracket thickness before plating. However, in some applications this assumption needs to be checked by an F test on the X variable. For our data from Table 16.4 such an F test would give

$$F_{2,9} = \frac{1771.2/2}{7469.5} = \frac{885.6}{829.9} = 1.07$$

which is obviously nonsignificant.

From this discussion of the assumptions, one notes that it is often a problem in the use of covariance analysis to satisfy all of the assumptions. This may be the reason many people avoid covariance although it is applicable whenever one suspects that another measured variable or variables affect the dependent variable.

The above techniques may be extended to handle several covariates, non-linear regression, and several factors and interactions of interest. In handling randomized blocks, Latin squares, or experiments with several factors, Ostle [22] points out that the proper technique is to add the sum of squares and sum of cross products of the term of interest to the sum of squares and sum of cross products of the error, adjust this total, then adjust the error and determine the adjusted treatment effect by subtraction.

When several covariates are involved, many questions arise concerning how many covariates can be handled, how many really affect the dependent variable, how a significant subset can be found, and so on. All of these questions are of concern in any multiple regression problem and some discussion of these and other problems can be found in [5].

SAS Programs

Example 16.1 can be easily analyzed by using SAS as shown below:

```
DATA A;
INPUT VENDOR $ X Y      @@;
CARDS;
A 110 40 A 75 38 A 93 30 A 97 47
B 60 25 B 75 32 B 38 13 B 140 25
C 62 27 C 90 24 C 45 20 C 59 13
;
TITLE'COVARIANCE EXAMPLE';
PROC PRINT;
PROC GLM;
CLASS VENDOR;
MODEL Y=VENDOR X/SOLUTION;
LSMEANS VENDOR/STDERR PDIFF;
```

COVARIANCE ANALYSIS

1. The Type I SS for VENDOR gives the between—lot sums of squares that would be obtained for the ANOVA model Y = VENDOR.

2. Type III SS for VENDOR gives the VENDOR SS ''adjusted'' for the covariate.

3. The LSMEANS printed are the same as adjusted means (means adjusted for the covariate).

4. The STDERR option on the LSMEANS statement causes the standard error of the least squares means and the probability of getting a large |t| value under the hypothesis Ho: LSM=0 to be printed.

5. Specifying the PDIFF option causes all probability values for the hypothesis Ho: LSM(i) = LSM(j) to be printed.

6. The Solution option to the model statement asks GLM to print a solution to the normal equations. GLM doesn't print a solution when a CLASS statement is used unless asked for by the Solution option.

7. The estimate for the covariate X is the ''average within VENDOR'' slope which is tested Ho: slope=0.

COVARIANCE EXAMPLE
GENERAL LINEAR MODELS PROCEDURE

DEPENDENT VARIABLE: Y

SOURCE	DF	SUM OF SQUARES	MEAN SQUARE	F VALUE	PR > F	R-SQUARE	C.V.
MODEL	3	881.26817949	293.75605983	7.18	0.00117	0.729124	22.3160
ERROR	8	327.39848718	40.92481090			ROOT MSE	Y MEAN
CORRECTED TOTAL	11	1208.66666667				6.39725026	28.66666667

SOURCE	DF	TYPE I SS	F VALUE	PR > F	TYPE III SS	F VALUE	PR > F
VENDOR	2	665.16666667	8.13	0.0118	292.42172781	3.57	0.0778
X	1	216.10151282	5.28	0.0506	216.10151282	5.28	0.0506

PARAMETER	ESTIMATE	T FOR H0: PARAMETER=0	PR > |T|	STD ERROR OF ESTIMATE
INTERCEPT	10.11413080 B	1.77	0.1148	5.71601862
VENDOR A	12.68977174 B	2.52	0.0357	5.03106293
B	2.82619319 B	0.61	0.5598	4.64488634
C	0.00000000 B	.	.	.
X	0.17009171	2.30	0.0506	0.07401974

COVARIANCE EXAMPLE
GENERAL LINEAR MODELS PROCEDURE
LEAST SQUARES MEANS

VENDOR	Y LSMEAN	STD ERR LSMEAN	PROB > |T| H0:LSMEAN=0	I/J	PROB > |T| HO: LSMEAN(I)=LSMEAN(J) 1	2	3
A	36.1844501	3.3878748	0.0001	1	.	0.0675	0.0357
B	26.3208715	3.1987738	0.0001	2	0.0675	.	0.5598
C	23.4946784	3.3778366	0.0001	3	0.0357	0.5598	.

NOTE: TO ENSURE OVERALL PROTECTION LEVEL, ONLY PROBABILITIES ASSOCIATED WITH PRE-PLANNED COMPARISONS SHOULD BE USED.

■ **EXAMPLE 16.2** If there is more than one factor of interest, a covariance adjustment can be made on each factor sum of squares plus the error sum of squares. Using this adjusted sum of squares, covariance is applied to main effects and interaction as shown in the SAS program below:

```
DATA A;
INPUT A B X Y @@;
CARDS;
1 1 22 5 1 2 30 8
1 1 19 3 1 2 25 5
1 1 20 4 1 2 35 8
2 1 27 3 2 2 40 9
2 1 30 4 2 2 29 7
2 1 32 6 2 2 30 6
;
TITLE'COVARIANCE EXAMPLE';
PROC PRINT;
PROC GLM;
CLASS A B;
MODEL Y=A|B X/SOLUTION;
LSMEANS A B/STDERR PDIFF;
```

          COVARIANCE EXAMPLE

OBS	A	B	X	Y
1	1	1	22	5
2	1	2	30	8
3	1	1	19	3
4	1	2	25	5
5	1	1	20	4
6	1	2	35	8
7	2	1	27	3
8	2	2	40	9
9	2	1	30	4
10	2	2	29	7
11	2	1	32	6
12	2	2	30	6

COVARIANCE EXAMPLE
GENERAL LINEAR MODELS PROCEDURE

DEPENDENT VARIABLE: Y

SOURCE	DF	SUM OF SQUARES	MEAN SQUARE	F VALUE	PR > F	R-SQUARE	C.V.
MODEL	4	40.01336478	10.00334119	15.05	0.0015	0.895822	14.3881
ERROR	7	4.65330189	0.66475741		ROOT MSE		Y MEAN
CORRECTED TOTAL	11	44.66666667			0.81532657		5.66666667

SOURCE	DF	TYPE I SS	F VALUE	PR > F	DF	TYPE III SS	F VALUE	PR > F
A	1	0.33333333	0.50	0.5018	1	3.80390312	5.72	0.0480
B	1	27.00000000	40.62	0.0004	1	1.75390980	2.64	0.1483
A*B	1	0.00000000	0.00	1.0000	1	2.22532394	3.35	0.1100
X	1	12.68003145	19.07	0.0033	1	12.68003145	19.07	0.0033

PARAMETER		ESTIMATE	T FOR HO: PARAMETER=0	PR > |T|	STD ERROR OF ESTIMATE
INTERCEPT		-2.55110063 B	-1.10	0.3063	2.31163732
A	1	0.56525157 B	0.81	0.4439	0.69678035
	2	0.00000000 B	.	.	.
B	1	-2.00157233 B	-2.84	0.0249	0.70386960
	2	0.00000000 B	.	.	.
A*B	1 1	1.89701258 B	1.83	0.1100	1.03682434
	1 2	0.00000000 B	.	.	.
	2 1	0.00000000 B	.	.	.
	2 2	0.00000000 B	.	.	.
X		0.29952830	4.37	0.0033	0.06858187

LEAST SQUARES MEANS

A	Y LSMEAN	STD ERR LSMEAN	PROB > |T| HO:LSMEAN=0	PROB > |T| HO: LSMEAN1=LSMEAN2
1	6.42354560	0.39434573	0.0001	0.0480
2	4.90978774	0.39434573	0.0001	

B	Y LSMEAN	STD ERR LSMEAN	PROB > |T| HO:LSMEAN=0	PROB > |T| HO: LSMEAN1=LSMEAN2
1	5.14013365	0.40059122	0.0001	0.1483
2	6.19319969	0.40059122	0.0001	

16.3 Response-Surface Experimentation

Philosophy

The concept of a response surface involves a dependent variable Y, called the response variable, and several independent or controlled variables, $X_1, X_2, \ldots,$ X_k. If all of these variables are assumed to be measurable, the response surface can be expressed as

$$Y = f(X_1, X_2, \ldots, X_k)$$

For the case of two independent variables such as temperature X_1 and time X_2, the yield Y of a chemical process can be expressed as

$$Y = f(X_1, X_2)$$

This surface can be plotted in three dimensions, with X_1 on the abscissa, X_2 on the ordinate, and Y perpendicular to the $X_1 X_2$ plane. If the values of X_1 and X_2 that yield the same Y are connected, we can picture the surface with a series of equal-yield lines, or contours. These are similar to the contours of equal height on topographic maps and the isobars on weather maps. Some response surfaces might be of the types in Figure 16.3. Two excellent references for understanding response-surface experimentation are [10] and [13].

The Twofold Problem

The problem involved in the use of response-surface experimentation is twofold: (1) to determine, on the basis of one experiment, where to move in the next experiment toward the optimal point on the underlying response surface; (2) having located the optimum or near optimum of the surface, to determine the equation of the response surface in an area near this optimum point.

One method of experimentation that seeks the optimal point of the response surface might be the traditional one-factor-at-a-time method. As shown in Figure 16.3(a), if X_2 is fixed and X_1 is varied, we find the X_1 optimal (or near optimal) value of response Y at the fixed value of X_2. Having found this X_1 value, experiments can now be run at this fixed X_1, and the X_2 for optimal response can be found. In the case of the mound in Figure 16.3(a), this method would lead eventually to the peak of the mound or near it. However, this same method, when applied to a surface such as the rising ridge in Figure 16.3(c), fails to lead to the maximum point on the response surface. In experimental work the type of response surface is usually unknown, so that a better method is necessary if the optimum set of conditions is to be found for any surface.

The method developed by those who have worked in this area is called the *path of steepest ascent* method. The idea here is to run a simple experiment over a small area of the response surface where, for all practical purposes, the surface may be regarded as a plane. We then determine the equation of this plane and from it the direction we should take from this experiment in order to

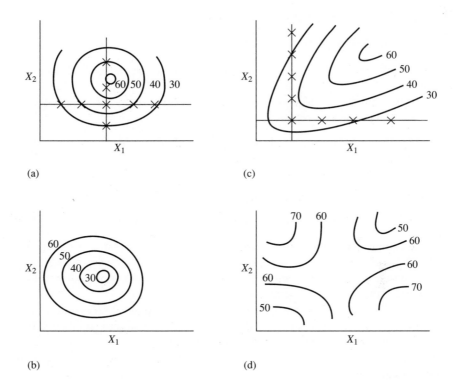

(a)

(c)

(b)

(d)

Figure 16.3 Some typical response surfaces in two dimensions: (a) mound; (b) depression; (c) rising ridge; (d) saddle.

move toward the optimum of the surface. Since the next experiment should be in a direction in which we hope to scale the height the fastest, this is referred to as the *path of steepest ascent*. This technique does not determine how far away from the original experiment succeeding sequential experiments should be run, but it does indicate to the experimenter the direction along which the next experiment should be performed. A simple example will illustrate this method.

To determine the equation of the response surface, several special experimental designs have been developed that attempt to approximate this equation using the smallest number of experiments possible. In two dimensions the simplest surface is a plane given by

$$Y = B_0 X_0 + B_1 X_1 + B_2 X_2 + \varepsilon \tag{16.6}$$

where Y is the observed response, X_0 is taken as unity, and estimates of the B's are to be determined by the method of least squares, which minimizes the sum of the squares of the errors ε. Such an equation is referred to as a first-order equation, since the power on each independent variable is unity.

If there is some evidence that the surface is not planar, a second-order equation in two dimensions may be a more suitable model

$$Y = B_0 X_0 + B_1 X_1 + B_2 X_2 + B_{11} X_1^2 + B_{12} X_1 X_2 + B_{22} X_2^2 + \varepsilon \qquad (16.7)$$

Here the $X_1 X_2$ term represents an interaction between the two variables X_1 and X_2.

If there are three independent or controlled variables, the first-order equation is again a plane or hyperplane.

$$Y = B_0 X_0 + B_1 X_1 + B_2 X_2 + B_3 X_3 + \varepsilon \qquad (16.8)$$

and the second-order equation is

$$Y = B_0 X_0 + B_1 X_1 + B_2 X_2 + B_3 X_3 + B_{11} X_1^2 + B_{22} X_2^2$$
$$+ B_{33} X_3^2 + B_{12} X_1 X_2 + B_{13} X_1 X_3 + B_{23} X_2 X_3 + \varepsilon \qquad (16.9)$$

As the complexity of the surface increases, more coefficients must be estimated, and the number of experimental points must necessarily increase. Several very clever designs have been developed that minimize the amount of work necessary to estimate these response-surface equations.

To determine the coefficients for these more complex surfaces and to interpret their geometric nature, both multiple regression techniques and the methods of solid analytical geometry are used. In the example that follows, only the simplest type of surface will be explored. The references quoted will give many more complex examples.

■ **EXAMPLE 16.3**   Consider an example in which an experimenter is seeking the proper values for both concentration of filler to epoxy resin X_1 and position in the mold X_2 to minimize the abrasion on a plastic die. This abrasion or wear is measured as a decrease in thickness of the material after 10,000 cycles of abrasion. Since the maximum thickness is being sought, the first experiment should attempt to discover the direction in which succeeding experiments should be run in order to approach this maximum by the steepest path. Assuming the surface to be a plane in a small area, the first experiment will be used to determine the equation of this plane. The response surface is then

$$Y = B_0 X_0 + B_1 X_1 + B_2 X_2 + \varepsilon$$

As there are three parameters to be estimated, B_0, B_1, and B_2, at least three experimental points must be taken to estimate these coefficients. Such a design might be an equilateral triangle, but since there are two factors, X_1 and X_2, each can be set at two levels and a 2^2 factorial may be used. Two concentrations were chosen, $\frac{1}{2}$:1 and 1:1 (ratio of filler to resin), and two positions, 1 inch and 2 inches from a reference point, with responses Y, the thickness of the material is 10^{-4} inches, as shown in Figure 16.4.

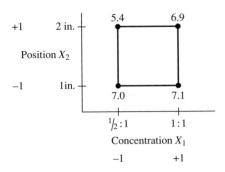

Figure 16.4 2^2 factorial example on response surface.

To determine the equation of the best fitting plane for these four points, consider the error in the prediction equation

$$\varepsilon = Y - B_0 X_0 - B_1 X_1 - B_2 X_2$$

The sum of the squares for this error is

$$\sum \varepsilon^2 = \sum (Y - B_0 X_0 - B_1 X_1 - B_2 X_2)^2$$

where the summation is over all points given in the design. To find the B's, we differentiate this expression with respect to each parameter and set these three expressions equal to zero. This provides three least squares normal equations, which can be solved for the best estimates of the B's. These estimates are designated as b's.

Differentiating the expression above gives

$$\frac{\partial (\sum \varepsilon^2)}{\partial B_0} = -2 \sum (Y - B_0 X_0 - B_1 X_1 - B_2 X_2) X_0 = 0$$

$$\frac{\partial (\sum \varepsilon^2)}{\partial B_1} = -2 \sum (Y - B_0 X_0 - B_1 X_1 - B_2 X_2) X_1 = 0$$

$$\frac{\partial (\sum \varepsilon^2)}{\partial B_2} = -2 \sum (Y - B_0 X_0 - B_1 X_1 - B_2 X_2) X_2 = 0$$

and from these results we get

$$\sum X_0 Y = b_0 \sum X_0^2 + b_1 \sum X_0 X_1 + b_2 \sum X_0 X_2$$
$$\sum X_1 Y = b_0 \sum X_0 X_1 + b_1 \sum X_1^2 + b_2 \sum X_1 X_2$$
$$\sum X_2 Y = b_0 \sum X_0 X_2 + b_1 \sum X_1 X_2 + b_2 \sum X_2^2 \qquad (16.10)$$

as the least squares normal equations.

By a proper choice of experimental variables, it is possible to reduce these equations considerably for simple solution. More complex models can be solved best by the use of matrix algebra. For the 2^2 factorial, the following coding scheme simplifies the solution of Equation (16.10). Set

$$X_1 = 4C - 3 \quad \text{where } C \text{ is the concentration}$$
$$X_2 = 2P - 3 \quad \text{where } P \text{ is the position}$$

X_0 is always taken as unity. For the experimental variables X_1 and X_2, the responses can be recorded as in Table 16.6. These experimental variables are also indicated on Figure 16.4.

Table 16.6 Orthogonal Layout for
Figure 16.4

Y	X_0	X_1	X_2
7.0	1	−1	−1
5.4	1	−1	1
7.1	1	1	−1
6.9	1	1	1

An examination of the data as presented in Table 16.6 shows that X_0, X_1, and X_2 are all orthogonal to each other as

$$\sum X_1 = \sum X_2 = 0 \quad \text{and} \quad \sum X_1 X_2 = 0$$

Hence the least squares normal equations become

$$\sum X_0 Y = b_0 n + b_1 \cdot 0 + b_2 \cdot 0$$
$$\sum X_1 Y = b_0 \cdot 0 + b_1 \cdot \sum X_1^2 + b_2 \cdot 0 \qquad (16.11)$$
$$\sum X_2 Y = b_0 \cdot 0 + b_1 \cdot 0 + b_2 \sum X_2^2$$

Solving gives

$$b_0 = \sum X_0 Y / n$$
$$b_1 = \sum X_1 Y / \sum X_1^2 \qquad (16.12)$$
$$b_2 = \sum X_2 Y / \sum X_2^2$$

For this problem

$$b_0 = \frac{26.4}{4} = 6.60$$

$$b_1 = \frac{1.6}{4} = 0.40$$

$$b_2 = \frac{-1.8}{4} = -0.45$$

and the response surface can be approximated as

$$\hat{Y} = 6.60 + 0.40X_1 - 0.45X_2$$

To determine the sum of squares due to each of these terms in the model, we can use the sum of squares due to b_i's

$$SS_{b_i} = b_i \cdot \sum X_i Y$$

Here

$$SS_{b_0} = (6.60)(26.4) = 174.24$$
$$SS_{b_1} = (0.40)(1.6) = 0.64$$
$$SS_{b_2} = (-0.45)(-1.8) = 0.81$$

each carrying 1 df. The ANOVA table is shown in Table 16.7.

Table 16.7 ANOVA for 2^2 Factorial Response-Surface Example

Source	df	SS
b_0	1	174.24
b_1	1	0.64
b_2	1	0.81
Residual	1	0.49
Totals	4	176.18

Here, all 4 df are shown, since the b_0 term represents the degree of freedom usually associated with the mean, that is, the correction term in many examples. The total sum of squares is $\sum Y^2$ of the responses and the residual is what is left over. With this 1-df residual, no good test is available on the significance of each term in the model, nor is there any way to assess the adequacy of the planar model to describe the surface.

To decide on the direction for the next experiment, plot contours of equal response using the equation of the plane determined above

$$\hat{Y} = 6.60 + 0.40X_1 - 0.45X_2$$

Solve for X_2,

$$X_2 = \frac{6.60 - \hat{Y} + 0.40X_1}{0.45}$$

If $\hat{Y} = 5.5$,

$$X_2 = \frac{1.10 + 0.40X_1}{0.45}$$

when

$$X_1 = -1 \qquad X_2 = 1.56$$
$$X_1 = 1 \qquad X_2 = 3.33$$

If $\hat{Y} = 6.0$,

$$X_2 = \frac{0.60 + 0.40X_1}{0.45}$$

when

$$X_1 = -1 \qquad X_2 = 0.44$$
$$X_1 = 1 \qquad X_2 = 2.22$$

If $\hat{Y} = 6.5$,

$$X_2 = \frac{0.10 + 0.40X_1}{0.45}$$

when

$$X_1 = -1 \qquad X_2 = -0.67$$
$$X_1 = 1 \qquad X_2 = 1.11$$

If $\hat{Y} = 7.0$,

$$X_2 = \frac{-0.40 + 0.40X_1}{0.45}$$

when

$$X_1 = -1 \qquad X_2 = -1.67$$
$$X_1 = 1 \qquad X_2 = 0$$

If $\hat{Y} = 7.5$,

$$X_2 = \frac{-0.90 + 0.40X_1}{0.45}$$

when

$$X_1 = -1 \qquad X_2 = -2.89$$
$$X_1 = 1 \qquad X_2 = -1.11$$

Plotting these five contours on the original diagram gives the pattern shown in Figure 16.5.

By moving in a direction normal to these contours and "up" the surface, we can anticipate larger values of response until the peak is reached. To decide on a possible set of conditions for the next experiment, consider the equation of a normal to these contours through the point (0, 0). The contours are

$$X_2 = \frac{6.60 - \hat{Y} + 0.40X_1}{0.45}$$

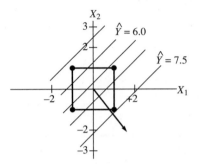

Figure 16.5 Contours on 2^2 factorial
response-surface example.

and their slope is $0.40 / 0.45 = 8 / 9$. The normal will have a slope $= -9/8$,
and its equation is

$$X_2 - 0 = \frac{-9}{8}(X_1 - 0)$$

$$X_2 = \frac{-9}{8}X_1 \qquad \text{(see arrow in Figure 16.5)}$$

This technique will not tell how far to go in this direction, but we might run the
next experiment with the center of the factorial at $(+1, -9/8)$. The four points
would be

X_1	X_2
0	$-\frac{1}{8}$
0	$-2\frac{1}{8}$
2	$-\frac{1}{8}$
2	$-2\frac{1}{8}$

These points are expressed in terms of the experimental variables, but must be
decoded to see where to set the concentration and position. Using the coding,

$$X_1 = 4C - 3 \qquad \text{or} \qquad C = \frac{X_1 + 3}{4}$$

$$X_2 = 2P - 3 \qquad \text{or} \qquad P = \frac{X_2 + 3}{2}$$

the four new points would be

X_1	X_2	C	P
0	$\frac{1}{8}$	$\frac{3}{4}{:}1$	$1\frac{7}{16}$
0	$-2\frac{1}{8}$	$\frac{3}{4}{:}1$	$\frac{7}{16}$
2	$-\frac{1}{8}$	$1\frac{1}{4}{:}1$	$1\frac{7}{16}$
2	$-2\frac{1}{8}$	$1\frac{1}{4}{:}1$	$\frac{7}{16}$

if these settings are possible. Responses may now be taken at these four points on a new 2^2 factorial, and after analysis we can again decide the direction of steepest ascent. This procedure continues until the optimum is obtained.

The above design is sufficient to indicate the direction for subsequent experiments, but it does not provide a good measure of experimental error to test the significance of b_0, b_1, and b_2, nor is there any test of how well the plane approximates the surface. One way to improve on this design is to take two or more points at the center of the square. By replication at the same point, an estimate of experimental error is obtained and the average of the center-point responses will provide an estimate of "goodness of fit" of the plane. If the experiment is near the maximum response, this center point might be somewhat above the four surrounding points, which would indicate the need for a more complex model.

To see how this design would help, consider two observations of response at the center of the example above. Using the same responses at the vertices of the square, the results might be those shown in Figure 16.6.

The coefficients in the model are still estimated by Equation (16.12):

$$b_0 = \frac{\sum X_0 Y}{n} = \frac{39.8}{6} = 6.63$$

$$b_1 = \frac{\sum X_1 Y}{\sum X_1^2} = \frac{1.6}{4} = 0.40$$

$$b_2 = \frac{\sum X_2 Y}{\sum X_2^2} = \frac{-1.8}{4} = -0.45$$

and

$$SS_{b_0} = 6.63(39.8) = 263.87$$
$$SS_{b_1} = 0.40(1.6) = 0.64$$
$$SS_{b_2} = -0.45(-1.8) = 0.81$$

For the sum of squares of the error at $(0, 0)$,

$$SS_e = (6.6)^2 + (6.8)^2 - \frac{(13.4)^2}{2} = 89.80 - 89.78 = 0.02$$

The analysis is shown in Table 16.8.

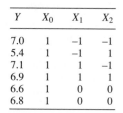

Y	X_0	X_1	X_2
7.0	1	−1	−1
5.4	1	−1	1
7.1	1	1	−1
6.9	1	1	1
6.6	1	0	0
6.8	1	0	0

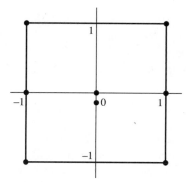

Figure 16.6 2^2 factorial with two points in center.

Table 16.8 ANOVA for 2^2+ Two Center Points

Source	df	SS	MS
Total	6	265.98	
b_0	1	263.87	236.87
b_1	1	0.64	0.64
b_2	1	0.81	0.81
Residual	3	0.66	—
Error	1	0.02	0.02
Lack of fit	2	0.64	0.32

Testing these effects against error gives

$$b_0: F_{1,1} = \frac{236.87}{0.02} = 11{,}843.5$$

$$b_1: F_{1,1} = \frac{0.64}{0.02} = 32$$

$$b_2: F_{1,1} = \frac{0.81}{0.02} = 40.5$$

$$\text{lack of fit}: F_{2,1} = \frac{0.32}{0.02} = 16$$

With such a small number of degrees of freedom, only b_0 shows significance at the 5 percent level. However, the tests on the b terms are larger than the test on lack of fit, which may indicate that the plane

$$\hat{Y} = 6.63 + 0.40X_1 - 0.45\dot{X}_2$$

is a fair approximation to the surface where this first experiment was run. ∎

More Complex Surfaces

For more factors—controlled variables—the model for the response variable is more complex, but several designs have been found useful in estimating the coefficients of these surfaces.

When three variables are involved, a first approximation is again a plane or hyperplane of the form

$$Y = B_0 X_0 + B_1 X_1 + B_2 X_2 + B_3 X_3 + \varepsilon$$

For this first-order surface, at least four points must be taken to estimate the four B's. Because three dimensions are involved, we might consider a 2^3 factorial. As this design has eight experimental conditions at its vertices, the design often used is a one-half replication of a 2^3 factorial with two or more points at the center of the cube. These six points (two at the center) are sufficient to estimate all four B's and to test for lack of fit of this plane to the surface in three dimensions. After this initial one-half replication of the 2^3 the other half might be run, giving more information for a better fit.

If a plane is not a good fit in two dimensions, a second-order model might be tried. Such a response surface has the form

$$Y = B_0 X_0 + B_1 X_1 + B_2 X_2 + B_{11} X_1^2 + B_{12} X_1 X_2 + B_{22} X_2^2 + \varepsilon$$

Here six B's are to be estimated. The simplest design for this model would be a pentagon (five points) plus center points. This would yield six or more responses, and all six B's could be estimated.

In developing these designs Box and others found that the calculations can be simplified if the design can be rotated. A *rotatable design* is one that has equal predictability in all directions from the center, and the points are at a constant distance from the center. All first-order designs are rotatable, as the square and one-half replication of the cube in the cases above. The simplest second-order design that is rotatable is the pentagon with a point at the center, as given above.

In three dimensions a second-order surface is given by

$$Y = B_0 X_0 + B_1 X_1 + B_2 X_2 + B_3 X_3 + B_{11} X_1^2 + B_{22} X_2^2$$
$$+ B_{33} X_3^2 + B_{12} X_1 X_2 + B_{13} X_1 X_3 + B_{23} X_2 X_3 + \varepsilon \qquad (16.13)$$

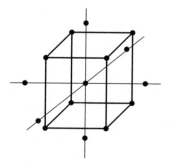

Figure 16.7 Central composite design.

which has 10 unknown coefficients. The cube for a three-dimensional model has only eight points, so a special design has been developed, called a *central composite* design. It is a 2^3 factorial with points along each axis at a distance from the center equal to the distance to each vertex. This gives $8 + 6 = 14$ points plus a point at the center makes 15, which is adequate for estimating the B's in Equation (16.13). This design is pictured in Figure 16.7.

In attempting to determine the equation of the response surface, some concepts in solid analytical geometry are often helpful. If a second-order model in two dimensions is

$$Y = B_0 X_0 + B_1 X_1 + B_2 X_2 + B_{11} X_1^2 + B_{12} X_1 X_2 + B_{22} X_2^2 + \varepsilon$$

the shape of this surface can be determined by reducing this equation to what is called *canonical form*. This would be

$$Y = B'_{11} X_1'^2 + B'_{22} X_2'^2$$

This is accomplished by a translation of axes to remove terms in X_1 and X_2 then a rotation of axes to remove the $X_1 X_2$ term. From B'_{11} and B'_{22} we can determine the shape of the surface, whether it is a sphere, ellipsoid, paraboloid, hyperboloid, or other. In higher dimensions this becomes more complicated, but it can still be useful in describing the response surface.

16.4 Evolutionary Operation (EVOP)

Philosophy

Evolutionary operation is a method of process operation that has a built-in procedure to increase productivity. The technique was developed by Box [6]. This book and Barnett's article [2] should be read to understand how the method works. Many chemical companies have reported considerable success using EVOP.

The procedure consists of running a simple experiment, usually a factorial, within the range of operability of a process as it is currently running. It is assumed that the variables to be controlled are measurable and can be set within a short distance of the current settings without disturbing production quality. The idea is to gather data on a response variable, usually yield, at the various points of an experimental design. When one set of data has been taken at all the points, one *cycle* is said to have been completed. One cycle is usually not sufficient to detect any shift in the response, so a second cycle is taken. This continues until the effect of one or more control variables, their interactions, or a change in the mean shows up as significant when compared with a measure of experimental error. This estimate of error is obtained from the cycle data, thus making the experiment self-contained. After a significant increase in yield has been detected, one *phase* is said to have been completed, and at this point a decision is usually made to change the basic operating conditions in a direction that should improve the yield. Several cycles may be necessary before a shift can be detected. The objective here, as with response surfaces, is to move in the direction of an optimum response. Response surface experimentation is primarily a laboratory or research technique; evolutionary operation is a production-line method.

To facilitate the EVOP procedure, a simple form has been developed to be used on the production line for each cycle of a 2^2 factorial with a point at the center. In the sections that follow an example will be run using these forms, and later the details of the form will be developed.

■ **EXAMPLE 16.4**   To illustrate the EVOP procedure, consider a chemical process in which temperature and pressure are varied over short ranges, and the resulting chemical yield is recorded. Since two controlled variables affect the yield, a 2^2 factorial should indicate the effect of each factor as well as a possible interaction between them. By taking a point at the center of a 2^2 factorial, we can also check on a change in the mean (CIM) by comparing this point at the center with the four points around the vertices of the square. If the process should be straddling a maximum, the center point should eventually (after several cycles) be significantly above the peripheral points at the vertices. The standard form for EVOP locates the five points in this design as indicated in Figure 16.8.

By comparing the responses (or average responses) at points 3 and 4 with those at 2 and 5, an effect of variable X_1 may be detected. Likewise, by comparing responses at 3 and 5 with those at 2 and 4, we may assess the X_2 effect. Comparing the responses at 2 and 3 with those at 4 and 5 will indicate an interaction effect, and comparing the responses at 2, 3, 4, and 5 with those at 1 will indicate a change in the mean, if present. The forms shown in Table 16.9 are filled in with data to show their use, which is self-explanatory.

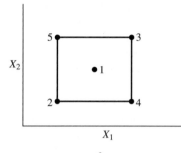

Figure 16.8 2^2 EVOP design.

Table 16.9 EVOP Work Sheet—First Cycle

	5	3		Project: 424
	1		Cycle: $n = 1$	Phase: 1
	2	4	Response: Yield	Date: 10/12

Calculation of Averages						Calculation of Standard Deviation

Operating Conditions	(1)	(2)	(3)	(4)	(5)	
(i) Previous cycle sum						Previous sum s =
(ii) Previous cycle average						Previous average s =
(iii) New observations	94.0	94.5	96.5	94.5	94.5	Range =
(iv) Differences [(ii) less (iii)]						New s = range $\times f_{k,n}$ =
(v) New sums	94.0	94.5	96.5	94.5	94.5	New sum s =
(vi) New averages $\hat{Y}_i$	94.0	94.5	96.5	94.5	94.5	New average $s = \dfrac{\text{new sum } s}{n-1}$

Calculation of Effects	Calculations of Error Limits

Temperature effect $= \frac{1}{2}(\bar{Y}_3 + \bar{Y}_4 - \bar{Y}_2 - \bar{Y}_5) = 1.00$ For new average $= \dfrac{2}{\sqrt{n}}s =$

Pressure effect $= \frac{1}{2}(\bar{Y}_3 + \bar{Y}_5 - \bar{Y}_2 - \bar{Y}_4) = 1.00$ For new effects $= \dfrac{2}{\sqrt{n}}s =$

$T \times P$ interaction effect $= \frac{1}{2}(\bar{Y}_2 + \bar{Y}_3 - \bar{Y}_4 - \bar{Y}_5) = 1.00$ For change in mean $= \dfrac{1.78}{\sqrt{n}}s =$

Change in mean effect $= \frac{1}{5}(\bar{Y}_2 + \bar{Y}_3 + \bar{Y}_4 + \bar{Y}_5 - 4\bar{Y}_1)$
$= 0.80$

Not much can be learned from the first cycle unless some separate estimate of standard deviation is available. The second cycle (Table 16.10) will begin to show the method for testing the effects.

Since, in the second cycle, none of the effects are numerically larger than their error limits, the true effect could easily be zero. In this case another cycle must be run. The only items that might need explaining are under the calculation of standard deviation. The range referred to here is the range of the differences (iv), and $f_{k,n}$ is found in a table where $k = 5$ for these five-point designs, and

Table 16.10 EVOP Work Sheet—Second Cycle

5	3		Project: 424
1		Cycle: $n = 2$	Phase: 1
2	4	Response: Yield	Date: 10/12

	Calculation of Averages					Calculation of Standard Deviation
Operating Conditions	*(1)*	*(2)*	*(3)*	*(4)*	*(5)*	
(i) Previous cycle sum	94.5	94.5	96.5	94.5	94.5	Previous sum s =
(ii) Previous cycle average	94.0	94.5	96.5	94.5	94.5	Previous average s =
(iii) New observations	96.0	95.0	95.0	96.5	94.0	Range = 3.5
(iv) Differences [(ii) less (iii)]	−2.0	−0.5	1.5	−2.0	−0.5	New s = range $\times f_{k,n}$ = 1.05
(v) New sums	190.0	189.5	191.5	191.0	188.5	New sum s = 1.05
(vi) New averages $\bar{Y}_i$	95.0	94.7	95.7	95.5	94.2	New average s = $\dfrac{\text{new sum } s}{n-1}$ = 1.05

	Calculation of Effects		Calculations of Error Limits

Temperature effect $= \frac{1}{2}(\bar{Y}_3 + \bar{Y}_4 - \bar{Y}_2 - \bar{Y}_5) = 1.15$ 　　For new average $= \dfrac{2}{\sqrt{n}}s$ = 1.48

Pressure effect $= \frac{1}{2}(\bar{Y}_3 + \bar{Y}_5 - \bar{Y}_2 - \bar{Y}_4) = -0.15$ 　　For new effects $= \dfrac{2}{\sqrt{n}}s$ = 1.48

$T \times P$ interaction effect $= \frac{1}{2}(\bar{Y}_2 + \bar{Y}_3 - \bar{Y}_4 - \bar{Y}_5)$ $= 0.35$ 　　For change in mean $= \dfrac{1.78}{\sqrt{n}}s$ = 1.32

Change in mean effect $= \frac{1}{5}(\bar{Y}_2 + \bar{Y}_3 + \bar{Y}_4 + \bar{Y}_5 - 4\bar{Y}_1)$ $= 0.02$

n is the cycle number. Part of such a table shows

$n =$	2	3	4	5	6	7	8
$f_{5,n} =$	0.30	0.35	0.37	0.38	0.39	0.40	0.40

The third cycle is shown in Table 16.11. Here the temperature effect is seen to be significant, with an increase in temperature giving a higher yield. Once a significant effect has been found, the first *phase* of the EVOP procedure has been completed. The average results at this point are usually displayed on an EVOP bulletin board as in Figure 16.9.

Table 16.11 EVOP Work Sheet—Third Cycle

```
 5      3
 ┌──────────┐          Project: 424
 │    1     │  Cycle: n = 3     Phase: 1
 └──────────┘  Response: Yield  Date: 10/12
 2      4
```

	Calculation of Averages					Calculation of Standard Deviation
Operating Conditions	*(1)*	*(2)*	*(3)*	*(4)*	*(5)*	
(i) Previous cycle sum	190.0	189.5	191.5	191.0	188.5	Previous sum $s = 1.05$
(ii) Previous cycle average	95.0	94.7	95.7	95.5	94.2	Previous average $s = 1.05$
(iii) New observations	94.5	93.5	96.0	97.0	94.0	Range $= 2.7$
(iv) Differences [(ii) less (iii)]	0.5	1.2	-0.3	-1.5	0.2	New $s = $ range $\times f_{k,n}$ $= 0.95$
(v) New sums	284.5	283.0	287.2	286.5	282.7	New sum $s = 2.00$
(vi) New averages $\bar{Y}_i$	94.8	94.3	95.7	95.5	94.2	New average $s = \dfrac{\text{new sum } s}{n-1}$ $= 1.00$

	Calculation of Effects	Calculations of Error Limits

Temperature effect $= \frac{1}{2}(\bar{Y}_3 + \bar{Y}_4 - \bar{Y}_2 - \bar{Y}_5) = 1.35^*$
For new average $= \dfrac{2}{\sqrt{n}}s$ $= 1.16$

Pressure effect $= \frac{1}{2}(\bar{Y}_3 + \bar{Y}_5 - \bar{Y}_2 - \bar{Y}_4) = 0.05$
For new effects $= \dfrac{2}{\sqrt{n}}s$ $= 1.16$

$T \times P$ interaction effect $= \frac{1}{2}(\bar{Y}_2 + \bar{Y}_3 - \bar{Y}_4 - \bar{Y}_5)$ $= -0.15$
For change in mean $= \dfrac{1.78}{\sqrt{n}}s$ $= 1.02$

Change in mean effect $= \frac{1}{5}(\bar{Y}_2 + \bar{Y}_3 + \bar{Y}_4 + \bar{Y}_5 - 4\bar{Y}_1)$ $= 0.10$

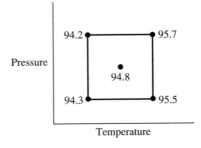

Figure 16.9　Final cycle of EVOP.

An EVOP committee usually reviews these data and decides whether or not to reset the operating conditions. If they change the operating conditions (point 1), then EVOP is reinstated around this new point and the second phase is begun. EVOP is continued again until significant changes are detected. In fact, it goes on continually, seeking to optimize a process. ■

This is a very simple example with just two independent variables. The references should be consulted for more complex situations.

EVOP Form Rationale

Most of the steps in the EVOP form above are quite clear. It may be helpful to see where some of the constants come from.

In estimating the standard deviation from the range of differences in step (iv), consider a general expression for these differences

$$D_p = \frac{X_{p1} + X_{p2} + \cdots + X_{p, n-1}}{n - 1} - X_{p, n}$$

where D_p represents the difference at any point p of the design. X_{pi} represents the observation at point p in the ith cycle. n is, of course, the number of cycles. Because the variance of a sum equals the sum of the variances for independent variables, and since the variance of a constant times a random variable equals the square of the constant times the variance of the random variable,

$$\sigma_{D_p}^2 = \frac{1}{(n - 1)^2} [\sigma_{xp1}^2 + \sigma_{xp2}^2 + \cdots + \sigma_{xp, n-1}^2] + \sigma_{xp, n}^2$$

Since all these x's represent the same population, their variances are all alike, and then

$$\sigma_D^2 = \frac{1}{(n - 1)^2} [(n - 1)\sigma_x^2] + \sigma_x^2 = \frac{n}{n - 1} \sigma_x^2$$

and

$$\sigma_D = \sqrt{\frac{n}{n-1}} \sigma_x$$

The standard deviation of the population can then be written in terms of the standard deviation of these differences

$$\sigma_x = \sqrt{\frac{n-1}{n}} \sigma_D$$

σ_D can now be estimated from the range of these differences R_d. From the quality control field

$$\sigma_D = \frac{R_d}{d_2}$$

where d_2 depends on the number of differences in the range, which is 5 on this form. Here $d_2 = 2.326$ for samples of 5,

$$\sigma_D = \frac{R_d}{2.326}$$

and the standard deviation of the population is estimated by

$$\sigma_x = \sqrt{\frac{n-1}{n}} \frac{R_d}{2.326}$$

The quantity

$$\sqrt{\frac{n-1}{n}} \frac{1}{2.326}$$

is called $f_{k,n}$ in the EVOP form where $k = 5$.

Note that

$$f_{5,2} = \sqrt{\frac{1}{2}} \frac{R_d}{2.326} = 0.30 R_d$$

$$f_{5,6} = \sqrt{\frac{5}{6}} \frac{R_d}{2.326} = 0.39 R_d$$

which tallies with the table values given.

To determine the error limits for the effects, two standard deviation limits are used, as they represent the approximate 95 percent confidence limits on the parameter being estimated.

For any effect such as

$$E = \frac{1}{2}(\bar{Y}_3 + \bar{Y}_4 - \bar{Y}_2 - \bar{Y}_5)$$

its variance would be

$$V_E = \frac{1}{4}(\sigma_{\bar{Y}_3}^2 + \sigma_{\bar{Y}_4}^2 + \sigma_{\bar{Y}_2}^2 + \sigma_{\bar{Y}_5}^2) = \frac{1}{4}(4\sigma_{\bar{Y}}^2) = \sigma_{\bar{Y}}^2 = \frac{\sigma_{\bar{Y}}^2}{n}$$

and two standard deviation limits on an effect would be

$$\pm 2 \frac{\sigma_Y}{\sqrt{n}} \quad \text{estimated by} \quad \pm 2 \frac{s}{\sqrt{n}}$$

For the change in mean effect

$$\text{CIM} = \frac{1}{5}(\bar{Y}_2 + \bar{Y}_3 + \bar{Y}_4 + \bar{Y}_5 - 4\bar{Y}_1)$$

$$V_{\text{CIM}} = \frac{1}{25}(\sigma_{\bar{Y}_2}^2 + \sigma_{\bar{Y}_3}^2 + \sigma_{\bar{Y}_4}^2 + \sigma_{\bar{Y}_5}^2 + 16\sigma_{\bar{Y}_1}^2)$$

$$= \frac{20}{25}\sigma_{\bar{Y}}^2 = \frac{20}{25}\frac{\sigma_Y^2}{n}$$

$$\sigma_{\text{CIM}} = \sqrt{\frac{4}{5}} \cdot \frac{\sigma_Y}{\sqrt{n}}$$

and two standard deviation limits would be

$$\pm 2 \cdot \sqrt{\frac{4}{5}} \frac{s}{\sqrt{n}} = \pm 1.78 \frac{s}{\sqrt{n}}$$

as given on the EVOP form.

■ **EXAMPLE 16.5** This EVOP procedure often has three factors, each at two levels. The $2^3 = 8$ experimental points are run as well as two points in the center of the cube, giving 10 experimental points per cycle. Usually this design is run in two blocks where the ABC interaction is confounded with blocks. The standard form for the location of the ten points is shown in Figure 16.10.

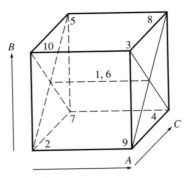

Figure 16.10 2^3 EVOP.

The first five points (1 through 5) are run as the first block and the last five points (6 through 10) are run as the second block. Tables 16.12 and 16.13 illustrate this EVOP technique for the second cycle of an example on the mean surface factor for grinder slivers where the factors are dropout temperature A, back-zone temperature B, and atmosphere C. The work sheets should be self-explanatory as they are based on the design principles of confounding a 2^3

Table 16.12 EVOP Work Sheet—Block 1 in 2^3

Cycle: $n = 2$ Project: 478

Response: Mean surface Phase: 1

factor for grinder slivers Date: 11/6

Factors: A = **Dropout temperature 2145° and 2165°**

B = **Back-zone temperature 2140° and 2160°**

C = **Atmosphere: Reducing and oxidizing**

	Calculation of Averages—Block 1					Calculation of Standard Deviation
Operating Conditions (Block 1)	*(1)*	*(2)*	*(3)*	*(4)*	*(5)*	
(i) Previous cycle sum	3.9	0.0	1.6	9.8	4.3	Previous sum s = (all blocks)
(ii) Previous cycle sum	3.9	0.0	1.6	9.8	4.3	Previous average s =
(iii) New observations	5.6	4.0	6.7	15.0	3.4	Range = 4.3
(iv) Differences [(ii) less (iii)]	−1.7	−4.0	−5.1	−5.2	−0.9	New s = range $\times f_{k,n}$ = 1.29
(v) New sums	9.5	4.0	8.3	24.8	7.7	New sum s = 1.29 (all blocks)
(vi) New averages $\bar{Y}_i$	4.8	2.0	4.2	12.4	3.8	New average $s = \dfrac{\text{new sum } s}{2n-3}$ = 1.29

Calculation of Effects—Block 1

$A - BC = \frac{1}{2}(\bar{Y}_3 + \bar{Y}_4 - \bar{Y}_2 - \bar{Y}_5) = 5.40$

$B - AC = \frac{1}{2}(\bar{Y}_3 + \bar{Y}_5 - \bar{Y}_2 - \bar{Y}_4) = -3.20$

$C - AB = \frac{1}{2}(\bar{Y}_4 + \bar{Y}_5 - \bar{Y}_2 - \bar{Y}_3) = 5.00$

Change in mean effect $= \frac{1}{5}(\bar{Y}_2 + \bar{Y}_3 + \bar{Y}_4 + \bar{Y}_5 - 4\bar{Y}_1) = 0.64$

factorial in two blocks of four observations. One could terminate this process with the second cycle as it shows dropout temperature A, back-zone temperature B, and their AB interaction to be significant in producing changes in the mean surface factor for grinder slivers.

Table 16.13 EVOP Work Sheet—Block 2 in 2^3

	Cycle: $n = 3$ Calculation of Averages—Block 2					Calculation of Standard Deviation
Operating Conditions (Block 2)	(6)	(7)	(8)	(9)	(10)	
(i) Previous cycle sum	4.2	4.4	2.8	10.1	0.0	Previous sum s = 1.29 (all blocks)
(ii) Previous cycle average	4.2	4.4	2.8	10.1	0.0	Previous average s = 1.29
(iii) New observations	3.2	3.4	6.6	15.0	1.9	Range = 5.9
(iv) Differences [(ii) less (iii)]	1.0	0.9	−3.8	−4.9	−1.9	New sum s = range $\times f_{k, n}$ = 1.77
(v) New sums	7.4	7.7	9.4	25.1	1.9	New sum s = 3.06 (all blocks)
(vi) New averages $\bar{Y}_i$	3.7	3.8	4.7	12.6	1.0	New average $s = \dfrac{\text{new sum } s}{2n - 2}$ = 1.53

Calculation of Effects—Block 2	Calculation of Error Limits
$A + BC = \frac{1}{2}(\bar{Y}_8 + \bar{Y}_9 - \bar{Y}_7 - \bar{Y}_{10}) = 6.25$	For new averages $= \dfrac{2s}{\sqrt{n}}$ $= \pm 2.16$
$B + AC = \frac{1}{2}(\bar{Y}_8 + \bar{Y}_{10} - \bar{Y}_7 - \bar{Y}_9) = -5.35$	For new effects $= \dfrac{1.42s}{\sqrt{n}}$ $= \pm 1.53$
$C + AB = \frac{1}{2}(\bar{Y}_7 + \bar{Y}_8 - \bar{Y}_9 - \bar{Y}_{10}) = -2.55$	For change in mean $= \dfrac{1.26s}{\sqrt{n}}$ $= \pm 1.36$
Change in mean effect $= \frac{1}{5}(\bar{Y}_7 + \bar{Y}_8 + \bar{Y}_9 + \bar{Y}_{10} - 4\bar{Y}_6) = 1.46$	

Calculation of Effects—Both Blocks	
$A = \frac{1}{2}[(A + BC) + (A - BC)] = 5.82^*$	$AB = \frac{1}{2}[(C + AB) - (C - AB)] = -3.78^*$
$B = \frac{1}{2}[(B + AC) + (B - AC)] = -4.28^*$	$AC = \frac{1}{2}[(B + AC) - (B - AC)] = -1.08$
$C = \frac{1}{2}[(C + AB) + (C - AB)] = 1.23$	$BC = \frac{1}{2}[(A + BC) - (A - BC)] = -0.42$
Change in mean effect $= \frac{1}{2}(\text{CIM}_1 + \text{CIM}_2) = 1.05$	

16.5 Analysis of Attribute Data

Philosophy

Two assumptions used in the application of the analysis of variance technique are that (1) the response variable is normally distributed and (2) the variances of the experimental errors are equal throughout the experiment. In practice it is often necessary to deal with attribute data where the response variable is either 0 or 1. In such cases one often records the number of occurrences of a particular phenomenon or the percentage of such occurrences.

It is well known that the number of occurrences per unit such as defects per piece, errors per page, customers per unit time, often follow a Poisson distribution where such response variables are not only nonnormal but variances and means are equal. When proportions or percentages are used as the response variable, the data are binomial and again variances are related to means and the basic assumptions of ANOVA do not hold. Some studies have shown the lack of normality is not too serious in applying the ANOVA technique, but most statisticians recommend that a transformation be made on the original data if it is known to be nonnormal. Recommendations for proper transformations are given in Davies [10].

Another technique, called *factorial chi square,* has been found very useful in treating attribute data in industrial problems. This technique is described by Batson [3], who gives several examples of its application to actual problems. Although this factorial chi-square technique may not be as precise as a regular ANOVA on transformed data, its simplicity makes it well worth consideration for many applied problems.

To illustrate the techniques mentioned above, consider the following example from a consulting-firm study.

■ **EXAMPLE 16.6** Interest centered on undesirable marks on steel samples from a grinding operation. The response variable was simply whether or not these marks occurred. Four factors were believed to affect these marks: blade size A, centering B, leveling C, and speed D. Each of these four factors was set at two levels, giving a 2^4 factorial experiment. Each of the 16 treatment combinations was run in a completely randomized order, and 20 steel samples were produced under each of the 16 experimental conditions. The number of damaged samples in each group of 20 is the recorded variable in Table 16.14.

If the responses are each divided by 20, the proportions of damaged samples can be determined. Since proportions follow a binomial distribution, an arc sine transformation is appropriate (see [12]). Taking the arc sin $\sqrt{p}$ as the response variable, Yates' method can be used to determine the sums of squares as shown in Table 16.15.

Table 16.14 Damaged Steel Samples in 2^4 Experiment

		Factor A			
		0		**1**	
		Factor B		Factor B	
Factor C	Factor D	0	1	0	1
0	0	0	0	16	20
	1	0	0	10	20
1	0	0	0	10	14
	1	1	0	12	20

Table 16.15 Yates' Method on Transformed Data of Steel Sample Experiment

Treatment	Proportion P	Transformed Variable $X \equiv$ arc sin $\sqrt{p}$	(1)	(2)	(3)	(4)	SS
(1)	0.00	0.00	1.11	2.68	4.46	9.51	—
a	0.80	1.11	1.57	1.78	5.05	9.05	5.12
b	0.00	0.00	0.79	2.36	4.46	1.89	0.22
ab	1.00	1.57	0.99	2.69	4.59	2.35	0.35
c	0.00	0.00	0.79	2.68	0.66	−1.23	0.09
ac	0.50	0.79	1.57	1.78	1.23	−1.03	0.06
bc	0.00	0.00	1.12	2.36	0.66	−0.59	0.03
abc	0.70	0.99	1.57	2.23	1.69	−0.13	0.01
d	0.00	0.00	1.11	0.46	−0.90	0.59	0.03
ad	0.50	0.79	1.57	0.20	−0.33	0.13	0.01
bd	0.00	0.00	0.79	0.78	−0.90	0.57	0.02
abd	1.00	1.57	0.99	0.45	−0.13	1.03	0.06
cd	0.05	0.23	0.79	0.46	−0.26	0.57	0.02
acd	0.60	0.89	1.57	0.20	−0.33	0.77	0.04
bcd	0.00	0.00	0.66	0.78	−0.26	−0.07	0.00
abcd	1.00	1.57	1.57	0.91	0.13	0.39	0.01

If the three- and four-way interactions are assumed to be nonexistent, the resulting ANOVA table is shown as Table 16.16.

This analysis shows the highly significant blade size effect, which is obvious from a glance at Table 16.14. It also shows a significant centering effect and a blade size–centering interaction that is not as obvious from Table 16.14.

Table 16.16 ANOVA for Steel Sample Experiment

Source	df	SS	MS
A	1	5.12	5.12***
B	1	0.22	0.22*
AB	1	0.35	0.35**
C	1	0.09	0.09
AC	1	0.06	0.06
BC	1	0.03	0.03
D	1	0.03	0.03
AD	1	0.01	0.01
BD	1	0.02	0.02
CD	1	0.02	0.02
Error, pooled	5	0.12	0.02

To apply the factorial chi-square technique to this same example, Table 16.17 is filled in. The plus and minus signs simply give the proper weights for the main effects and two-way interactions and T is the value of the corresponding contrast. D is the total number of samples in the total experiment; here 20 per cell or $20 \times 2^4 = 320$ samples. The chi-square (χ^2) value is determined by the formula

$$\chi^2_{[1]} = \frac{N^2}{(S)(F)} \times \frac{T^2}{D} \qquad (16.14)$$

Here N is the total number of items in the whole experiment and S is the number of occurrences (successes) in the N. F is the number of nonoccurrences (or failures). In the example $N = 320$ samples, $S = 123$ with undesirable marks. For factor A, then

$$\chi^2_{[1]} = \frac{(320)^2}{(123)(197)}(45.75)$$

$$= (4.23)(45.75) = 193.52$$

The 4.23 is constant for these data. The resulting chi-square values are then compared with χ^2 with 1 df at the significance level desired. For the 5 percent level, $\chi^2_{[1]} = 3.84$ and again factors A, B, and AB are significant.

As seen in Table 16.17, the calculations for this technique are quite simple and the results are consistent with those using a more precise method. Extensions to problems other than 2^f factorials can be found in Batson [3]. ∎

Table 16.17 Factorial Chi-Square Work Sheet for Table 16.14

														$\Sigma +$	$\Sigma -$	T	T^2	D	T^2/D	χ^2
Blade A						0						1								
Center B			0					0					1								
Level C	0		1		0		1		0		1		0	1							
Speed D	0	1	0	1	0	1	0	1	0	1	0	1	0	1							
$S/20$	0	0	0	0	16	10	10	12	20	20	20	20	14	20							
Main effects																					
Blade A	−	−	−	−	+	+	+	+	+	+	+	+	+	+	122	−1	121	14,641	320	45.75	193.52*
Center B	−	−	−	+	+	−	−	+	+	+	+	+	−	+	74	−49	25	625	320	1.95	8.25*
Level C	−	−	+	−	+	−	+	−	+	−	+	−	+	+	57	−66	−9	81	320	0.25	1.06
Speed D	−	+	−	+	−	+	−	+	−	+	−	+	−	+	63	−60	3	9	320	0.03	0.12
Interactions																					
AB	+	+	+	−	−	−	−	+	+	+	+	+	−	+	75	−48	27	729	320	2.28	9.65*
AC	+	+	−	+	−	−	+	−	+	+	−	−	−	+	56	−67	−9	81	320	0.25	1.06
AD	+	−	+	−	−	+	−	+	−	+	−	+	−	+	62	−61	1	1	320	0.01	0.04
BC	+	+	−	−	+	+	−	−	+	−	+	−	+	+	60	−63	−3	9	320	0.03	0.12
BD	+	−	+	−	+	−	+	−	−	+	+	−	+	+	66	−57	9	81	320	0.25	1.06
CD	+	−	−	+	−	−	+	+	−	+	+	−	−	+	69	−54	+15	225	320	0.70	2.96

16.6 Randomized Incomplete Blocks—Restriction on Experimentation

Method for Balanced Blocks

In some randomized block designs it may not be possible to apply all treatments in every block. If there were, for example, six brands of tires to test, only four could be tried on a given car (not using the spare), and such a block would be incomplete, having only four out of the six treatments in it.

■ **EXAMPLE 16.7**   Take the problem of determining the effect of current flow of four treatments applied to the coils of TV tube filaments. As each treatment application requires some time, it is not possible to run several observations of these treatments in one day. If days are taken as blocks, all four treatments must be run in random order on each of several days in order to have a randomized block design. After checking it is found that even four treatments cannot be completed in a day; three are the most that can be run. The question then is: Which treatments are to be run on the first day, which on the second, and so forth, if information is desired on all four treatments?

The solution to this problem is to use a balanced incomplete block design. An *incomplete block design* is simply one in which there are more treatments than can be put in a single block. A *balanced incomplete block design* is an incomplete block design in which every pair of treatments occurs the same number of times in the experiment. Tables of such designs may be found in Fisher and Yates [12]. The number of blocks necessary for balancing will depend upon the number of treatments that can be run in a single block.

For the example mentioned there are four treatments and only three treatments can be run in a block. The balanced design for this problem requires four blocks (days) as shown in Table 16.18.

In this design only treatments A, C, and D are run on the first day; B, C, and D on the second day, and so forth. Note that each pair of treatments, such

Table 16.18 Balanced Incomplete Block Design for TV Filament Example

Block (days)	Treatment A	B	C	D	$T_{i.}$
1	2	—	20	7	29
2	—	32	14	3	49
3	4	13	31	—	48
4	0	23	—	11	34
$T_{.j}$	6	68	65	21	$160 = T_{..}$

as AB, occurs together twice in the experiment. A and B occur together on days 3 and 4; C and D occur together on days 1 and 2; and so on. As in randomized complete block designs, the order in which the three treatments are run on a given day is completely randomized.

The analysis of such a design is easier if some new notation is introduced. Let

b = number of blocks in the experiment (b = 4)

t = number of treatments in the experiment (t = 4)

k = number of treatments per block (k = 3)

r = number of replications of a given treatment throughout the experiment (r = 3)

N = total number of observations

  = $bk = tr(N = 12)$

λ = number of times each pair of treatments appears together throughout the experiment

  = $r(k - 1)/(t - 1)$ (λ = 2)

Table 16.18 has the current readings after they have been coded by subtracting 513 milliamperes and the block and treatment totals. The analysis for a balanced incomplete block design proceeds as follows:

1. Calculate the total sum of squares as usual:

$$SS_{total} = \sum_i \sum_j Y_{ij}^2 - \frac{T_{..}^2}{N}$$

$$= 3478 - \frac{(160)^2}{12} = 1344.67$$

2. Calculate the block sum of squares, ignoring treatments:

$$SS_{block} = \sum_{i=1}^{b} \frac{T_{i.}^2}{k} - \frac{T_{..}^2}{N}$$

$$= \frac{(29)^2 + (49)^2 + (48)^2 + (34)^2}{3} - \frac{(160)^2}{12} = 100.67$$

3. Calculate treatment effects, adjusting for blocks:

$$SS_{treatment} = \frac{\sum_{j=1}^{t} Q_j^2}{k\lambda t} \tag{16.15}$$

where

$$Q_j = kT_{.j} - \sum_i n_{ij} T_{i.} \tag{16.16}$$

where $n_{ij} = 1$ if treatment j appears in block i, and $n_{ij} = 0$ if treatment j does *not* appear in block i. Note that $\sum_i n_{ij}T_i$. is merely the sum of all block totals that contain treatment j.

For the data given,

$$Q_1 = 3(6) - (29 + 48 + 34) = 18 - 111 = -93$$
$$Q_2 = 3(68) - 131 = 73$$
$$Q_3 = 3(65) - 126 = 69$$
$$Q_4 = 3(21) - 112 = \underline{-49}$$
$$0$$

Note that

$$\sum_{j=1}^{t} Q_j = 0$$

which is always true. Then

$$SS_{treatment} = \frac{(-93)^2 + (73)^2 + (69)^2 + (-49)^2}{3(2)4} = 880.83$$

4. Calculate the error sum of squares by subtraction

$$SS_{error} = SS_{total} - SS_{block} - SS_{treatment}$$
$$= 1344.67 - 100.67 - 880.83 = 363.17$$

Table 16.19 summarizes in an ANOVA table.

Table 16.19 ANOVA for Incomplete Block Design Example

Source	df	SS	MS
Blocks (days)	3	100.67	—
Treatments (adjusted)	3	880.83	293.61
Error	5	363.17	72.63
Totals	11	1344.67	

An F test gives $F_{3,5} = 293.61 / 72.63 = 4.04$, which is not significant at the 5 percent level (Statistical Table D).

In Table 16.19 the error degrees of freedom are determined by subtraction rather than as the product of the block and treatment degrees of freedom. However, this error degrees of freedom is seen to be the product of treatment and block degrees of freedom (9) if 4 is subtracted for the four missing values in the design.

In some incomplete block designs it may be desirable to test for a block effect. The mean square for blocks was not computed in Table 16.19 since it had not been adjusted for treatments. In the case of a *symmetrical balanced incomplete randomized block design*, where $b = t$, the block sum of squares may be adjusted in the same manner as the treatment sums of squares

$$Q_1' = rT_{i.} - \sum_j n_{ij}T_{.j}$$

$$Q_1' = 3(29) - 92 = -5$$
$$Q_2' = 3(49) - 154 = -7$$
$$Q_3' = 3(48) - 139 = +5$$
$$Q_4' = 3(34) - 95 = +7$$
$$\sum_i Q_i' = 0$$

$$SS_{block} = \sum_{i=1}^{b} (Q_i')^2 / r\lambda b = \frac{(-5)^2 + (-7)^2 + (5)^2 + (7)^2}{3(2)4} = 6.17$$

and

$$SS_{treatment \ (unadjusted)} = \frac{6^2 + 68^2 + 65^2 + 21^2}{3} - \frac{(160)^2}{12} = 975.34$$

The results of this adjustment and the one in treatments may now be summarized for this symmetrical case as shown in Table 16.20.

Table 16.20 ANOVA for Incomplete Block Design Example for Both Treatments and Blocks

Source	df	SS	MS
Blocks (adjusted)	3	6.17	2.06
Blocks	(3)	(100.67)	—
Treatments (adjusted)	3	880.83	293.61
Treatments	(3)	(975.34)	—
Error	5	363.17	72.63
Totals	11	1344.67	

The terms in parentheses are inserted only to show how the error term was computed for one adjusted effect and one unadjusted effect

$$SS_{error} = SS_{total} - SS_{treatment \ (adjusted)} - SS_{block}$$
$$= 1344.67 - 880.83 - 100.67 = 363.17$$

or

$$SS_{error} = SS_{total} - SS_{treatment} - SS_{block \ (adjusted)}$$
$$= 1344.67 - 975.34 - 6.17 = 363.17$$

It should be noted also that the final sum of squares values used in Table 16.20 to get the mean square values do not add up to the total sum of squares. This is characteristic of a nonorthogonal design. The F test for blocks was not run because its value is obviously extremely small, which indicates no day-to-day effect on current flow.

For nonsymmetrical or unbalanced designs, the general regression method may be a useful alternative. If contrasts are to be computed for an incomplete block design, it can be shown that the sum of squares for a contrast is given by

$$SS_{C_m} = \frac{(C_m)^2}{\left(\sum_{j=1}^{t} c_{jm}^2 \right) k\lambda t} \tag{16.17}$$

where contrasts C_m are made on the Q_j's rather than $T_{.j}$'s.

As an example consider the following orthogonal contrasts on the data of Table 16.18:

$$C_1 = Q_1 - Q_2 \qquad\qquad = -166$$
$$C_2 = Q_1 + Q_2 - 2Q_3 \qquad = -158$$
$$C_3 = Q_1 + Q_2 + Q_3 - 3Q_4 = 196$$

The corresponding sums of squares are

$$SS_{C_1} = \frac{(-166)^2}{(2)(3)(2)(4)} = 574.08$$

$$SS_{C_2} = \frac{(-158)^2}{(6)(3)(2)(4)} = 173.36$$

$$SS_{C_3} = \frac{(196)^2}{(12)(24)} = \underline{133.39}$$

$$= 880.83$$

which checks with the adjusted sums of squares for treatments. Comparing each of the above sums of squares with their 1 df against the error mean square in Table 16.19 indicates that contrast C_1 is the only one of these three that is significant at the 5 percent level of significance. This indicates a real difference in current flow between treatments A and B, even though treatments in general showed no significant difference in current flow. ■

Example 16.7 can be handled with an SAS program as shown below:

```
DATA A;
INPUT BLK TRT $ RDG @@;CARDS;
1 A 2 1 C 20 1 D 7
2 B 32 2 C 14 2 D 3
3 A 4 3 B 13 3 C 31
4 A 0 4 B 23 4 D 11
;
TITLE'INC BLOCK EXAMPLE';
PROC PRINT;
PROC GLM;
CLASS BLK TRT;
MODEL RDG=BLK TRT;
```

```
       INC BLOCK EXAMPLE
OBS      BLK      TRT     RDG
  1       1        A        2
  2       1        C       20
  3       1        D        7
  4       2        B       32
  5       2        C       14
  6       2        D        3
  7       3        A        4
  8       3        B       13
  9       3        C       31
 10       4        A        0
 11       4        B       23
 12       4        D       11
```

INC BLOCK EXAMPLE
GENERAL LINEAR MODELS PROCEDURE

DEPENDENT VARIABLE: RDG

SOURCE	DF	SUM OF SQUARES	MEAN SQUARE	F VALUE	PR > F	R-SQUARE	C.V.
MODEL	6	981.50000000	163.58333333	2.25	0.1954	0.729921	63.9189
ERROR	5	363.16666667	72.63333333		ROOT MSE		RDG MEAN
CORRECTED TOTAL	11	1344.66666667			8.52251919		13.33333333

SOURCE	DF	TYPE I SS	F VALUE	PR > F	DF	TYPE III SS	F VALUE	PR > F
BLK	3	100.6666667	0.46	0.7211	3	6.16666667	0.03	0.9928
TRT	3	880.83333333	4.04	0.0834	3	880.83333333	4.04	0.0834

Because this is a nonorthogonal design, the Type III column should be used to evaluate the effects of interest. The above SAS results check with Table 16.20.

16.7 Youden Squares

When the conditions for a Latin square are met except for the fact that only three treatments are possible (for example, because in one block only three positions are available) and where there are four blocks altogether, the design is an incomplete Latin square. This design is called a *Youden square*. One such Youden square is illustrated in Table 16.21.

Table 16.21 Youden Square Design

	Position		
Block	1	2	3
I	A	B	C
II	D	A	B
III	B	C	D
IV	C	D	A

Note that the addition of a column (*D*, *C*, *A*, *B*) would make this a Latin square if another position were available. A situation calling for a Youden square might occur if four materials were to be tested on four machines but there were only three heads on each machine whose orientation might affect the results.

■ **EXAMPLE 16.8**   The analysis of a Youden square proceeds like the incomplete block analysis. Assuming hypothetical values for some measured variable Y_{ijk} where

$$Y_{ijk} = \mu + \beta_i + \tau_j + \gamma_k + \varepsilon_{ijk}$$

with

$$i = 1, \ldots, 4 \qquad j = 1, 2, \ldots, 4 \qquad k = 1, 2, 3$$

we might have the data of Table 16.22.

Table 16.22 Youden Square Design Data

	Position			
Block	1	2	3	$T_{i..}$
I	A = 2	B = 1	C = 0	3
II	D = −2	A = 2	B = 2	2
III	B = −1	C = −1	D = −3	−5
IV	C = 0	D = −4	A = 2	−2
$T_{..k}$	−1	−2	+1	−2 = $T_{...}$

In Table 16.22

$$t = b = 4$$
$$r = k = 3$$
$$\lambda = 2$$

Treatment totals of A, B, C, D are

$$T_{.j}: 6, 2, -1, -9$$

From this,

$$SS_{total} = \sum_i \sum_j \sum_k Y_{ijk}^2 - \frac{T_{...}^2}{N} = 48 - \frac{(-2)^2}{12} = 47.67$$

Position effect may first be ignored, since every position occurs once and only once in each block and once with each treatment, so that positions are orthogonal to blocks and treatments:

$$SS_{block\ (ignoring\ treatments)} = \frac{(3)^2 + (2)^2 + (-5)^2 + (-2)^2}{3} - \frac{(-2)^2}{12}$$

$$= \frac{42}{3} - \frac{1}{3} = \frac{41}{3} = 13.67$$

For treatment sum of squares adjusted for blocks we get

$$Q_1 = 3(+6) - 3 = 15$$
$$Q_2 = 3(2) - 0 = 6$$
$$Q_3 = 3(-1) - (-4) = 1$$
$$Q_4 = 3(-9) - (-5) = -22$$

and

$$\sum_i Q_i = 0$$

$$SS_{treatment} = \frac{(15)^2 + (6)^2 + (1)^2 + (-22)^2}{(4)(6)} = 31.08$$

$$SS_{position} = \frac{(-1)^2 + (-2)^2 + (1)^2}{4} - \frac{(-2)^2}{12} = 1.17$$

$$SS_{error} = SS_{total} - SS_{block} - SS_{treatment\ (adjusted)} - SS_{position}$$

$$= 47.67 - 13.67 - 31.08 - 1.17$$

$$= 1.75$$

The analysis appears in Table 16.23.

Table 16.23 Youden Square ANOVA

Source	df	SS	MS
Treatment τ_j (adjusted)	3	31.08	10.36
Blocks β_i	3	13.67	—
Position γ_k	2	1.17	0.58
Error ε_{ijk}	3	1.75	0.58
Totals	11	47.67	

The position effect here is not significant and it might be desirable to pool it with the error, getting

$$s_{\varepsilon}^2 = \frac{1.17 + 1.75}{2 + 3} = \frac{2.92}{5} = 0.58$$

as an estimate of error variance, with 5 df. Then the treatment effect is highly significant. The block mean square is not given, as blocks must be adjusted by treatments if block effects are to be assessed. The procedure is the same as shown for incomplete block designs. ∎

Problems

16.1 One study reports on the breaking strength Y of samples of seven different starch films where the thickness of each film X was also recorded as a covariate. Results rounded off considerably are as follows:

Source	df	Σy^2	Σxy	Σx^2	Adjusted Σy^2	df	MS
Between starches		6,000,000	50,000	500			
Within starches	100						
Totals		8,000,000	60,000	600			

a. Complete this table and set up a test for the effect of starch film or breaking strength both before and after adjustment for thickness.
b. Write the mathematical model for this experiment, briefly describing each term.
c. State the assumptions implied in the tests of hypotheses made above and describe briefly how you would test their validity.

16.2 In studying the effect of two different teaching methods the gain score of each pupil in a public school test was used as the response variable. The pretest scores are also available. Run a covariance analysis on the data below to determine

whether or not the new teaching method did show an improvement in average
achievement as measured by this public school test.

	Method I			Method II			Method I			Method II	
Pupil	Pretest Score	Gain	Pupil	Pretest Score	Gain	Pupil	Pretest Score	Gain	Pupil	Pretest Score	Gain
1	28	5	1	24	7	15	25	7	15	43	−2
2	43	−8	2	40	9	16	43	1	16	34	2
3	36	4	3	41	4	17	34	3	17	44	7
4	35	1	4	33	4	18	40	5	18	43	4
5	31	4	5	45	5	19	36	4	19	28	3
6	34	1	6	41	−2	20	30	9	20	46	−1
7	33	4	7	41	6	21	35	−6	21	43	2
8	38	8	8	33	11	22	38	0	22	46	−4
9	39	1	9	41	−1	23	31	−14	23	21	4
10	44	−1	10	30	15	24	41	4			
11	36	−7	11	45	4	25	35	10			
12	21	2	12	50	2	26	27	10			
13	34	−1	13	42	9	27	45	1			
14	33	1	14	47	3						

16.3 Data below are for five individuals who have been subjected to four conditions or
treatments represented by the groups or lots 1–4. Y represents some measure of
an individual supposedly affected by the variations in the treatments of the four
lots. X represents another measure of an individual that may affect the value of Y
even in the absence of the four treatments. The problem is to determine whether
or not the Y means of the four groups differ significantly from each other after
the effects of the X variable have been removed.

Groups or Lots							
1		2		3		4	
X	Y	X	Y	X	Y	X	Y
29	22	15	30	16	12	5	23
20	22	9	32	31	8	25	25
14	20	1	26	26	13	16	28
21	24	6	25	35	25	10	26
6	12	19	37	12	7	24	23

16.4   Consider a two-way classified design with more than one observation per cell.
  a. Show the F test for one of the main effects in a covariance analysis if both factors are fixed.
  b. Show the F test for the interaction in a covariance analysis if one factor is fixed and the other one is random.

16.5   For Example 16.5 in the text, another response variable was measured—the mean surface factor for heat slivers. Results of the first three cycles at the 10 experimental points gave

Cycle	(1)	(2)	(3)	(4)	(5)	(6)	(7)	(8)	(9)	(10)
1	4.0	6.8	5.9	4.5	4.3	4.2	7.8	8.3	11.0	5.9
2	5.2	7.2	6.1	3.7	4.0	4.5	7.0	8.1	12.5	6.3
3	4.8	8.0	7.0	3.6	3.8	4.3	6.5	7.8	12.7	7.0

Set up EVOP work sheets for these data and comment on your results after three cycles.

16.6   Data on yield in pounds for varying percent concentration at levels 29, 30, and 31 percent, and varying power at levels 400, 450, and 500 watts gave the following results in four cycles at the five points in a 2^2 design.

Cycle	(1)	(2)	(3)	(4)	(5)
1	477	472	411	476	372
2	469	452	430	468	453
3	465	396	375	468	292
4	451	469	363	432	460

Analyze by EVOP methods.

16.7   The response variable in the data below is the number of defective knife handles in samples of 72 knives in which the factors are machines, lumber grades, and replications. Analyze after making a suitable transformation of the data.

Lumber Grade	Replication	Machine 1	2
A	I	8	4
	II	4	6
B	I	9	3
	II	10	4

16.8 Analyze the data of Problem 16.7 using factorial chi square. Compare with the results of Problem 16.7.

16.9 Data on screen color difference on a television tube measured in degrees Kelvin are to be compared for four operators. On a given day only three operators can be used in the experiment. A balanced incomplete block design gave results as follows:

Day	Operator			
	A	B	C	D
Monday	780	820	800	—
Tuesday	950	—	920	940
Wednesday	—	880	880	820
Thursday	840	780	—	820

Do a complete analysis of these data and discuss your findings with regard to differences between operators.

16.10 Run orthogonal contrasts on the operators in Problem 16.9.

16.11 At the Bureau of Standards an experiment was to be run on the wear resistance of a new experimental shoe leather as compared with the standard leather in use by the Army. It was decided to equip several men with one shoe of the experimental-type leather and the other shoe of standard leather and after many weeks in the field compare wear on the two types. Considering each man as a block, suggest a reasonable number of men to be used in this experiment and outline its possible analysis.

16.12 If three experimental types of leather were to be tested along with the standard in Problem 16.11, it is obvious that only two of the four types can be tested on one man. Set up a balanced incomplete block design for this situation. Insert some arbitrary numerical values and complete an ANOVA table for your data.

Summary and Special Problems

Throughout this book the three phases of an experiment have been emphasized: the experiment, the design, and the analysis. At the end of most chapters an outline of the designs considered up to that point has been presented. It may prove useful to the reader to have a complete outline as a summary of the designs presented in Chapters 1–15.

This is not the only way such an outline could be constructed, but it represents an attempt to see each design as part of the overall picture of designed experiments. A look at the outline will reveal that the same design is often used for an entirely different experiment, which simply illustrates the fact that much care must be taken to spell out the experiment and its design or method of randomization before an experiment is performed.

For each experiment, the chapter reference in this book is given.

After the summary several special problems are presented. Many do not fit any one pattern or design but may be of interest as they represent the kinds of problems actually encountered in practice. These problems have no "pat" solutions and the way they are set up may depend a great deal upon assumptions the reader may make. They should serve to evoke some good discussion about designing an experiment to solve a given practical problem.

Experiment	Design	Analysis	Chapter Reference
I. Single factor			
	1. Completely randomized $Y_{ij} = \mu + \tau_j + \varepsilon_{ij}$	1. One-way Anova	Chapter 3
	2. Randomized block $Y_{ij} = \mu + \beta_i + \tau_j + \varepsilon_{ij}$	2.	
	a. Complete	a. Two-way ANOVA	Chapter 4
	b. Incomplete, balanced	b. Special ANOVA	Chapter 16
	c. Incomplete, general	c. General regression method	
	3. Latin square $Y_{ijk} = \mu + \beta_i + \tau_j + \gamma_k + \varepsilon_{ijk}$	3.	
	a. Complete	a. Three-way ANOVA	Chapter 4
	b. Incomplete, Youden square	b. Special ANOVA (like 2b)	Chapter 16
	4. Graeco–Latin square $Y_{ijkm} = \mu + \beta_i + \tau_j + \gamma_k + \omega_m + \varepsilon_{ijkm}$	4. Four-way ANOVA	Chapter 4
II. Two or more factors A. Factorial (crossed)			
	1. Completely randomized $Y_{ijk} = \mu + A_i + B_j + AB_{ij} + \varepsilon_{k(ij)}, \ldots$ for more factors	1.	
	a. General case	a. ANOVA with interactions	Chapter 5
	b. 2^f case	b. Yates method or general ANOVA; use: $(1), a, b, ab, \ldots$	Chapter 9
	c. 3^f case	c. General ANOVA; use $00, 10, 20, 01, 011, \ldots$ and $A \times B = AB + AB^2, \ldots$ for interaction	Chapter 10

Experiment	Design	Analysis	Chapter Reference
	2. Randomized block	2.	
	a. Complete $Y_{ijk} = \mu + R_k + A_i + B_j + AB_{ij} + \varepsilon_{ijk}$	a. Factorial ANOVA with replications R_k	Chapter 8
	b. Incomplete, confounding:	b.	
	i. Main effect—split plot $Y_{ijk} = \mu + \underbrace{R_i + A_j + RA_{ij}}_{\text{whole plot}}$ $\underbrace{+B_k + RB_{ik} \\ +AB_{jk} + RAB_{ijk}}_{\text{split plot}}$	i. Split-plot ANOVA	Chapter 11
	ii. Interactions in 2^f and 3^f	ii.	Chapter 12
	(1) Several replications $Y_{ijkq} = \mu + R_i + \beta_j + R\beta_{ij} + A_k + B_q + AB_{kq} + AB_{kq} + \varepsilon_{ijkq}$	(1) Factorial ANOVA with replications R_i and blocks β_j or confounded interaction	
	(2) One replication only $Y_{ijk} = \mu + \beta_i + A_j + B_k + AB_{jk}, \ldots$	(2) Factorial ANOVA with blocks β_i or confounded interaction	Chapter 12
	(3) Fractional replication— aliases $Y_{ijk} = \mu + A_i + B_j + AB_{ij}, \ldots$ (a) in 2^f (b) in 3^f; $f = 3$: Latin square	(3) Factorial ANOVA with aliases or aliased interactions	Chapter 13
	3. Latin square	3.	
	a. Complete $Y_{ijkm} = \mu + R_k + \gamma_m + A_i + B_j + AB_{ij} + \varepsilon_{ijkm}$	a. Factorial ANOVA with replications and positions	Chapter 8

Experiment	Design	Analysis	Chapter Reference
B. Nested (hierarchical)			
	1. Completely randomized $Y_{ijk} = \mu + A_i +$ $\quad B_{j(i)} + \varepsilon_{k(ij)}$	1. Nested ANOVA	Chapter 7
	2. Randomized block a. Complete $Y_{ijk} = \mu + R_k + A_i$ $\quad + B_{j(i)} + \varepsilon_{ijk}$	2. a. Nested ANOVA with blocks R_k	Chapters 4 and 7
	3. Latin square a. Complete $X_{ijkm} = \mu + R_k + \gamma_m$ $\quad + A_i + B_{j(i)}$ $\quad + \varepsilon_{ijkm}$	3. a. Nested ANOVA with blocks and positions	Chapters 4 and 7
C. Nested factorial			
	1. Completely randomized $Y_{ijkm} = \mu + A_i + B_{j(i)}$ $\quad + C_k + AC_{ik}$ $\quad + BC_{kj(i)} + \varepsilon_{m(ijk)}$	1. Nested-factorial ANOVA	Chapter 7
	2. Randomized block a. Complete $Y_{ijkm} = \mu + R_k + A_i$ $\quad + B_{j(i)} + C_m$ $\quad + AC_{im} + BC_{mj(i)}$ $\quad + \varepsilon_{ijkm}$	2. a. Nested-factorial ANOVA with blocks R_k	Chapters 4 and 7
	3. Latin square a. Complete $Y_{ijkmq} = \mu + R_k + \gamma_m$ $\quad + A_i + B_{j(i)}$ $\quad + C_q + AC_{iq}$ $\quad + BC_{qj(i)} + \varepsilon_{ijkmq}$	3. a. Nested-factorial ANOVA with blocks and positions	Chapters 4 and 7

Special Problems

1. Three different time periods are chosen in which to plant a certain vegetable: early May, late May, and early June. Plots are selected in early May and five different fertilizers are randomly tried in each plot. This is repeated in late May and again in early June. The criterion is the number of bushels per acre at harvest time. Assuming r plots for each fertilizer can be used at each planting:
   a. Discuss the nature of this experiment.
   b. Recommend a value for r to make this an acceptable study. (Time periods and fertilizers are fixed.)

2. In the study of the factors affecting the surface finish of steel in problem 13.19 there was a sixth factor—factor A (the flux type)—also set at two levels. It was possible to run all 64 (2^6) treatment combinations except for the fact that the experimenter wanted a complete replication of the experiment that was to be run. Thus, it was decided to do a one-half replicate of the 2^6 and then replicate the same block at another time. It was also learned that first a flux type had to be chosen, and then a slab width, and then the other four factors randomized. After this the experimenter would run the other slab width and then repeat the experiment with the other flux type.

   Outline a data format for this problem and show what its ANOVA table might look like.

3. In Problem 11.18 regarding the 81 engines, it took about a year to gather all the data. When the data were ready they were found to be incomplete at the low-temperature level so the experiment was really a one-third replication of a $3^4 \times 2$ factorial instead of the 3^5 factorial. How would you analyze the data if one whole level in your answer to this problem were missing?

4. In a food industry there was a strong feeling that viscosity readings on a slurry were not consistent. Some claimed that different laboratories gave different readings. Others said it was the fault of the instruments in these laboratories. And still others claimed it was the fault of the operators who read the instruments. To determine where the main sources of variability were, slurries from two products with vastly different viscosities were taken. Each sample of slurry was mixed well and then divided into four jars, one of which was sent to each of four laboratories for analysis. In each laboratory the jar was divided into eight samples in order for each of two randomly chosen operators within that laboratory to test two samples on each of two randomly chosen test instruments. Considering the laboratories as fixed, the two products as fixed, operators as random, and instruments as random, show a data layout for this problem. Show its model and its ANOVA outline and discuss any concerns about the tests to be made.

5. A total of 256 seventh grade students with equal numbers of boys and girls and representing both high and low socioeconomic status (SES) were subjected to a mathematics lesson that was believed to be either interesting or uninteresting under one of four reinforcement conditions: no knowledge of results, immediate verbal knowledge of results, delayed verbal knowledge of results, and delayed written knowledge of results. Task interest and type of reinforcement emerged as significant at the 1 percent significance level. The delayed verbal condition was significantly

superior to the other three, which did not differ from each other. Also, low SES children learned the high-interest task as quickly as high SES children, and more quickly than they learned the low-interest task. One the basis of this information answer the following.

a. What is the implied criterion Y?

b. List each independent variable and state the levels of each.

c. Set up an ANOVA table for this problem and indicate which results are significant.

d. Show a data layout indicating the number of children in each treatment.

e. For the significant results set some hypothetical means to show possible interpretation of such results.

6. In testing for a significant effect of treatments on some measured variable Y the resulting F test was almost, but not quite, significant at the 5 percent level (say, $F = 4.00$ with 3 and 8 df). Now the experimenter repeats the whole experiment on another set of independent observations under the same treatments and again the F fails to reach the 5 percent level (say, $F = 4.01$ with 3 and 8 df). What conclusion—if any—would you reach with regard to the hypotheses of equal treatment effects? Briefly explain your reasoning.

7. Six different formulas or mixes of concrete are to be purchased from five competing suppliers. Tests of crushing strength are to be made on two blocks constructed according to each formula–supplier combination. One block will be poured, hardened, and tested for each combination before the second block is formed, thus making two replications of the whole experiment. Assuming mixes and suppliers fixed and replications random, set up a mathematical model for this situation, outline its ANOVA properties, and show what F tests are appropriate.

8. Four chemical treatments A, B, C, and D are prepared in three concentrations (3 to 1, 2 to 1, and 1 to 1) and two samples of each combination are poured; then three determinations of percent of pH are made from each sample. Considering treatments and concentrations fixed and samples and determinations random, set up the appropriate ANOVA table and comment on statistical tests to be made. Also explain in some detail how you would proceed to set 95 percent confidence limits on the average percent of pH for treatment B, concentration 2 to 1.

9. A graduate student carried out an experiment involving six experimental conditions. From an available group of subjects she assigned five men and five women at random to each of the six conditions, using 60 subjects in all. Upon completion of the experiment, she carried out the following analysis to test the differences among the condition means:

Source	df	
Between conditions	5	$F = \dfrac{MS_b}{MS_w}$ with 5 and 54 df
Within conditions	54	
Total	59	

At this point a statistical consultant was called in.

a. The consultant blew his top. Why?

b. You are now the consultant. Set up the analysis the student should use, with sources of variation, df, and F tests, and explain to her why she must use your analysis and why hers is incorrect.

c. As consultant you must point out the possibility that one of the significance tests in the appropriate analysis may affect the interpretation of the other. How? What should be done in such a case?

10. For the problem (Problem 6.6) on the effect of four factors on chemical yield, the original data had an unequal number of observations at each treatment combination. The numbers in each treatment combination below are the n's for each particular treatment. Explain how you might handle the analysis of this problem.

		A_1		A_2	
		B_1	B_2	B_1	B_2
C_1	D_1	2	2	7	5
	D_2	2	1	2	3
C_2	D_1	4	2	5	1
	D_2	1	1	2	2

Glossary of Terms

Alias.   An effect in a fractionally replicated design that "looks like" or cannot be distinguished from another effect.

Alpha (α).   Size of the type I error or probability of rejecting a hypothesis when true.

Autocorrelation Coefficient.   A correlation between measures within the same sample in the order taken.

Beta (β).   Size of the type II error or probability of accepting a hypothesis when some alternative hypothesis is true.

Canonical Form.   Form of a second-degree response-surface equation that allows determination of the type of surface.

Completely Randomized Design.   A design in which all treatments are assigned to the experimental units in a completely random manner.

Confidence Limits.   Two values between which a parameter is said to lie with a specified degree of confidence.

Confounding.   An experimental arrangement in which certain effects cannot be distinguished from others. One such effect is usually blocks.

Consistent Estimator.   An estimator of a parameter whose value comes closer to the parameter as the sample size is increased.

Contrast.   A linear combination of treatment totals or averages in which the sum of the coefficients is zero.

Correlation Coefficient (Pearson r).   The square root of the proportion of total variation accounted for by linear regression.

Correlation Index R.   The square root of the proportion of total variation accounted for by the regression equation of the degree being fitted to the data.

Covariance Analysis.   Technique for removing the effect of one or more variables on the response variable before assessing the effect of treatments on the response variable.

Critical Region.   A set of values of a test statistic in which the hypothesis under test is rejected.

Defining Contrast.   An expression that indicates which effects are confounded with blocks in a factorial design that is confounded.

Effect of a Factor.   The change in response produced by a change in level of the factor.

Errors.   Type I: Rejecting a hypothesis when true. Type II: Accepting a hypothesis when false.

Eta Squared (η^2).   Proportion of total variation accounted for by a regression line through all average responses for each value of the independent variable.

Evolutionary Operation.   An experimental procedure for collection information to improve a process without disturbing production.

Expected Value of a Statistic.   The average value of a statistic if it were calculated from an infinite number of equal-sized samples from a given population.

Ex-post-facto Research.   Research in which the variables have already acted and one is seeking some possible relationships between the variables.

Factorial Chi Square.   Technique for handling attribute data in an analysis of variance.

Factorial Experiment.   An experiment in which all levels of each factor in the experiment are combined with all levels of every other factor.

Fixed Effect.   Effect of a factor whose levels are set at specific values.

Fixed Model.   A model that includes fixed effects only.

Fractional Replication.   An experimental design in which only a fraction of a complete factorial is run.

Graeco–Latin Square.   An experimental design in which four factors are so arranged that each level of each factor is combined only once with each level of the other three factors.

Incomplete Block Design.   A randomized block design in which not all treatment combinations can be included in one block. Such a design is called a *balanced* design if each pair of treatments occurs together the same number of times.

Interaction.   An interaction between two factors means that a change in response between levels of one factor is not the same for all levels of the other factor.

Latin Square.   An experimental design in which each level of each factor is combined only once with each level of two other factors.

Least Squares Method.   Method of assigning estimates of parameters in a model such that the sum of the squares of the errors is minimized.

Linear Graphs.   Graphs of effects and interactions in a Taguchi design.

Max-min-con Principle.   A design principle in which the experimenter attempts to maximize the variance due to the manipulated variables, minimize the error variance, and control other possible sources of variance.

Mean.   Of a sample:

$$\bar{Y} = \sum_{i=1}^{n} Y_i / n$$

Of a population:

$$\mu = E(Y).$$

Mean Square.   An unbiased estimate of a population variance. Determined by dividing a sum of squares by its degree of freedom.

Minimum Variance Estimate.   If two or more estimates $u_1, u_2, \ldots, u_k$ are made of the same parameter θ, the estimate with the smallest variance is called the minimum variance estimate.

Mixed Model.   The model of a factorial experiment in which one or more factors are at fixed levels and at least one factor is at random levels.

Nested Experiment.   An experiment in which the levels of one factor are chosen within the levels of another factor.

Nested-Factorial Experiment.   An experiment in which some factors are factorial or crossed with others and some factors are nested within others.

Operating Characteristic Curve.   A plot of the probability of acceptance of a hypothesis under test versus the true parameter of the population.

Orthogonal Array.   A special design used in Taguchi procedures.

Orthogonal Contrasts.   Two contrasts are said to be orthogonal if the products of their corresponding coefficients add to zero.

Orthogonal Polynomials.   Sets of polynomials used in regression analysis such that each polynomial in the set is orthogonal to all other polynomials in the set.

Parameter.   A characteristic of a population, such as the population mean or variance.

Point Estimate.   A single statistic used to estimate a parameter.

Power of a Test.   A plot of the probability of rejection of a hypothesis under test versus the true parameter. It is the complement of the operating characteristic curve.

Principal Block.   The block in a confounded design that contains the treatment combination in which all factors are at their lowest level.

Random Effect.   Effect of a factor whose levels are set at random from many possible levels.

Random Model.   A model of a factorial experiment in which the levels of the factor are chosen at random.

Random Sample.   A sample in which each member of the population sampled has an equal chance of being selected in this sample.

Randomized Block Design.   An experimental design in which the treatment combinations are randomized within a block and several blocks are run.

Repeated-Measures Design.   A design in which measurements are taken on the experimental units more than once.

Research.   A systematic quest for undiscovered truth.

Residual Analysis.   Analysis of what remains of response variation after effects in the model have been removed.

Resolution.   A scheme for determining which effects and interactions can be assessed in a given fractional factorial experiment.

Rotatable Design.   An experimental design that has equal predictive power in all directions from a center point, and in which all experimental points are equidistant from this center point.

SAS. Statistical Analysis System.   A procedure used to analyze statistical data on a computer.

Split-Plot Design.   An experimental design in which a main effect is confounded with blocks due to the practical necessities of the order of experimentation.

Standard Error of Estimate.   The standard deviation of errors of estimate around a least squares fitted regression model.

Statistic u.   A measure computed from a sample.

Statistical Hypothesis H_0.   An assumption about a population being sampled.

Statistical Inference.   Inferring something about a population of measures from a sample of that population.

Statistics.   A tool for decision making in the light of uncertainty.

Steepest Ascent Method.   A method of sequential experiments that will direct the experimenter toward the optimum of a response surface.

Sum of Squares (SS).   The sum of the squares of deviations of a random variable from its mean,

$$SS = \sum_{i=1}^{n} (Y_i - \bar{Y})^2$$

Test of a Hypothesis.   A rule by which a hypothesis is accepted or rejected.

Test Statistic.   A statistic used to test a hypothesis.

Treatment Combination.   A given combination showing the levels of all factors to be run for that set of experimental conditions.

True Experiment.   A study in which certain independent variables are manipulated, their effect on one or more dependent variables is observed, and the levels of the independent variables are assigned at random to the experimental units of the study.

Unbiased Statistic.   A statistic whose expected value equals the parameter it is estimating.

Variance.   Of a sample:

$$s^2 = \frac{\sum_{i=1}^{n} (Y_i - \bar{Y})^2}{n-1} = \frac{SS}{df}$$

Of a population:

$$\sigma^2 = E(Y - \mu)^2.$$

Youden Square.   An incomplete Latin square.

References

1. Anderson, V. L., and R. A. McLean, "Restriction Errors: Another Dimension in Teaching Experimental Statistics," *The American Statistician* (Nov. 1974).

2. Barnett, E. H., "Introduction to Evolutionary Operation," *Industrial and Engineering Chemistry,* vol. 52 (June 1960), p. 500.

3. Batson, H. C., "Applications of Factorial Analysis to Experiments in Chemistry," *National Convention Transactions of the American Society for Quality Control* (1956), pp. 9–23.

4. Bennett, C. A., and N. L. Franklin, *Statistical Analysis in Chemistry and the Chemical Industry.* New York: John Wiley & Sons, Inc. (1954).

5. *Biometrics,* vol. 13 (Sept. 1957), pp. 261–405.

6. Box, G. E. P., and N. R. Draper, *Evolutionary Operation: A Statistical Method for Process Improvement.* New York: John Wiley & Sons, Inc. (1969).

7. Box, G. E. P., Hunter, J. S., and W. G. Hunter, *Statistics for Experimenters: An Introduction to Design, Data Analysis and Model Building.* New York: John Wiley & Sons, Inc. (1978).

8. Burr, I. W., *Statistical Quality Control Methods.* New York: Marcel Dekker (1976).

9. Burr, I. W., *Applied Statistical Methods.* New York: Academic Press (1974).

10. Davies, O. L., *Design and Analysis of Industrial Experiments.* New York: Hafner Publishing Company (1954).

11. Dixon, W. J., and F. J. Massey, *An Introduction to Statistical Analysis* (2nd Ed.). New York: McGraw-Hill Book Company (1957).

12. Fisher, R. A., and F. Yates, *Statistical Tables for Biological, Agricultural and Medical Research* (4th Ed.). Edinburgh and London: Oliver & Boyd, Ltd. (1953).

13. Hunter, J. S., "Determination of Optimum Operating Conditions by Experimental Methods," *Industrial Quality Control* (Dec.–Feb. 1958–1959).

14. Kempthorne, O., *The Design and Analysis of Experiments.* New York: John Wiley & Sons, Inc. (1952).

15. Kerlinger, F. N., *Foundations of Behavioral Research* (2nd Ed.). New York: Holt, Rinehart and Winston (1973).

16. Keuls, M., "The Use of the Studentized Range in Connection with an Analysis of Variance," *Euphytica,* vol. 1 (1952), pp. 112–122.

17. Leedy, P. D., *Practical Research: Planning and Design.* New York: Macmillan (1974).

18. McCall, C. H., Jr., "Linear Contrasts, Parts I, II, and III," *Industrial Quality Control* (July–Sept., 1960).

19. Miller, L. D., "An Investigation of the Machinability of Malleable Iron Using Ceramic Tools," unpublished MSIE thesis, Purdue University (1959).

20. National Bureau of Standards Applied Mathematics Series. Nos. 48 (1957) and 54 (1959).

21. Nelson, L. S. "Extreme Screening Designs," *Journal of Quality Technology,* vol. 14 (April 1982).

22. Ostle, B., *Statistics in Research* (2nd Ed.). Ames, Iowa: Iowa State University Press (1963).

23. Owen, D. B., *Handbook of Statistical Tables.* Boston: Addison-Wesley Publishing Company, Inc. (1962)

24. Statistical Analysis System–*SAS User's Guide: Statistics Version 5 Edition.* SAS Institute Inc., Box 8000, Cary, NC 27511-8000 (1985).

25. Scheffe, H., "A Method for Judging All Contrasts in the Analysis of Variance," *Biometrics,* vol. XL (June, 1953).

26. Snedecor, G. W., and W. C. Cochran, *Statistical Methods* (6th Ed.). Ames, Iowa: Iowa State University Press (1967).

27. Taguchi, G. *Introduction to Quality Engineering.* Tokyo: Asian Productivity Organization (1986).

28. Tanur, Judith M., editor, and others, *Statistics: A Guide to the Unknown* (2nd Ed.). San Francisco: Holden-Day (1978).

29. Winer, B. J., *Statistical Principles in Experimental Design* (2nd Ed.). New York: McGraw-Hill Book Company (1971).

30. Wortham, A. W., and T. E. Smith, *Practical Statistics in Experimental Design.* Dallas: Dallas Publishing House (1960).

31. Yates, F., *Design and Analysis of Factorial Experiments.* London: Imperial Bureau of Soil Sciences (1937).

Statistical Tables

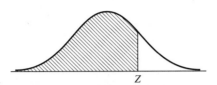

Z

Table A Areas under the Normal Curve* (Proportion of Total Area under the Curve from $-\infty$ to Designated Z Value)

Z	0.09	0.08	0.07	0.06	0.05	0.04	0.03	0.02	0.01	0.00
−3.5	0.00017	0.00017	0.00018	0.00019	0.00019	0.00020	0.00021	0.00022	0.00022	0.00023
−3.4	0.00024	0.00025	0.00026	0.00027	0.00028	0.00029	0.00030	0.00031	0.00033	0.00034
−3.3	0.00035	0.00036	0.00038	0.00039	0.00040	0.00042	0.00043	0.00045	0.00047	0.00048
−3.2	0.00050	0.00052	0.00054	0.00056	0.00058	0.00060	0.00062	0.00064	0.00066	0.00069
−3.1	0.00071	0.00074	0.00076	0.00079	0.00082	0.00085	0.00087	0.00090	0.00094	0.00097
−3.0	0.00100	0.00104	0.00107	0.00111	0.00114	0.00118	0.00122	0.00126	0.00131	0.00135
−2.9	0.0014	0.0014	0.0015	0.0015	0.0016	0.0016	0.0017	0.0017	0.0018	0.0019
−2.8	0.0019	0.0020	0.0021	0.0021	0.0022	0.0023	0.0023	0.0024	0.0025	0.0026
−2.7	0.0026	0.0027	0.0028	0.0029	0.0030	0.0031	0.0032	0.0033	0.0034	0.0035
−2.6	0.0036	0.0037	0.0038	0.0039	0.0040	0.0041	0.0043	0.0044	0.0045	0.0047
−2.5	0.0048	0.0049	0.0051	0.0052	0.0054	0.0055	0.0057	0.0059	0.0060	0.0062
−2.4	0.0064	0.0066	0.0068	0.0069	0.0071	0.0073	0.0075	0.0078	0.0080	0.0082
−2.3	0.0084	0.0087	0.0089	0.0091	0.0094	0.0096	0.0099	0.0102	0.0104	0.0107
−2.2	0.0110	0.0113	0.0116	0.0119	0.0122	0.0125	0.0129	0.0132	0.0136	0.0139
−2.1	0.0143	0.0146	0.0150	0.0154	0.0158	0.0162	0.0166	0.0170	0.0174	0.0179
−2.0	0.0183	0.0188	0.0192	0.0197	0.0202	0.0207	0.0212	0.0217	0.0222	0.0228
−1.9	0.0233	0.0239	0.0244	0.0250	0.0256	0.0262	0.0268	0.0274	0.0281	0.0287
−1.8	0.0294	0.0301	0.0307	0.0314	0.0322	0.0329	0.0336	0.0344	0.0351	0.0359
−1.7	0.0367	0.0375	0.0384	0.0392	0.0401	0.0409	0.0418	0.0427	0.0436	0.0446
−1.6	0.0455	0.0465	0.0475	0.0485	0.0495	0.0505	0.0516	0.0526	0.0537	0.0548
−1.5	0.0559	0.0571	0.0582	0.0594	0.0606	0.0618	0.0630	0.0643	0.0655	0.0668
−1.4	0.0681	0.0694	0.0708	0.0721	0.0735	0.0749	0.0764	0.0778	0.0793	0.0808
−1.3	0.0823	0.0838	0.0853	0.0869	0.0885	0.0901	0.0918	0.0934	0.0951	0.0968
−1.2	0.0985	0.1003	0.1020	0.1038	0.1057	0.1075	0.1093	0.1112	0.1131	0.1151
−1.1	0.1170	0.1190	0.1210	0.1230	0.1251	0.1271	0.1292	0.1314	0.1335	0.1357
−1.0	0.1379	0.1401	0.1423	0.1446	0.1469	0.1492	0.1515	0.1539	0.1562	0.1587
−0.9	0.1611	0.1635	0.1660	0.1685	0.1711	0.1736	0.1762	0.1788	0.1814	0.1841
−0.8	0.1867	0.1894	0.1922	0.1949	0.1977	0.2005	0.2033	0.2061	0.2090	0.2119
−0.7	0.2148	0.2177	0.2207	0.2236	0.2266	0.2297	0.2327	0.2358	0.2389	0.2420
−0.6	0.2451	0.2483	0.2514	0.2546	0.2578	0.2611	0.2643	0.2676	0.2709	0.2743
−0.5	0.2776	0.2810	0.2843	0.2877	0.2912	0.2946	0.2981	0.3015	0.3050	0.3085
−0.4	0.3121	0.3156	0.3192	0.3228	0.3264	0.3300	0.3336	0.3372	0.3409	0.3446
−0.3	0.3483	0.3520	0.3557	0.3594	0.3632	0.3669	0.3707	0.3745	0.3783	0.3821
−0.2	0.3859	0.3897	0.3936	0.3974	0.4013	0.4052	0.4090	0.4129	0.4168	0.4207
−0.1	0.4247	0.4286	0.4325	0.4364	0.4404	0.4443	0.4483	0.4522	0.4562	0.4602
−0.0	0.4641	0.4681	0.4721	0.4761	0.4801	0.4840	0.4880	0.4920	0.4960	0.5000

*Adapted from E. L. Grant, *Statistical Quality Control* (2nd Ed.). New York: McGraw-Hill Book Company, Inc. (1952), Table A, pp. 510–511. Reproduced by permission of the publisher.

Table A (Continued)

Z	0.00	0.01	0.02	0.03	0.04	0.05	0.06	0.07	0.08	0.09
+0.0	0.5000	0.5040	0.5080	0.5120	0.5160	0.5199	0.5239	0.5279	0.5319	0.5359
+0.1	0.5398	0.5438	0.5478	0.5517	0.5557	0.5596	0.5636	0.5675	0.5714	0.5753
+0.2	0.5793	0.5832	0.5871	0.5910	0.5948	0.5987	0.6026	0.6064	0.6103	0.6141
+0.3	0.6179	0.6217	0.6255	0.6293	0.6331	0.6368	0.6406	0.6443	0.6480	0.6517
+0.4	0.6554	0.6591	0.6628	0.6664	0.6700	0.6736	0.6772	0.6808	0.6844	0.6879
+0.5	0.6915	0.6950	0.6985	0.7019	0.7054	0.7088	0.7123	0.7157	0.7190	0.7224
+0.6	0.7257	0.7291	0.7324	0.7357	0.7389	0.7422	0.7454	0.7486	0.7517	0.7549
+0.7	0.7580	0.7611	0.7642	0.7673	0.7704	0.7734	0.7764	0.7794	0.7823	0.7852
+0.8	0.7881	0.7910	0.7939	0.7967	0.7995	0.8023	0.8051	0.8079	0.8106	0.8133
+0.9	0.8159	0.8186	0.8212	0.8238	0.8264	0.8289	0.8315	0.8340	0.8365	0.8389
+1.0	0.8413	0.8438	0.8461	0.8485	0.8508	0.8531	0.8554	0.8577	0.8599	0.8621
+1.1	0.8643	0.8665	0.8686	0.8708	0.8729	0.8749	0.8770	0.8790	0.8810	0.8830
+1.2	0.8849	0.8869	0.8888	0.8907	0.8925	0.8944	0.8962	0.8980	0.8997	0.9015
+1.3	0.9032	0.9049	0.9066	0.9082	0.9099	0.9115	0.9131	0.9147	0.9162	0.9177
+1.4	0.9192	0.9207	0.9222	0.9236	0.9251	0.9265	0.9279	0.9292	0.9306	0.9319
+1.5	0.9332	0.9345	0.9357	0.9370	0.9382	0.9394	0.9406	0.9418	0.9429	0.9441
+1.6	0.9452	0.9463	0.9474	0.9484	0.9495	0.9505	0.9515	0.9525	0.9535	0.9545
+1.7	0.9554	0.9564	0.9573	0.9582	0.9591	0.9599	0.9608	0.9616	0.9625	0.9633
+1.8	0.9641	0.9649	0.9656	0.9664	0.9671	0.9678	0.9686	0.9693	0.9699	0.9706
+1.9	0.9713	0.9719	0.9726	0.9732	0.9738	0.9744	0.9750	0.9756	0.9761	0.9767
+2.0	0.9773	0.9778	0.9783	0.9788	0.9793	0.9798	0.9803	0.9808	0.9812	0.9817
+2.1	0.9821	0.9826	0.9830	0.9834	0.9838	0.9842	0.9846	0.9850	0.9854	0.9857
+2.2	0.9861	0.9864	0.9868	0.9871	0.9875	0.9878	0.9881	0.9884	0.9887	0.9890
+2.3	0.9893	0.9896	0.9898	0.9901	0.9904	0.9906	0.9909	0.9911	0.9913	0.9916
+2.4	0.9918	0.9920	0.9922	0.9925	0.9927	0.9929	0.9931	0.9932	0.9934	0.9936
+2.5	0.9938	0.9940	0.9941	0.9943	0.9945	0.9946	0.9948	0.9949	0.9951	0.9952
+2.6	0.9953	0.9955	0.9956	0.9957	0.9959	0.9960	0.9961	0.9962	0.9963	0.9964
+2.7	0.9965	0.9966	0.9967	0.9968	0.9969	0.9970	0.9971	0.9972	0.9973	0.9974
+2.8	0.9974	0.9975	0.9976	0.9977	0.9977	0.9978	0.9979	0.9979	0.9980	0.9981
+2.9	0.9981	0.9982	0.9983	0.9983	0.9984	0.9984	0.9985	0.9985	0.9986	0.9986
+3.0	0.99865	0.99869	0.99874	0.99878	0.99882	0.99886	0.99889	0.99893	0.99896	0.99900
+3.1	0.99903	0.99906	0.99910	0.99913	0.99915	0.99918	0.99921	0.99924	0.99926	0.99929
+3.2	0.99931	0.99934	0.99936	0.99938	0.99940	0.99942	0.99944	0.99946	0.99948	0.99950
+3.3	0.99952	0.99953	0.99955	0.99957	0.99958	0.99960	0.99961	0.99962	0.99964	0.99965
+3.4	0.99966	0.99967	0.99969	0.99970	0.99971	0.99972	0.99973	0.99974	0.99975	0.99976
+3.5	0.99977	0.99978	0.99978	0.99979	0.99980	0.99981	0.99981	0.99982	0.99983	0.99983

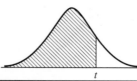

Table B Student's t Distribution

df	Percentile point						
	70	80	90	95	97.5	99	99.5
1	.73	1.38	3.08	6.31	12.71	31.82	63.66
2	.62	1.06	1.89	2.92	4.30	6.96	9.92
3	.58	.98	1.64	2.35	3.18	4.54	5.84
4	.57	.94	1.53	2.13	2.78	3.75	4.60
5	.56	.92	1.48	2.01	2.57	3.36	4.03
6	.55	.91	1.44	1.94	2.45	3.14	3.71
7	.55	.90	1.42	1.90	2.36	3.00	3.50
8	.55	.89	1.40	1.86	2.31	2.90	3.36
9	.54	.88	1.38	1.83	2.26	2.82	3.25
10	.54	.88	1.37	1.81	2.23	2.76	3.17
11	.54	.88	1.36	1.80	2.20	2.72	3.11
12	.54	.87	1.36	1.78	2.18	2.68	3.06
13	.54	.87	1.35	1.77	2.16	2.65	3.01
14	.54	.87	1.34	1.76	2.14	2.62	2.98
15	.54	.87	1.34	1.75	2.13	2.60	2.95
16	.54	.86	1.34	1.75	2.12	2.58	2.92
17	.53	.86	1.33	1.74	2.11	2.57	2.90
18	.53	.86	1.33	1.73	2.10	2.55	2.88
19	.53	.86	1.33	1.73	2.09	2.54	2.86
20	.53	.86	1.32	1.72	2.09	2.53	2.84
21	.53	.86	1.32	1.72	2.08	2.52	2.83
22	.53	.86	1.32	1.72	2.07	2.51	2.82
23	.53	.86	1.32	1.71	2.07	2.50	2.81
24	.53	.86	1.32	1.71	2.06	2.49	2.80
25	.53	.86	1.32	1.71	2.06	2.48	2.79
26	.53	.86	1.32	1.71	2.06	2.48	2.78
27	.53	.86	1.31	1.70	2.05	2.47	2.77
28	.53	.86	1.31	1.70	2.05	2.47	2.76
29	.53	.85	1.31	1.70	2.04	2.46	2.76
30	.53	.85	1.31	1.70	2.04	2.46	2.75
40	.53	.85	1.30	1.68	2.02	2.42	2.70
50	.53	.85	1.30	1.67	2.01	2.40	2.68
60	.53	.85	1.30	1.67	2.00	2.39	2.66
80	.53	.85	1.29	1.66	1.99	2.37	2.64
100	.53	.84	1.29	1.66	1.98	2.36	2.63
200	.52	.84	1.29	1.65	1.97	2.34	2.60
500	.52	.84	1.28	1.65	1.96	2.33	2.59
∞	.52	.84	1.28	1.64	1.96	2.33	2.58

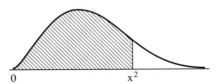

0   x^2

Table C Cumulative Chi-Square Distribution

Entry is χ_p^2 where $F(\chi_p^2) = P(\chi^2 \le \chi_p^2) = p$ and $\chi^2 \sim \chi_{(\nu)}^2$

ν	0.005	0.01	0.025	0.05	0.10	0.90	0.95	0.975	0.99	0.995
1	0.000	0.000	0.001	0.004	0.016	2.706	3.841	5.024	6.635	7.879
2	0.010	0.020	0.051	0.103	0.211	4.605	5.991	7.378	9.210	10.597
3	0.072	0.115	0.216	0.352	0.584	6.251	7.815	9.348	11.345	12.838
4	0.207	0.297	0.484	0.711	1.064	7.779	9.488	11.143	13.277	14.860
5	0.412	0.554	0.831	1.145	1.610	9.236	11.070	12.832	15.086	16.750
6	0.676	0.872	1.237	1.635	2.204	10.645	12.592	14.449	16.812	18.548
7	0.989	1.239	1.690	2.167	2.833	12.017	14.067	16.013	18.475	20.278
8	1.344	1.646	2.180	2.733	3.490	13.362	15.507	17.535	20.090	21.955
9	1.735	2.088	2.700	3.325	4.168	14.684	16.919	19.023	21.666	23.589
10	2.156	2.558	3.247	3.940	4.865	15.987	18.307	20.483	23.209	25.188
11	2.603	3.053	3.816	4.575	5.578	17.275	19.675	21.920	24.725	26.757
12	3.074	3.571	4.404	5.226	6.304	18.549	21.026	23.337	26.217	28.299
13	3.565	4.107	5.009	5.892	7.042	19.812	22.362	24.736	27.688	29.819
14	4.075	4.660	5.629	6.571	7.790	21.064	23.685	26.119	29.141	31.319
15	4.601	5.229	6.262	7.261	8.547	22.307	24.996	27.488	30.578	32.801
16	5.142	5.812	6.908	7.962	9.312	23.542	26.296	28.845	32.000	34.267
17	5.697	6.408	7.564	8.672	10.085	24.769	27.587	30.191	33.409	35.718
18	6.265	7.015	8.231	9.390	10.865	25.989	28.869	31.526	34.805	37.156
19	6.844	7.633	8.907	10.117	11.651	27.204	30.143	32.852	36.191	38.582
20	7.434	8.260	9.591	10.851	12.443	28.412	31.410	34.170	37.566	39.997
21	8.034	8.897	10.283	11.591	13.240	29.615	32.671	35.479	38.932	41.401
22	8.643	9.542	10.982	12.338	14.042	30.813	33.924	36.781	40.289	42.796
23	9.260	10.196	11.689	13.090	14.848	32.007	35.172	38.076	41.638	44.181
24	9.886	10.856	12.401	13.848	15.659	33.196	36.415	39.364	42.980	45.558
25	10.520	11.524	13.120	14.611	16.473	34.382	37.652	40.647	44.314	46.928
26	11.160	12.198	13.844	15.379	17.292	35.563	38.885	41.923	45.642	48.290
27	11.808	12.878	14.573	16.151	18.114	36.741	40.113	43.195	46.963	49.645
28	12.461	13.565	15.308	16.928	18.939	37.916	41.337	44.461	48.278	50.993
29	13.121	14.257	16.047	17.708	19.768	39.087	42.557	45.722	49.588	52.336
30	13.787	14.953	16.791	18.493	20.599	40.256	43.773	46.979	50.892	53.672
35	17.192	18.509	20.596	22.465	24.797	46.059	49.801	53.203	57.342	60.274
40	20.707	22.164	24.433	26.509	29.051	51.805	55.759	59.342	63.691	66.766
45	24.314	25.901	28.365	30.612	33.350	57.505	61.658	65.411	69.956	73.166
50	27.991	29.707	32.357	34.764	37.689	63.167	67.505	71.420	76.154	79.490
55	31.734	33.570	36.398	38.958	42.060	68.796	73.311	77.380	82.292	85.749
60	35.535	37.485	40.482	43.188	46.459	74.397	79.082	83.298	88.379	91.952
70	43.275	45.442	48.758	51.739	55.329	85.527	90.531	95.023	100.425	104.215
80	51.172	53.540	57.153	60.392	64.278	96.578	101.879	106.629	112.329	116.321
90	59.196	61.754	65.647	69.126	73.291	107.565	113.145	118.136	124.116	128.299
100	67.328	70.065	74.222	77.930	82.358	118.498	124.342	129.561	135.807	140.169

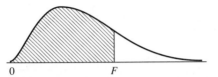

0   F

Table D Cumulative F Distribution

Tabled value is F_p where $p = P(F \le F_p)$ and $F \sim F_{(\nu_1, \nu_2)}$.

ν_2	p	ν_1									
		1	2	3	4	5	6	7	8	9	10
1	0.005	0.00006	0.0051	0.018	0.032	0.044	0.054	0.062	0.068	0.073	0.078
	0.01	0.00025	0.010	0.029	0.047	0.062	0.073	0.082	0.089	0.095	0.100
	0.025	0.0015	0.026	0.057	0.082	0.100	0.113	0.124	0.132	0.139	0.144
	0.05	0.0062	0.054	0.099	0.130	0.151	0.167	0.179	0.188	0.195	0.201
	0.10	0.025	0.117	0.181	0.220	0.246	0.265	0.279	0.289	0.298	0.304
	0.90	39.9	49.5	53.6	55.8	57.2	58.2	58.9	59.4	59.9	60.2
	0.95	161	200	216	225	230	234	237	239	241	242
	0.975	648	800	864	900	922	937	948	957	963	969
	0.99	4050	5000	5400	5620	5760	5860	5930	5980	6020	6060
	0.995	16200	20000	21600	22500	23100	23400	23700	23900	24100	24200
2	0.005	0.00005	0.0050	0.020	0.038	0.055	0.069	0.081	0.091	0.099	0.106
	0.01	0.00020	0.010	0.032	0.056	0.075	0.092	0.105	0.116	0.125	0.132
	0.025	0.0013	0.026	0.062	0.094	0.119	0.138	0.153	0.165	0.175	0.183
	0.05	0.0050	0.053	0.105	0.144	0.173	0.194	0.211	0.224	0.235	0.244
	0.10	0.020	0.111	0.183	0.231	0.265	0.289	0.307	0.321	0.333	0.342
	0.90	8.53	9.00	9.16	9.24	9.29	9.33	9.35	9.37	9.38	9.39
	0.95	18.5	19.0	19.2	19.2	19.3	19.3	19.4	19.4	19.4	19.4
	0.975	38.5	39.0	39.2	39.2	39.3	39.3	39.4	39.4	39.4	39.4
	0.99	98.5	99.0	99.2	99.2	99.3	99.3	99.4	99.4	99.4	99.4
	0.995	198	199	199	199	199	199	199	199	199	199
3	0.005	0.00005	0.0050	0.021	0.041	0.060	0.077	0.092	0.104	0.115	0.124
	0.01	0.00019	0.010	0.034	0.060	0.083	0.102	0.118	0.132	0.143	0.153
	0.025	0.0012	0.026	0.065	0.100	0.129	0.152	0.170	0.185	0.197	0.207
	0.05	0.0046	0.052	0.108	0.152	0.185	0.210	0.230	0.246	0.259	0.270
	0.10	0.019	0.109	0.185	0.239	0.276	0.304	0.325	0.342	0.356	0.367
	0.90	5.54	5.46	5.39	5.34	5.31	5.28	5.27	5.25	5.24	5.23
	0.95	10.1	9.55	9.28	9.12	9.01	8.94	8.89	8.85	8.81	8.79
	0.975	17.4	16.0	15.4	15.1	14.9	14.7	14.6	14.5	14.5	14.4
	0.99	34.1	30.8	29.5	28.7	28.2	27.9	27.7	27.5	27.3	27.2
	0.995	55.6	49.8	47.5	46.2	45.4	44.8	44.4	44.1	43.9	43.7

Table D (*Continued*)

ν_2	p	11	12	15	20	24	30	40	60	120	∞
							ν_1				
1	0.005	0.082	0.085	0.093	0.101	0.105	0.109	0.113	0.118	0.122	0.127
	0.01	0.104	0.107	0.115	0.124	0.128	0.132	0.137	0.141	0.146	0.151
	0.025	0.149	0.153	0.161	0.170	0.175	0.180	0.184	0.189	0.194	0.199
	0.05	0.207	0.211	0.220	0.230	0.235	0.240	0.245	0.250	0.255	0.261
	0.10	0.310	0.315	0.325	0.336	0.342	0.347	0.353	0.358	0.364	0.370
	0.90	60.5	60.7	61.2	61.7	62.0	62.3	62.5	62.8	63.1	63.3
	0.95	243	244	246	248	249	250	251	252	253	254
	0.975	973	977	985	993	997	1000	1010	1010	1010	1020
	0.99	6080	6110	6160	6210	6230	6260	6290	6310	6340	6370
	0.995	24300	24400	24600	24800	24900	25000	25100	25300	25400	25500
2	0.005	0.112	0.118	0.130	0.143	0.150	0.157	0.165	0.173	0.181	0.189
	0.01	0.139	0.144	0.157	0.171	0.178	0.186	0.193	0.201	0.209	0.217
	0.025	0.190	0.196	0.210	0.224	0.232	0.239	0.247	0.255	0.263	0.271
	0.05	0.251	0.257	0.272	0.286	0.294	0.302	0.309	0.317	0.326	0.334
	0.10	0.350	0.356	0.371	0.386	0.394	0.402	0.410	0.418	0.426	0.434
	0.90	9.40	9.41	9.42	9.44	9.45	9.46	9.47	9.47	9.48	9.49
	0.95	19.4	19.4	19.4	19.4	19.5	19.5	19.5	19.5	19.5	19.5
	0.975	39.4	39.4	39.4	39.4	39.5	39.5	39.5	39.5	39.5	39.5
	0.99	99.4	99.4	99.4	99.4	99.5	99.5	99.5	99.5	99.5	99.5
	0.995	199	199	199	199	199	199	199	199	199	200
3	0.005	0.132	0.138	0.154	0.172	0.181	0.191	0.201	0.211	0.222	0.234
	0.01	0.161	0.168	0.185	0.203	0.212	0.222	0.232	0.242	0.253	0.264
	0.025	0.216	0.224	0.241	0.259	0.269	0.279	0.289	0.299	0.310	0.321
	0.05	0.279	0.287	0.304	0.323	0.332	0.342	0.352	0.363	0.373	0.384
	0.10	0.376	0.384	0.402	0.420	0.430	0.439	0.449	0.459	0.469	0.480
	0.90	5.22	5.22	5.20	5.18	5.18	5.17	5.16	5.15	5.14	5.13
	0.95	8.76	8.74	8.70	8.66	8.63	8.62	8.59	8.57	8.55	8.53
	0.975	14.4	14.3	14.3	14.2	14.1	14.1	14.0	14.0	13.9	13.9
	0.99	27.1	27.1	26.9	26.7	26.6	26.5	26.4	26.3	26.2	26.1
	0.995	43.5	43.4	43.1	42.8	42.6	42.5	42.3	42.1	42.0	41.8

Table D (Continued)   *Numerator*

						ν_1					
ν_2	p	1	2	3	4	5	6	7	8	9	10
	0.005	0.00004	0.0050	0.022	0.043	0.064	0.083	0.100	0.114	0.126	0.137
	0.01	0.00018	0.010	0.035	0.063	0.088	0.109	0.127	0.143	0.156	0.167
	0.025	0.0011	0.026	0.066	0.104	0.135	0.161	0.181	0.198	0.212	0.224
	0.05	0.0044	0.052	0.110	0.157	0.193	0.221	0.243	0.261	0.275	0.288
	0.10	0.018	0.108	0.187	0.243	0.284	0.314	0.338	0.356	0.371	0.384
4											
	0.90	4.54	4.32	4.19	4.11	4.05	4.01	3.98	3.95	3.94	3.92
	0.95	7.71	6.94	6.59	6.39	6.26	6.16	6.09	6.04	6.00	5.96
	0.975	12.2	10.6	9.98	9.60	9.36	9.20	9.07	8.98	8.90	8.84
	0.99	21.2	18.0	16.7	16.0	15.5	15.2	15.0	14.8	14.7	14.5
	0.995	31.3	26.3	24.3	23.2	22.5	22.0	21.6	21.4	21.1	21.0
	0.005	0.00004	0.0050	0.022	0.045	0.067	0.087	0.105	0.120	0.134	0.146
	0.01	0.00017	0.010	0.035	0.064	0.091	0.114	0.134	0.151	0.165	0.177
	0.025	0.0011	0.025	0.067	0.107	0.140	0.167	0.189	0.208	0.223	0.236
	0.05	0.0043	0.052	0.111	0.160	0.198	0.228	0.252	0.271	0.287	0.301
	0.10	0.017	0.108	0.188	0.247	0.290	0.322	0.347	0.367	0.383	0.397
5											
	0.90	4.06	3.78	3.62	3.52	3.45	3.40	3.37	3.34	3.32	3.30
	0.95	6.61	5.79	5.41	5.19	5.05	4.95	4.88	4.82	4.77	4.74
	0.975	10.0	8.43	7.76	7.39	7.15	6.98	6.85	6.76	6.68	6.62
	0.99	16.3	13.3	12.1	11.4	11.0	10.7	10.5	10.3	10.2	10.1
	0.995	22.8	18.3	16.5	15.6	14.9	14.5	14.2	14.0	13.8	13.6
	0.005	0.00004	0.0050	0.022	0.045	0.069	0.090	0.109	0.126	0.140	0.153
	0.01	0.00017	0.010	0.036	0.066	0.094	0.118	0.139	0.157	0.172	0.186
	0.025	0.0011	0.025	0.068	0.109	0.143	0.172	0.195	0.215	0.231	0.246
	0.05	0.0043	0.052	0.112	0.162	0.202	0.233	0.259	0.279	0.296	0.311
	0.10	0.017	0.107	0.189	0.249	0.294	0.327	0.354	0.375	0.392	0.406
6											
	0.90	3.78	3.46	3.29	3.18	3.11	3.05	3.01	2.98	2.96	2.94
	0.95	5.99	5.14	4.76	4.53	4.39	4.28	4.21	4.15	4.10	4.06
	0.975	8.81	7.26	6.60	6.23	5.99	5.82	5.70	5.60	5.52	5.46
	0.99	13.7	10.9	9.78	9.15	8.75	8.47	8.26	8.10	7.98	7.87
	0.995	18.6	14.5	12.9	12.0	11.5	11.1	10.8	10.6	10.4	10.2
	0.005	0.00004	0.0050	0.023	0.046	0.070	0.093	0.113	0.130	0.145	0.159
	0.01	0.00017	0.010	0.036	0.067	0.096	0.121	0.143	0.162	0.178	0.192
	0.025	0.0010	0.025	0.068	0.110	0.146	0.176	0.200	0.221	0.238	0.253
	0.05	0.0042	0.052	0.113	0.164	0.205	0.238	0.264	0.286	0.304	0.319
	0.10	0.017	0.107	0.190	0.251	0.297	0.332	0.359	0.381	0.399	0.414
7											
	0.90	3.59	3.26	3.07	2.96	2.88	2.83	2.78	2.75	2.72	2.70
	0.95	5.59	4.74	4.35	4.12	3.97	3.87	3.79	3.73	3.68	3.64
	0.975	8.07	6.54	5.89	5.52	5.29	5.12	4.99	4.90	4.82	4.76
	0.99	12.2	9.55	8.45	7.85	7.46	7.19	6.99	6.84	6.72	6.62
	0.995	16.2	12.4	10.9	10.0	9.52	9.16	8.89	8.68	8.51	8.38

Table D *(Continued)*

ν_2	p	11	12	15	20	24	30	40	60	120	∞
						ν_1					
4	0.005	0.145	0.153	0.172	0.193	0.204	0.216	0.229	0.242	0.255	0.269
	0.01	0.176	0.185	0.204	0.226	0.237	0.249	0.261	0.274	0.287	0.301
	0.025	0.234	0.243	0.263	0.284	0.296	0.308	0.320	0.332	0.346	0.359
	0.05	0.298	0.307	0.327	0.349	0.360	0.372	0.384	0.396	0.409	0.422
	0.10	0.394	0.403	0.424	0.445	0.456	0.467	0.478	0.490	0.502	0.514
	0.90	3.91	3.90	3.87	3.84	3.83	3.82	3.80	3.79	3.78	3.76
	0.95	5.94	5.91	5.86	5.80	5.77	5.75	5.72	5.69	5.66	5.63
	0.975	8.79	8.75	8.66	8.56	8.51	8.46	8.41	8.36	8.31	8.26
	0.99	14.4	14.4	14.2	14.0	13.9	13.8	13.7	13.7	13.6	13.5
	0.995	20.8	20.7	20.4	20.2	20.0	19.9	19.8	19.6	19.5	19.3
5	0.005	0.156	0.165	0.186	0.210	0.223	0.237	0.251	0.266	0.282	0.299
	0.01	0.188	0.197	0.219	0.244	0.257	0.270	0.285	0.299	0.315	0.331
	0.025	0.248	0.257	0.280	0.304	0.317	0.330	0.344	0.359	0.374	0.390
	0.05	0.313	0.322	0.345	0.369	0.382	0.395	0.408	0.422	0.437	0.452
	0.10	0.408	0.418	0.440	0.463	0.476	0.488	0.501	0.514	0.527	0.541
	0.90	3.28	3.27	3.24	3.21	3.19	3.17	3.16	3.14	3.12	3.10
	0.95	4.71	4.68	4.62	4.56	4.53	4.50	4.46	4.43	4.40	4.36
	0.975	6.57	6.52	6.43	6.33	6.28	6.23	6.18	6.12	6.07	6.02
	0.99	9.96	9.89	9.72	9.55	9.47	9.38	9.29	9.20	9.11	9.02
	0.995	13.5	13.4	13.1	12.9	12.8	12.7	12.5	12.4	12.3	12.1
6	0.005	0.164	0.174	0.197	0.224	0.238	0.253	0.269	0.286	0.304	0.324
	0.01	0.197	0.207	0.232	0.258	0.273	0.288	0.304	0.321	0.338	0.357
	0.025	0.258	0.268	0.293	0.320	0.334	0.349	0.364	0.381	0.398	0.415
	0.05	0.324	0.334	0.358	0.385	0.399	0.413	0.428	0.444	0.460	0.476
	0.10	0.418	0.429	0.453	0.478	0.491	0.505	0.519	0.533	0.548	0.564
	0.90	2.92	2.90	2.87	2.84	2.82	2.80	2.78	2.76	2.74	2.72
	0.95	4.03	4.00	3.94	3.87	3.84	3.81	3.77	3.74	3.70	3.67
	0.975	5.41	5.37	5.27	5.17	5.12	5.07	5.01	4.96	4.90	4.85
	0.99	7.79	7.72	7.56	7.40	7.31	7.23	7.14	7.06	6.97	6.88
	0.995	10.1	10.0	9.81	9.59	9.47	9.36	9.24	9.12	9.00	8.88
7	0.005	0.171	0.181	0.206	0.235	0.251	0.267	0.285	0.304	0.324	0.345
	0.01	0.205	0.216	0.241	0.270	0.286	0.303	0.320	0.339	0.358	0.379
	0.025	0.266	0.277	0.304	0.333	0.348	0.364	0.381	0.399	0.418	0.437
	0.05	0.332	0.343	0.369	0.398	0.413	0.428	0.445	0.461	0.479	0.498
	0.10	0.427	0.438	0.463	0.491	0.504	0.519	0.534	0.550	0.566	0.582
	0.90	2.68	2.67	2.63	2.59	2.58	2.56	2.54	2.51	2.49	2.47
	0.95	3.60	3.57	3.51	3.44	3.41	3.38	3.34	3.30	3.27	3.23
	0.975	4.71	4.67	4.57	4.47	4.42	4.36	4.31	4.25	4.20	4.14
	0.99	6.54	6.47	6.31	6.16	6.07	5.99	5.91	5.82	5.74	5.65
	0.995	8.27	8.18	7.97	7.75	7.65	7.53	7.42	7.31	7.19	7.08

Table D *(Continued)*

ν_2	p					ν_1					
		1	2	3	4	5	6	7	8	9	10
8	0.005	0.00004	0.0050	0.027	0.047	0.072	0.095	0.115	0.133	0.149	0.164
	0.01	0.00017	0.010	0.036	0.068	0.097	0.123	0.146	0.166	0.183	0.198
	0.025	0.0010	0.025	0.069	0.111	0.148	0.179	0.204	0.226	0.244	0.259
	0.05	0.0042	0.052	0.113	0.166	0.208	0.241	0.268	0.291	0.310	0.326
	0.10	0.017	0.107	0.190	0.253	0.299	0.335	0.363	0.386	0.405	0.421
	0.90	3.46	3.11	2.92	2.81	2.73	2.67	2.62	2.59	2.56	2.54
	0.95	5.32	4.46	4.07	3.84	3.69	3.58	3.50	3.44	3.39	3.35
	0.975	7.57	6.06	5.42	5.05	4.82	4.65	4.53	4.43	4.36	4.30
	0.99	11.3	8.65	7.59	7.01	6.63	6.37	6.18	6.03	5.91	5.81
	0.995	14.7	11.0	9.60	8.81	8.30	7.95	7.69	7.50	7.34	7.21
9	0.005	0.00004	0.0050	0.023	0.047	0.073	0.096	0.117	0.136	0.153	0.168
	0.01	0.00017	0.010	0.037	0.068	0.098	0.125	0.149	0.169	0.187	0.202
	0.025	0.0010	0.025	0.069	0.112	0.150	0.181	0.207	0.230	0.248	0.265
	0.05	0.0040	0.052	0.113	0.167	0.210	0.244	0.272	0.296	0.315	0.331
	0.10	0.017	0.107	0.191	0.254	0.302	0.338	0.367	0.390	0.410	0.426
	0.90	3.36	3.01	2.81	2.69	2.61	2.55	2.51	2.47	2.44	2.42
	0.95	5.12	4.26	3.86	3.63	3.48	3.37	3.29	3.23	3.18	3.14
	0.975	7.21	5.71	5.08	4.72	4.48	4.32	4.20	4.10	4.03	3.96
	0.99	10.6	8.02	6.99	6.42	6.06	5.80	5.61	5.47	5.35	5.26
	0.995	13.6	10.1	8.72	7.96	7.47	7.13	6.88	6.69	6.54	6.42
10	0.005	0.00004	0.0050	0.023	0.048	0.073	0.098	0.119	0.139	0.156	0.171
	0.01	0.00017	0.010	0.037	0.069	0.100	0.127	0.151	0.172	0.190	0.206
	0.025	0.0010	0.025	0.069	0.113	0.151	0.183	0.210	0.233	0.252	0.269
	0.05	0.0041	0.052	0.114	0.168	0.211	0.246	0.275	0.299	0.319	0.336
	0.10	0.017	0.106	0.191	0.255	0.303	0.340	0.370	0.394	0.414	0.430
	0.90	3.28	2.92	2.73	2.61	2.52	2.46	2.41	2.38	2.35	2.32
	0.95	4.96	4.10	3.71	3.48	3.33	3.22	3.14	3.07	3.02	2.98
	0.975	6.94	5.46	4.83	4.47	4.24	4.07	3.95	3.85	3.78	3.72
	0.99	10.0	7.56	6.55	5.99	5.64	5.39	5.20	5.06	4.94	4.85
	0.995	12.8	9.43	8.08	7.34	6.87	6.54	6.30	6.12	5.97	5.85
11	0.005	0.00004	0.0050	0.023	0.048	0.074	0.099	0.121	0.141	0.158	0.174
	0.01	0.00016	0.010	0.037	0.069	0.100	0.128	0.153	0.175	0.193	0.210
	0.025	0.0010	0.025	0.069	0.114	0.152	0.185	0.212	0.236	0.256	0.273
	0.05	0.0041	0.052	0.114	0.168	0.212	0.248	0.278	0.302	0.323	0.340
	0.10	0.017	0.106	0.192	0.256	0.305	0.342	0.373	0.397	0.417	0.435
	0.90	3.23	2.86	2.66	2.54	2.45	2.39	2.34	2.30	2.27	2.25
	0.95	4.84	3.98	3.59	3.36	3.20	3.09	3.01	2.95	2.90	2.85
	0.975	6.72	5.26	4.63	4.28	4.04	3.88	3.76	3.66	3.59	3.53
	0.99	9.65	7.21	6.22	5.67	5.32	5.07	4.89	4.74	4.63	4.54
	0.995	12.2	8.91	7.60	6.88	6.42	6.10	5.86	5.68	5.54	5.42

Table D *(Continued)*

ν_2	p	11	12	15	20	24	30	40	60	120	∞
						ν_1					
8	0.005	0.176	0.187	0.214	0.244	0.261	0.279	0.299	0.319	0.341	0.364
	0.01	0.211	0.222	0.250	0.281	0.297	0.315	0.334	0.354	0.376	0.398
	0.025	0.273	0.285	0.313	0.343	0.360	0.377	0.395	0.415	0.435	0.456
	0.05	0.339	0.351	0.379	0.409	0.425	0.441	0.459	0.477	0.496	0.516
	0.10	0.435	0.445	0.472	0.500	0.515	0.531	0.547	0.563	0.581	0.599
	0.90	2.52	2.50	2.46	2.42	2.40	2.38	2.36	2.34	2.32	2.29
	0.95	3.31	3.28	3.22	3.15	3.12	3.08	3.04	3.01	2.97	2.93
	0.975	4.24	4.20	4.10	4.00	3.95	3.89	3.84	3.78	3.73	3.67
	0.99	5.73	5.67	5.52	5.36	5.28	5.20	5.12	5.03	4.95	4.86
	0.995	7.10	7.01	6.81	6.61	6.50	6.40	6.29	6.18	6.06	5.95
9	0.005	0.181	0.192	0.220	0.253	0.271	0.290	0.310	0.332	0.356	0.382
	0.01	0.216	0.228	0.257	0.289	0.307	0.326	0.346	0.368	0.391	0.415
	0.025	0.279	0.291	0.320	0.352	0.370	0.388	0.408	0.428	0.450	0.473
	0.05	0.345	0.358	0.386	0.418	0.435	0.452	0.471	0.490	0.510	0.532
	0.10	0.441	0.452	0.479	0.509	0.525	0.541	0.558	0.575	0.594	0.613
	0.90	2.40	2.38	2.34	2.30	2.28	2.25	2.23	2.21	2.18	2.16
	0.95	3.10	3.07	3.01	2.94	2.90	2.86	2.83	2.79	2.75	2.71
	0.975	3.91	3.87	3.77	3.67	3.61	3.56	3.51	3.45	3.39	3.33
	0.99	5.18	5.11	4.96	4.81	4.73	4.65	4.57	4.48	4.40	4.31
	0.995	6.31	6.23	6.03	5.83	5.73	5.62	5.52	5.41	5.30	5.19
10	0.005	0.185	0.197	0.226	0.260	0.279	0.299	0.321	0.344	0.370	0.397
	0.01	0.220	0.233	0.263	0.297	0.316	0.336	0.357	0.380	0.405	0.431
	0.025	0.283	0.296	0.327	0.360	0.379	0.398	0.419	0.441	0.464	0.488
	0.05	0.351	0.363	0.393	0.426	0.444	0.462	0.481	0.502	0.523	0.546
	0.10	0.444	0.457	0.486	0.516	0.532	0.549	0.567	0.586	0.605	0.625
	0.90	2.30	2.28	2.24	2.20	2.18	2.16	2.13	2.11	2.08	2.06
	0.95	2.94	2.91	2.85	2.77	2.74	2.70	2.66	2.62	2.58	2.54
	0.975	3.66	3.62	3.52	3.42	3.37	3.31	3.26	3.20	3.14	3.08
	0.99	4.77	4.71	4.56	4.41	4.33	4.25	4.17	4.08	4.00	3.91
	0.995	5.75	5.66	5.47	5.27	5.17	5.07	4.97	4.86	4.75	4.64
11	0.005	0.188	0.200	0.231	0.266	0.286	0.308	0.330	0.355	0.382	0.412
	0.01	0.224	0.237	0.268	0.304	0.324	0.344	0.366	0.391	0.417	0.444
	0.025	0.288	0.301	0.332	0.368	0.386	0.407	0.429	0.450	0.476	0.503
	0.05	0.355	0.368	0.398	0.433	0.452	0.469	0.490	0.513	0.535	0.559
	0.10	0.448	0.461	0.490	0.524	0.541	0.559	0.578	0.595	0.617	0.637
	0.90	2.23	2.21	2.17	2.12	2.10	2.08	2.05	2.03	2.00	1.97
	0.95	2.82	2.79	2.72	2.65	2.61	2.57	2.53	2.49	2.45	2.40
	0.975	3.47	3.43	3.33	3.23	3.17	3.12	3.06	3.00	2.94	2.88
	0.99	4.46	4.40	4.25	4.10	4.02	3.94	3.86	3.78	3.69	3.60
	0.995	5.32	5.24	5.05	4.86	4.76	4.65	4.55	4.45	4.34	4.23

Table D *(Continued)*

ν_2	p	ν_1									
		1	2	3	4	5	6	7	8	9	10
12	0.005	0.00004	0.0050	0.023	0.048	0.075	0.100	0.122	0.143	0.161	0.177
	0.01	0.00016	0.010	0.037	0.070	0.101	0.130	0.155	0.176	0.196	0.212
	0.025	0.0010	0.025	0.070	0.114	0.153	0.186	0.214	0.238	0.259	0.276
	0.05	0.0041	0.052	0.114	0.169	0.214	0.250	0.280	0.305	0.325	0.343
	0.10	0.016	0.106	0.192	0.257	0.306	0.344	0.375	0.400	0.420	0.438
	0.90	3.18	2.81	2.61	2.48	2.39	2.33	2.28	2.24	2.21	2.19
	0.95	4.75	3.89	3.49	3.26	3.11	3.00	2.91	2.85	2.80	2.75
	0.975	6.55	5.10	4.47	4.12	3.89	3.73	3.61	3.51	3.44	3.37
	0.99	9.33	6.93	5.95	5.41	5.06	4.82	4.64	4.50	4.39	4.30
	0.995	11.8	8.51	7.23	6.52	6.07	5.76	5.52	5.35	5.20	5.09
15	0.005	0.00004	0.0050	0.023	0.049	0.076	0.102	0.125	0.147	0.166	0.183
	0.01	0.00016	0.010	0.037	0.070	0.103	0.132	0.158	0.181	0.202	0.219
	0.025	0.0010	0.025	0.070	0.116	0.156	0.190	0.219	0.244	0.265	0.284
	0.05	0.0041	0.051	0.115	0.170	0.216	0.254	0.285	0.311	0.333	0.351
	0.10	0.016	0.106	0.192	0.258	0.309	0.348	0.380	0.406	0.427	0.446
	0.90	3.07	2.70	2.49	2.36	2.27	2.21	2.16	2.12	2.09	2.06
	0.95	4.54	3.68	3.29	3.06	2.90	2.79	2.71	2.64	2.59	2.54
	0.975	6.20	4.76	4.15	3.80	3.58	3.41	3.29	3.20	3.12	3.06
	0.99	8.68	6.36	5.42	4.89	4.56	4.32	4.14	4.00	3.89	3.80
	0.995	10.8	7.70	6.48	5.80	5.37	5.07	4.85	4.67	4.54	4.42
20	0.005	0.00004	0.0050	0.023	0.050	0.077	0.104	0.129	0.151	0.171	0.190
	0.01	0.00016	0.010	0.037	0.071	0.105	0.135	0.162	0.187	0.208	0.227
	0.025	0.0010	0.025	0.071	0.117	0.158	0.193	0.224	0.250	0.273	0.292
	0.05	0.0040	0.051	0.115	0.172	0.219	0.258	0.290	0.318	0.340	0.360
	0.10	0.016	0.106	0.193	0.260	0.312	0.353	0.385	0.412	0.435	0.454
	0.90	2.97	2.59	2.38	2.25	2.16	2.09	2.04	2.00	1.96	1.94
	0.95	4.35	3.49	3.10	2.87	2.71	2.60	2.51	2.45	2.39	2.35
	0.975	5.87	4.46	3.86	3.51	3.29	3.13	3.01	2.91	2.81	2.77
	0.99	8.10	5.85	4.94	4.43	4.10	3.87	3.70	3.56	3.46	3.37
	0.995	9.94	6.99	5.82	5.17	4.76	4.47	4.26	4.09	3.96	3.85
24	0.005	0.00004	0.0050	0.023	0.050	0.078	0.106	0.131	0.154	0.175	0.193
	0.01	0.00016	0.010	0.038	0.072	0.106	0.137	0.165	0.189	0.211	0.231
	0.025	0.0010	0.025	0.071	0.117	0.159	0.195	0.227	0.253	0.277	0.297
	0.05	0.0040	0.051	0.116	0.173	0.221	0.260	0.293	0.321	0.345	0.365
	0.10	0.016	0.106	0.193	0.261	0.313	0.355	0.388	0.416	0.439	0.459
	0.90	2.93	2.54	2.33	2.19	2.10	2.04	1.98	1.94	1.91	1.88
	0.95	4.26	3.40	3.01	2.78	2.62	2.51	2.42	2.36	2.30	2.25
	0.975	5.72	4.32	3.72	3.38	3.15	2.99	2.87	2.78	2.70	2.64
	0.99	7.82	5.61	4.72	4.22	3.90	3.67	3.50	3.36	3.26	3.17
	0.995	9.55	6.66	5.52	4.89	4.49	4.20	3.99	3.83	3.69	3.59

Table D *(Continued)*

ν_2	p	ν_1									
		11	12	15	20	24	30	40	60	120	∞
12	0.005	0.191	0.204	0.235	0.272	0.292	0.315	0.339	0.365	0.393	0.424
	0.01	0.227	0.241	0.273	0.310	0.330	0.352	0.375	0.401	0.428	0.458
	0.025	0.292	0.305	0.337	0.374	0.394	0.416	0.437	0.461	0.487	0.514
	0.05	0.358	0.372	0.404	0.439	0.458	0.478	0.499	0.522	0.545	0.571
	0.10	0.452	0.466	0.496	0.528	0.546	0.564	0.583	0.604	0.625	0.647
	0.90	2.17	2.15	2.11	2.06	2.04	2.01	1.99	1.96	1.93	1.90
	0.95	2.72	2.69	2.62	2.54	2.51	2.47	2.43	2.38	2.34	2.30
	0.975	3.32	3.28	3.18	3.07	3.02	2.96	2.91	2.85	2.79	2.72
	0.99	4.22	4.16	4.01	3.86	3.78	3.70	3.62	3.54	3.45	3.36
	0.995	4.99	4.91	4.72	4.53	4.43	4.33	4.23	4.12	4.01	3.90
15	0.005	0.198	0.212	0.246	0.286	0.308	0.333	0.360	0.389	0.422	0.457
	0.01	0.235	0.249	0.284	0.324	0.346	0.370	0.397	0.425	0.456	0.490
	0.025	0.300	0.315	0.349	0.389	0.410	0.433	0.458	0.485	0.514	0.546
	0.05	0.368	0.382	0.416	0.454	0.474	0.496	0.519	0.545	0.571	0.600
	0.10	0.461	0.475	0.507	0.542	0.561	0.581	0.602	0.624	0.647	0.672
	0.90	2.04	2.02	1.97	1.92	1.90	1.87	1.85	1.82	1.79	1.76
	0.95	2.51	2.48	2.40	2.33	2.39	2.25	2.20	2.16	2.11	2.07
	0.975	3.01	2.96	2.86	2.76	2.70	2.64	2.59	2.52	2.46	2.40
	0.99	3.73	3.67	3.52	3.37	3.29	3.21	3.13	3.05	2.96	2.87
	0.995	4.33	4.25	4.07	3.88	3.79	3.69	3.59	3.48	3.37	3.26
20	0.005	0.206	0.221	0.258	0.301	0.327	0.354	0.385	0.419	0.457	0.500
	0.01	0.244	0.259	0.297	0.340	0.365	0.392	0.422	0.455	0.491	0.532
	0.025	0.310	0.325	0.363	0.406	0.430	0.456	0.484	0.514	0.548	0.585
	0.05	0.377	0.393	0.430	0.471	0.493	0.518	0.544	0.572	0.603	0.637
	0.10	0.472	0.485	0.520	0.557	0.578	0.600	0.623	0.648	0.675	0.704
	0.90	1.91	1.89	1.84	1.79	1.77	1.74	1.71	1.68	1.64	1.61
	0.95	2.31	2.28	2.20	2.12	2.08	2.04	1.99	1.95	1.90	1.84
	0.975	2.72	2.68	2.57	2.46	2.41	2.35	2.29	2.22	2.16	2.09
	0.99	3.29	3.23	3.09	2.94	2.86	2.78	2.69	2.61	2.52	2.42
	0.995	3.76	3.68	3.50	3.32	3.22	3.12	3.02	2.92	2.81	2.69
24	0.005	0.210	0.226	0.264	0.310	0.337	0.367	0.400	0.437	0.479	0.527
	0.01	0.249	0.264	0.304	0.350	0.376	0.405	0.437	0.473	0.513	0.558
	0.025	0.315	0.331	0.370	0.415	0.441	0.468	0.498	0.531	0.568	0.610
	0.05	0.383	0.399	0.437	0.480	0.504	0.530	0.558	0.588	0.622	0.659
	0.10	0.476	0.491	0.527	0.566	0.588	0.611	0.635	0.662	0.691	0.723
	0.90	1.85	1.83	1.78	1.73	1.70	1.67	1.64	1.61	1.57	1.53
	0.95	2.21	2.18	2.11	2.03	1.98	1.94	1.89	1.84	1.79	1.73
	0.975	2.59	2.54	2.44	2.33	2.27	2.21	2.15	2.08	2.01	1.94
	0.99	3.09	3.03	2.89	2.74	2.66	2.58	2.49	2.40	2.31	2.21
	0.995	3.50	3.42	3.25	3.06	2.97	2.87	2.77	2.66	2.55	2.43

Table D *(Continued)*

ν_2	p	ν_1									
		1	2	3	4	5	6	7	8	9	10
30	0.005	0.00004	0.0050	0.024	0.050	0.079	0.107	0.133	0.156	0.178	0.197
	0.01	0.00016	0.010	0.038	0.072	0.107	0.138	0.167	0.192	0.215	0.235
	0.025	0.0010	0.025	0.071	0.118	0.161	0.197	0.229	0.257	0.281	0.302
	0.05	0.0040	0.051	0.116	0.174	0.222	0.263	0.296	0.325	0.349	0.370
	0.10	0.016	0.106	0.193	0.262	0.315	0.357	0.391	0.420	0.443	0.464
	0.90	2.88	2.49	2.28	2.14	2.05	1.98	1.93	1.88	1.85	1.82
	0.95	4.17	3.32	2.92	2.69	2.53	2.42	2.33	2.27	2.21	2.16
	0.975	5.57	4.18	3.59	3.25	3.03	2.87	2.75	2.65	2.57	2.51
	0.99	7.56	5.39	4.51	4.02	3.70	3.47	3.30	3.17	3.07	2.98
	0.995	9.18	6.35	5.24	4.62	4.23	3.95	3.74	3.58	3.45	3.34
40	0.005	0.00004	0.0050	0.024	0.051	0.080	0.108	0.135	0.159	0.181	0.201
	0.01	0.00016	0.010	0.038	0.073	0.108	0.140	0.169	0.195	0.219	0.240
	0.025	0.00099	0.025	0.071	0.119	0.162	0.199	0.232	0.260	0.285	0.307
	0.05	0.0040	0.051	0.116	0.175	0.224	0.265	0.299	0.329	0.354	0.376
	0.10	0.016	0.106	0.194	0.263	0.317	0.360	0.394	0.424	0.448	0.469
	0.90	2.84	2.44	2.23	2.09	2.00	1.93	1.87	1.83	1.79	1.76
	0.95	4.08	3.23	2.84	2.61	2.45	2.34	2.25	21.8	2.12	2.08
	0.975	5.42	4.05	3.46	3.13	2.90	2.74	2.62	2.53	2.45	2.39
	0.99	7.31	5.18	4.31	3.83	3.51	3.29	3.12	2.99	2.89	2.80
	0.995	8.83	6.07	4.98	4.37	3.99	3.71	3.51	3.35	3.22	3.12
60	0.005	0.00004	0.0050	0.024	0.051	0.081	0.110	0.137	0.162	0.185	0.206
	0.01	0.00016	0.010	0.038	0.073	0.109	0.142	0.172	0.199	0.223	0.245
	0.025	0.00099	0.025	0.071	0.120	0.163	0.202	0.235	0.264	0.290	0.313
	0.05	0.0040	0.051	0.116	0.176	0.226	0.267	0.303	0.333	0.359	0.382
	0.10	0.016	0.106	0.194	0.264	0.318	0.362	0.398	0.428	0.453	0.475
	0.90	2.79	2.39	2.18	2.04	1.95	1.87	1.82	1.77	1.74	1.71
	0.95	4.00	3.15	2.76	2.53	2.37	2.25	2.17	2.10	2.04	1.99
	0.975	5.29	3.93	3.34	3.01	2.79	2.63	2.51	2.41	2.33	2.27
	0.99	7.08	4.98	4.13	3.65	3.34	3.12	2.95	2.82	2.72	2.63
	0.995	8.49	5.80	4.73	4.14	3.76	3.49	3.29	3.13	3.01	2.90
120	0.005	0.00004	0.0050	0.024	0.051	0.081	0.111	0.139	0.165	0.189	0.211
	0.01	0.00016	0.010	0.038	0.074	0.110	0.143	0.174	0.202	0.227	0.250
	0.025	0.00099	0.025	0.072	0.120	0.165	0.204	0.238	0.268	0.295	0.318
	0.05	0.0039	0.051	0.117	0.177	0.227	0.270	0.306	0.337	0.364	0.388
	0.10	0.016	0.105	0.194	0.265	0.320	0.365	0.401	0.432	0.458	0.480
	0.90	2.75	2.35	2.13	1.99	1.90	1.82	1.77	1.72	1.68	1.65
	0.95	3.92	3.07	2.68	2.45	2.29	2.18	2.09	2.02	1.96	1.91
	0.975	5.15	3.80	3.23	2.89	2.67	2.52	2.39	2.30	2.22	2.16
	0.99	6.85	4.79	3.95	3.48	3.17	2.96	2.79	2.66	2.56	2.47
	0.995	8.18	5.54	4.50	3.92	3.55	3.28	3.09	2.93	2.81	2.71

Table D *(Continued)*

ν_2	p	ν_1									
		11	12	15	20	24	30	40	60	120	∞
30	0.005	0.215	0.231	0.271	0.320	0.349	0.381	0.416	0.457	0.504	0.559
	0.01	0.254	0.270	0.311	0.360	0.388	0.419	0.454	0.493	0.538	0.590
	0.025	0.321	0.337	0.378	0.426	0.453	0.482	0.515	0.551	0.592	0.639
	0.05	0.389	0.406	0.445	0.490	0.516	0.543	0.573	0.606	0.644	0.685
	0.10	0.481	0.497	0.534	0.575	0.598	0.623	0.649	0.678	0.710	0.746
	0.90	1.79	1.77	1.72	1.67	1.64	1.61	1.57	1.54	1.50	1.46
	0.95	2.13	2.09	2.01	1.93	1.89	1.84	1.79	1.74	1.68	1.62
	0.975	2.46	2.41	2.31	2.20	2.14	2.07	2.01	1.94	1.87	1.79
	0.99	2.91	2.84	2.70	2.55	2.47	2.39	2.30	2.21	2.11	2.01
	0.995	3.25	3.18	3.01	2.82	2.73	2.63	2.52	2.42	2.30	2.18
40	0.005	0.220	0.237	0.279	0.331	0.362	0.396	0.436	0.481	0.534	0.599
	0.01	0.259	0.276	0.319	0.371	0.401	0.435	0.473	0.516	0.567	0.628
	0.025	0.327	0.344	0.387	0.437	0.466	0.498	0.533	0.573	0.620	0.674
	0.05	0.395	0.412	0.454	0.502	0.529	0.558	0.591	0.627	0.669	0.717
	0.10	0.488	0.504	0.542	0.585	0.609	0.636	0.644	0.696	0.731	0.772
	0.90	1.73	1.71	1.66	1.61	1.57	1.54	1.51	1.47	1.42	1.38
	0.95	2.04	2.00	1.92	1.84	1.79	1.74	1.69	1.64	1.58	1.51
	0.975	2.33	2.29	2.18	2.07	2.01	1.94	1.88	1.80	1.72	1.64
	0.99	2.73	2.66	2.52	2.37	2.29	2.20	2.11	2.02	1.92	1.80
	0.995	3.03	2.95	2.78	2.60	2.50	2.40	2.30	2.18	2.06	1.93
60	0.005	0.225	0.243	0.287	0.343	0.376	0.414	0.458	0.510	0.572	0.652
	0.01	0.265	0.283	0.328	0.383	0.416	0.453	0.495	0.545	0.604	0.679
	0.025	0.333	0.351	0.396	0.450	0.481	0.515	0.555	0.600	0.654	0.720
	0.05	0.402	0.419	0.463	0.514	0.543	0.575	0.611	0.652	0.700	0.759
	0.10	0.493	0.510	0.550	0.596	0.622	0.650	0.682	0.717	0.758	0.806
	0.90	1.68	1.66	1.60	1.54	1.51	1.48	1.44	1.40	1.35	1.29
	0.95	1.95	1.92	1.84	1.75	1.70	1.65	1.59	1.53	1.47	1.39
	0.975	2.22	2.17	2.06	1.94	1.88	1.82	1.74	1.67	1.58	1.48
	0.99	2.56	2.50	2.35	2.20	2.12	2.03	1.94	1.84	1.73	1.60
	0.995	2.82	2.74	2.57	2.39	2.29	2.19	2.08	1.96	1.83	1.69
120	0.005	0.230	0.249	0.297	0.356	0.393	0.434	0.484	0.545	0.623	0.733
	0.01	0.271	0.290	0.338	0.397	0.433	0.474	0.522	0.579	0.652	0.755
	0.025	0.340	0.359	0.406	0.464	0.498	0.536	0.580	0.633	0.698	0.789
	0.05	0.408	0.427	0.473	0.527	0.559	0.594	0.634	0.682	0.740	0.819
	0.10	0.500	0.518	0.560	0.609	0.636	0.667	0.702	0.742	0.791	0.855
	0.90	1.62	1.60	1.55	1.48	1.45	1.41	1.37	1.32	1.26	1.19
	0.95	1.87	1.83	1.75	1.66	1.61	1.55	1.50	1.43	1.35	1.25
	0.975	2.10	2.05	1.95	1.82	1.76	1.69	1.61	1.53	1.43	1.31
	0.99	2.40	2.34	2.19	2.03	1.95	1.86	1.76	1.66	1.53	1.38
	0.995	2.62	2.54	2.37	2.19	2.09	1.98	1.87	1.75	1.61	1.43

Table E.1 Upper 5 Percent of Studentized Range q^*

p^{**} Number of Groups Being Compared

n_2	2	3	4	5	6	7	8	9	10	11	12	13	14	15	16	17	18	19	20
1	18.0	26.7	32.8	37.2	40.5	43.1	45.4	47.3	49.1	50.6	51.9	53.2	54.3	55.4	56.3	57.2	58.0	58.8	59.6
2	6.09	8.28	9.80	10.89	11.73	12.43	13.03	13.54	13.99	14.39	14.75	15.08	15.38	15.65	15.91	16.14	16.36	16.57	16.77
3	4.50	5.88	6.83	7.51	8.04	8.47	8.85	9.18	9.46	9.72	9.95	10.16	10.35	10.52	10.69	10.84	10.98	11.12	11.24
4	3.93	5.00	5.76	6.31	6.73	7.06	7.35	7.60	7.83	8.03	8.21	8.37	8.52	8.67	8.80	8.92	9.03	9.14	9.24
5	3.61	4.54	5.18	5.64	5.99	6.28	6.52	6.74	6.93	7.10	7.25	7.39	7.52	7.64	7.75	7.86	7.95	8.04	8.13
6	3.46	4.34	4.90	5.31	5.63	5.89	6.12	6.32	6.49	6.65	6.79	6.92	7.04	7.14	7.24	7.34	7.43	7.51	7.59
7	3.34	4.16	4.68	5.06	5.35	5.59	5.80	5.99	6.15	6.29	6.42	6.54	6.65	6.75	6.84	6.93	7.01	7.08	7.16
8	3.26	4.04	4.53	4.89	5.17	5.40	5.60	5.77	5.92	6.05	6.18	6.29	6.39	6.48	6.57	6.65	6.73	6.80	6.87
9	3.20	3.95	4.42	4.76	5.02	5.24	5.43	5.60	5.74	5.87	5.98	6.09	6.19	6.28	6.36	6.44	6.51	6.58	6.65
10	3.15	3.88	4.33	4.66	4.91	5.12	5.30	5.46	5.60	5.72	5.83	5.93	6.03	6.12	6.20	6.27	6.34	6.41	6.47
11	3.11	3.82	4.26	4.58	4.82	5.03	5.20	5.35	5.49	5.61	5.71	5.81	5.90	5.98	6.06	6.14	6.20	6.27	6.33
12	3.08	3.77	4.20	4.51	4.75	4.95	5.12	5.27	5.40	5.51	5.61	5.71	5.80	5.88	5.95	6.02	6.09	6.15	6.21
13	3.06	3.73	4.15	4.46	4.69	4.88	5.05	5.19	5.32	5.43	5.53	5.63	5.71	5.79	5.86	5.93	6.00	6.06	6.11
14	3.03	3.70	4.11	4.41	4.64	4.83	4.99	5.13	5.25	5.36	5.46	5.56	5.64	5.72	5.79	5.86	5.92	5.98	6.03
15	3.01	3.67	4.08	4.37	4.59	4.78	4.94	5.08	5.20	5.31	5.40	5.49	5.57	5.65	5.72	5.79	5.85	5.91	5.96
16	3.00	3.65	4.05	4.34	4.56	4.74	4.90	5.03	5.15	5.26	5.35	5.44	5.52	5.59	5.66	5.73	5.79	5.84	5.90
17	2.98	3.62	4.02	4.31	4.52	4.70	4.86	4.99	5.11	5.21	5.31	5.39	5.47	5.55	5.61	5.68	5.74	5.79	5.84
18	2.97	3.61	4.00	4.28	4.49	4.67	4.83	4.96	5.07	5.17	5.27	5.35	5.43	5.50	5.57	5.63	5.69	5.74	5.79
19	2.96	3.59	3.98	4.26	4.47	4.64	4.79	4.92	5.04	5.14	5.23	5.32	5.39	5.46	5.53	5.59	5.65	5.70	5.75
20	2.95	3.58	3.96	4.24	4.45	4.62	4.77	4.90	5.01	5.11	5.20	5.28	5.36	5.43	5.50	5.56	5.61	5.66	5.71
24	2.92	3.53	3.90	4.17	4.37	4.54	4.68	4.81	4.92	5.01	5.10	5.18	5.25	5.32	5.38	5.44	5.50	5.55	5.59
30	2.89	3.48	3.84	4.11	4.30	4.46	4.60	4.72	4.83	4.92	5.00	5.08	5.15	5.21	5.27	5.33	5.38	5.43	5.48
40	2.86	3.44	3.79	4.04	4.23	4.39	4.52	4.63	4.74	4.82	4.90	4.98	5.05	5.11	5.17	5.22	5.27	5.32	5.36
60	2.83	3.40	3.74	3.98	4.16	4.31	4.44	4.55	4.65	4.73	4.81	4.88	4.94	5.00	5.06	5.11	5.15	5.20	5.24
120	2.80	3.36	3.69	3.92	4.10	4.24	4.36	4.47	4.56	4.64	4.71	4.78	4.84	4.90	4.95	5.00	5.04	5.09	5.13
∞	2.77	3.32	3.63	3.86	4.03	4.17	4.29	4.39	4.47	4.55	4.62	4.68	4.74	4.80	4.84	4.89	4.93	4.97	5.01

*From J. M. May, "Extended and Corrected Tables of the Upper Percentage Points of the Studentized Range," *Biometrika*, vol. 39 (1952), pp. 192–193. Reproduced by permission of the trustees of *Biometrika*.

** p is the number of quantities (for example, means) whose range is involved. n_2 is the degrees of freedom in the error estimate.

Table E.2 Upper 1 Percent Points of Studentized Range q

n_2*	2	3	4	5	6	7	8	9	10	11	12	13	14	15	16	17	18	19	20
1	90.0	135	164	186	202	216	227	237	246	253	260	266	272	227	282	286	290	294	298
2	14.0	19.0	22.3	24.7	26.6	28.2	29.5	30.7	31.7	32.6	33.4	34.1	34.8	35.4	36.0	36.5	37.0	37.5	37.9
3	8.26	10.6	12.2	13.3	14.2	15.0	15.6	16.2	16.7	17.1	17.5	17.9	18.2	18.5	18.8	19.1	19.3	19.5	19.8
4	6.51	8.12	9.17	9.96	10.6	11.1	11.5	11.9	12.3	12.6	12.8	13.1	13.3	13.5	13.7	13.9	14.1	14.2	14.4
5	5.70	6.97	7.80	8.42	8.91	9.32	9.67	9.97	10.24	10.48	10.70	10.89	11.08	11.24	11.40	11.55	11.68	11.81	11.93
6	5.24	6.33	7.03	7.56	7.97	8.32	8.61	8.87	9.10	9.30	9.49	9.65	9.81	9.95	10.08	10.21	10.32	10.43	10.54
7	4.95	5.92	6.54	7.01	7.37	7.68	7.94	8.17	8.37	8.55	8.71	8.86	9.00	9.12	9.24	9.35	9.46	9.55	9.65
8	4.74	5.63	6.20	6.63	6.96	7.24	7.47	7.68	7.87	8.03	8.18	8.31	8.44	8.55	8.66	8.76	8.85	8.94	9.03
9	4.60	5.43	5.96	6.35	6.66	6.91	7.13	7.32	7.49	7.65	7.78	7.91	8.03	8.13	8.23	8.32	8.41	8.49	8.57
10	4.48	5.27	5.77	6.14	6.43	6.67	6.87	7.05	7.21	7.36	7.48	7.60	7.71	7.81	7.91	7.99	8.07	8.15	8.22
11	4.39	5.14	5.62	5.97	6.25	6.48	6.67	6.84	6.99	7.13	7.25	7.36	7.46	7.56	7.65	7.73	7.81	7.88	7.95
12	4.32	5.04	5.50	5.84	6.10	6.32	6.51	6.67	6.81	6.94	7.06	7.17	7.26	7.36	7.44	7.52	7.59	7.66	7.73
13	4.26	4.96	5.40	5.73	5.98	6.19	6.37	6.53	6.67	6.79	6.90	7.01	7.10	7.19	7.27	7.34	7.42	7.48	7.55
14	4.21	4.89	5.32	5.63	5.88	6.08	6.26	6.41	6.54	6.66	6.77	6.87	6.96	7.05	7.12	7.20	7.27	7.33	7.39
15	4.17	4.83	5.25	5.56	5.80	5.99	6.16	6.31	6.44	6.55	6.66	6.76	6.84	6.93	7.00	7.07	7.14	7.20	7.26
16	4.13	4.78	5.19	5.49	5.72	5.92	6.08	6.22	6.35	6.46	6.56	6.66	6.74	6.82	6.90	6.97	7.03	7.09	7.15
17	4.10	4.74	5.14	5.43	5.66	5.85	6.01	6.15	6.27	6.38	6.48	6.57	6.66	6.73	6.80	6.87	6.94	7.00	7.05
18	4.07	4.70	5.09	5.38	5.60	5.79	5.94	6.08	6.20	6.31	6.41	6.50	6.58	6.65	6.72	6.79	6.85	6.91	6.96
19	4.05	4.67	5.05	5.33	5.55	5.73	5.89	6.02	6.14	6.25	6.34	6.43	6.51	6.58	6.65	6.72	6.78	6.84	6.89
20	4.02	4.64	5.02	5.29	5.51	5.69	5.84	5.97	6.09	6.19	6.29	6.37	6.45	6.52	6.59	6.65	6.71	6.76	6.82
24	3.96	4.54	4.91	5.17	5.37	5.54	5.69	5.81	5.92	6.02	6.11	6.19	6.26	6.33	6.39	6.45	6.51	6.56	6.61
30	3.89	4.45	4.80	5.05	5.24	5.40	5.54	5.65	5.76	5.85	5.93	6.01	6.08	6.14	6.20	6.26	6.31	6.36	6.41
40	3.82	4.37	4.70	4.93	5.11	5.27	5.39	5.50	5.60	5.69	5.77	5.84	5.90	5.96	6.02	6.07	6.12	6.17	6.21
60	3.76	4.28	4.60	4.82	4.99	5.13	5.25	5.36	5.45	5.53	5.60	5.67	5.73	5.79	5.84	5.89	5.93	5.98	6.02
120	3.70	4.20	4.50	4.71	4.87	5.01	5.12	5.21	5.30	5.38	5.44	5.51	5.56	5.61	5.66	5.71	5.75	5.79	5.83
∞	3.64	4.12	4.40	4.60	4.76	4.88	4.99	5.08	5.16	5.23	5.29	5.35	5.40	5.45	5.49	5.54	5.57	5.61	5.65

* p is the number of quantities (for example, means) whose range is involved. n_2 is the degrees of freedom in the error estimate.

Table F Coefficients of Orthogonal Polynomials

k	Polynomial	X										$\sum \xi_j^2$	λ
		1	2	3	4	5	6	7	8	9	10		
3	Linear	-1	0	1								2	1
	Quadratic	1	-2	1								6	3
	Linear	-3	-1	1	3							20	2
4	Quadratic	1	-1	-1	1							4	1
	Cubic	-1	3	-3	1							20	$\frac{10}{3}$
	Linear	-2	-1	0	1	2						10	1
5	Quadratic	2	-1	-2	-1	2						14	1
	Cubic	-1	2	0	-2	1						10	$\frac{5}{6}$
	Quartic	1	-4	6	-4	1						70	$\frac{35}{12}$
	Linear	-5	-3	-1	1	3	5					70	2
6	Quadratic	5	-1	-4	-4	-1	5					84	$\frac{3}{2}$
	Cubic	-5	7	4	-4	-7	5					180	$\frac{5}{3}$
	Quartic	1	-3	2	2	-3	1					28	$\frac{7}{12}$
	Linear	-3	-2	-1	0	1	2	3				28	1
7	Quadratic	5	0	-3	-4	-3	0	5				84	1
	Cubic	-1	1	1	0	-1	-1	1				6	$\frac{1}{6}$
	Quartic	3	-7	1	6	1	-7	3				154	$\frac{7}{12}$
	Linear	-7	-5	-3	-1	1	3	5	7			168	2
	Quadratic	7	1	-3	-5	-5	-3	1	7			168	1
8	Cubic	-7	5	7	3	-3	-7	-5	7			264	$\frac{2}{3}$
	Quartic	7	-13	-3	9	9	-3	-13	7			616	$\frac{7}{12}$
	Quintic	-7	23	-17	-15	15	17	-23	7			2184	$\frac{7}{10}$
	Linear	-4	-3	-2	-1	0	1	2	3	4		60	1
	Quadratic	28	7	-8	-17	-20	-17	-8	7	28		2772	3
9	Cubic	-14	7	13	9	0	-9	-13	-7	14		990	$\frac{5}{6}$
	Quartic	14	-21	-11	9	18	9	-11	-21	14		2002	$\frac{7}{12}$
	Quintic	-4	11	-4	-9	0	9	4	-11	4		468	$\frac{3}{20}$
	Linear	-9	-7	-5	-3	-1	1	3	5	7	9	330	2
	Quadratic	6	2	-1	-3	-4	-4	-3	-1	2	6	132	$\frac{1}{2}$
10	Cubic	-42	14	35	31	12	-12	-31	-35	-14	42	8580	$\frac{5}{3}$
	Quartic	18	-22	-17	3	18	18	3	-17	-22	18	2860	$\frac{5}{12}$
	Quintic	-6	14	-1	-11	-6	6	11	1	-14	6	780	$\frac{1}{10}$

Answers to Odd-Numbered Problems

Chapter 2

2.1 $\bar{Y} = 18{,}472.9$, $s^2 = 41.7$.

2.3 For a two-sided alternative, some points are

$\mu = 18{,}473$; 18,472; 18,471; 18,470; 18,469; 18,468; 18,467

$P_a = \beta = 0.08$, 0.39, 0.80, 0.95, 0.80, 0.39, 0.08.

2.5 $n = 7$

2.7 Do not reject hypothesis, since $\chi^2 = 17.05$ with 11 df.

2.9 Do not reject hypothesis, since $F = 1.8$.

2.11 Reject hypothesis at 1 percent level, since $t = 3.56$.

2.13 a. Reject hypothesis, since $z = 4.0$.
b. Do not reject hypothesis, since $z = 1.71$.
c. Do not reject hypothesis z, since $z < 1$.

2.15 For $\quad\quad \mu = \quad$ 13 $\quad$ 15 $\quad$ 17 $\quad$ 19
Power $(1 - \beta) = \quad$ 0.04 $\;$ 0.27 $\;$ 0.70 $\;$ 0.95

2.17 Reject hypothesis at $\alpha = 0.05$, since $t = 8.32$.

2.19 Reject hypothesis of equal variances at $\alpha = 0.05$, since $F = 17.4$. Do not reject hypothesis of equal means at $\alpha = 0.05$, since t' is < 1.

2.21 Reject hypothesis at $\alpha = 0.05$, since $t = 2.68$.

2.23 Do not reject hypothesis at $\alpha = 0.05$, since $t = 1.62$.

2.25 Reject hypothesis at $\alpha = 0.05$, since $z = -1.96$.

Chapter 3

3.1

Source	df	SS	MS
Between A levels	4	253.04	63.26
Error	20	76.80	3.84
Totals	24	329.84	

Significant at the 1 percent level.

3.3   Two such contrasts might be

$$
\begin{array}{lr}
 & \text{SS} \\
\hline
C_1 = T_A - T_C = 31 & 60.06 \\
C_2 = T_A - 2T_B + T_C = -79 & 130.02 \\
\hline
 & 190.08
\end{array}
$$

Neither is significant at the 5 percent level.

3.5
$$
\begin{array}{cccc}
 & B & A & C \\
\bar{Y}_{.j} = & 25.4 & 22.4 & 18.5
\end{array}
$$

None significantly different.

3.7   Two sets and their sums of squares might be

$$
\begin{array}{lr}
 & \text{SS} \\
\hline
\text{Set 1: } C_1 = \quad 2T_1 \qquad\qquad\qquad - \quad 2T_5 = \quad -28 & 49.0 \\
C_2 = \qquad\quad 4T_2 \quad - \quad 6T_4 \qquad\qquad = -108 & 48.6 \\
C_3 = 11T_1 + 11T_2 - 14T_3 + 11T_4 + 11T_5 = \quad 71 & 1.3 \\
C_4 = 10T_1 - \quad 4T_2 \qquad\qquad - 4T_4 + 10T_5 = \quad 8 & 0.1 \\
\hline
 & 99.0
\end{array}
$$

$$
\begin{array}{lr}
 & \text{SS} \\
\hline
\text{Set 2: } C_1 = \quad 6T_1 - \quad 2T_2 \qquad\qquad\qquad\qquad = \quad -18 & 3.4 \\
C_2 = 11T_1 + 11T_2 - 8T_3 \qquad\qquad\quad = -237 & 33.6 \\
C_3 = \quad 4T_1 + \quad 4T_2 + 4T_3 - 19T_4 \qquad = -252 & 36.3 \\
C_4 = \quad 2T_1 + \quad 2T_2 + 2T_3 + \quad 2T_4 - 23T_5 = -172 & 25.7 \\
\hline
 & 99.0
\end{array}
$$

3.9   75.6 percent due to the levels of A, 24.4 percent due to error.

3.11   Two might be $C_1 = T_{.1} - T_{.2}$ and $C_2 = 2T_{.1} - T_{.2} - T_{.3}$ and $C_1 = -24$, which is less than $A_{S_{C_1}} = 60.65$, so is not significant at 5 percent, and $C_2 = 7$, which is less than $A_{S_{C_2}} = 105.07$, so is not significant at 5 percent.

3.13

Source	df	SS	MS	F
Between temperatures	4	1268.54	317.14	70.3^*
Error	25	112.83	4.51	

*Significant at the .1 percent level.

3.15

Source	df	SS	MS	F
Between coatings	3	1135.0	378.3	29.8^*
Error	16	203.2	12.7	

3.17 Newman–Keuls tests show:

Coating	IV	III	II	I
Means	42.0	43.6	57.2	58.4

3.19

Source	df	SS	MS	F
Bonder	3	7822	2607	1.95
Error	36	48,110	1336	

All bonders do about the same job.

3.21 One set of orthogonal contrasts might be

$$C_1 = 4T_{.1} - T_{.2} - T_{.3} - T_{.4} - T_{.5}$$
$$C_2 = \qquad\ T_{.2} - T_{.3}$$
$$C_3 = \qquad\qquad\qquad\ T_{.4} - T_{.5}$$
$$C_4 = \qquad\ T_{.2} + T_{.3} - T_{.4} - T_{.5}$$

3.23 For C_1 in Problem 3.21, $s_C = 24.5$.

3.25 Litter to litter differences account for 28.6 percent of the total variance.

3.27 Logical contrasts are $2T_A - T_B - T_C = -32.3$ and $T_B - T_C = -2.5$, whose SS are 28.98 and 0.52, respectively. The first is highly significant.

Chapter 4

4.1

Source	df	SS	MS
Between coater types	3	1.53	0.51[*]
Between days	2	0.21	0.10
Error	6	0.54	0.09
Totals	11	2.28	

[*]Significant at the 5 percent level.

4.3

	K	A	L	M
$\bar{Y}_{.j}$:	5.27	4.87	4.73	4.27

K and M are significantly different at the 5 percent level.

4.5 Four contrasts might be

$C_1 = T_M - T_A$	Nonsignificant
$C_2 = T_M - T_K$	Significant at 5 percent by Scheffé
$C_3 = T_M - T_L$	Nonsignificant
$C_4 = T_A - T_K$	Nonsignificant

4.7

Source	df	SS	MS
Electrodes	4	4.2966	1.0742^*
Positions	4	0.5696	0.1424
Strips	4	0.3136	0.0784
Error	12	0.8058	0.0672

*Significant at the 1 percent level.

4.9 Proof.

4.11 Error df $= 0$.

4.13 Three groups (treatments) and six lessons (blocks) ANOVA:

Source	df
Groups	2
Lessons	5
Error	10

4.15 Newman–Keuls on means:

Group	C	B	A
Means	12.33	19.17	22.17

4.17 a. $Y_{ijk} = \mu + S_i + t_j + D_k + \varepsilon_{ijk}$ and ANOVA:

Source	df
Scales	4
Times	4
Days	4
Error	12

b. ANOVA:

Source	df
Scales	4
Days	4
Error	16

4.19 $H_0: F_j = 0$ for all j. Model: $Y_{ij} = \mu + F_j + \beta_i + \varepsilon_{ij}$. $F = 8.75$, which is significant at the 1 percent level.

4.21 From -0.38 to 0.88.

4.23 For example,

$$C_1 = T_{.1} - T_{.2}$$
$$C_2 = T_{.1} + T_{.2} - 2T_{.3}$$
$$C_3 = \qquad\qquad T_{.3} - T_{.4}$$

4.25

Source	df	SS	MS	F
Vendors	2	10.89	5.44	49.5
Scales	2	32.89	16.44	149.5
Inspectors	2	0.22	0.11	1.0
Error	2	0.22	0.11	

Vendors significant at the 5 percent level and scales at the 1 percent.

4.27

Source	df	SS	MS	F
Curriculum	2	6.22	3.11	28.3*
Schools	2	3.56	1.78	16.2
Grades	2	0.89	0.44	4.0
Error	2	0.22	0.11	

Only curricula significant at 5 percent level.

Newman–Keuls on curricula means give

A	B	C
29.0	30.33	31.00

B and C better than A

4.29

Source	df	SS	MS	F
Machines	4	3027	757	2.11
Replications	2	3370	1685	4.71*
Error	8	2863	358	

No significant differences between machines.

Chapter 5

5.1

Source	df	SS	MS	F
Phosphors	2	1244.44	622.22	8.96**
Glass	1	13,338.89	13,338.89	192.09***
$P \times G$ interaction	2	44.45	22.22	< 1
Error	12	833.33	69.44	

5.3 G_2 better than G_1—lower current. For phosphors, Newman–Keuls:

Phosphor: *C*   *A*   *B*
Means: 253.33   260.00   273.33

so *A* or *C* is better than *B*.

5.5

Source	df	SS	MS
Humidity	2	9.07	4.53
Temperature	2	8.66	4.33
$T \times H$ interaction	4	6.07	1.52
Error	27	28.50	1.06
Totals	35	52.30	

5.7   At the 5 percent significance level reject H_2 and H_3 but not H_1.

5.9   Plot of cell totals versus feed for the two material lines are not parallel (interaction), and both material and feed effect are obvious.

5.11

Source	df	SS	MS
Temperature	2	600.09	300.04**
Mix	2	1.59	0.80*
$T \times M$	4	0.98	0.25*
Laboratory	3	3.85	1.28**
$T \times L$	6	1.54	0.26*
$M \times L$	6	0.76	0.13*
$T \times M \times L$	12	0.90	0.07*
Error	36	0.67	0.02
Totals	71	610.38	

5.13

Source	df	SS	MS
Soil	4	41.47	10.37
Fertilizer	2	68.87	34.44
Interaction	8	16.13	2.02
Error	15	52.50	3.50

5.15   Newman–Keuls on fertilizers:

Types: 3   2   1
Means: 3.6   5.2   7.3

Hence type 1 gives best yield with any of the five soil types.

5.17 a. No.
 b. Yes.
 c. Yes.

5.19 Two-dimensional stimulus more variable than three dimensional.

J by A means:

| 1.04 | 1.14 | 1.17 | 1.20 | 1.24 | 1.42 | 1.58 | 1.59 | 1.70 | 1.70 |

5.21

Source	df	SS	MS
Thickness A	1	846.81	846.81**
Temperature B	1	5041.00	5041.00**
AB	1	509.63	509.63**
Drying condition C	1	5.88	5.88
AC	1	1.44	1.44
BC	1	15.21	15.21*
ABC	1	0.14	0.14
Length of wash L	3	69.76	23.25*
AL	3	15.79	5.26
BL	3	3.04	1.01
ABL	3	11.45	3.82
CL	3	9.65	3.22
ACL	3	7.57	2.52
BCL	3	6.43	2.14
$ABCL$	3	5.66	1.89
Error	32	87.21	2.73
Totals	63	66,336.67	

5.23 a.

Source	df
Waxes	3
Times	2
$W \times T$ interaction	6
Error	12

 b. Changes in gloss index due to the four waxes are different for the three polishing times.

5.25 a. Significants are A, B, C, and the ABC interaction.
 b. Run Newman–Keuls on 24 cell means
 c. $4 \times 3 \times 2$ layout with five observations per treatment.
 d. 7.0 to 12.2.

5.27

Source	df	SS	MS	F
Cool	4	219.93	54.98	24.01***
Preheat	1	0.52	0.52	< 1
$C*P$	4	29.00	7.25	3.17*
Error	20	45.78	2.29	

As interaction significant at 5 percent level, run N-K on 10 "cell" means and recommend $C = 10$, $P = 5$.

3.71   4.08   4.68   5.35   5.78   5.94   7.88   9.98   10.94   12.28

Chapter 6

6.1

Source	EMS
O_i	$\sigma_\varepsilon^2 + 10\sigma_O^2$
A_j	$\sigma_\varepsilon^2 + 2\sigma_{OA}^2 + 6\phi_A$
OA_{ij}	$\sigma_\varepsilon^2 + 2\sigma_{OA}^2$
$\varepsilon_{k(ij)}$	σ_ε^2

Tests are indicated by arrows. None significant at the 5 percent level.

6.3

Source	EMS
A_i	$\sigma_\varepsilon^2 + nc\sigma_{AB}^2 + nbc\sigma_A^2$
B_j	$\sigma_\varepsilon^2 + nc\sigma_{AB}^2 + nac\sigma_B^2$
AB_{ij}	$\sigma_\varepsilon^2 + nc\sigma_{AB}^2$
C_k	$\sigma_\varepsilon^2 + n\sigma_{ABC}^2 + na\sigma_{BC}^2 + nb\sigma_{AC}^2 + nab\phi_C$
AC_{ik}	$\sigma_\varepsilon^2 + n\sigma_{ABC}^2 + nb\sigma_{AC}^2$
BC_{jk}	$\sigma_\varepsilon^2 + n\sigma_{ABC}^2 + na\sigma_{BC}^2$
ABC_{ijk}	$\sigma_\varepsilon^2 + n\sigma_{ABC}^2$
$\varepsilon_{m(ijk)}$	σ_ε^2

Tests are obvious.
No direct test on C.

6.5

Source	EMS
A_i	$\sigma_\varepsilon^2 + nb\sigma_{ACD}^2 + nbc\sigma_{AD}^2 + nbd\sigma_{AC}^2 + nbcd\phi_A$
B_j	$\sigma_\varepsilon^2 + na\sigma_{BCD}^2 + nac\sigma_{BD}^2 + nad\sigma_{BC}^2 + nacd\phi_B$
AB_{ij}	$\sigma_\varepsilon^2 + n\sigma_{ABCD}^2 + nc\sigma_{ABD}^2 + nd\sigma_{ABC}^2 + ncd\phi_{AB}$
C_k	$\sigma_\varepsilon^2 + nab\sigma_{CD}^2 + nabd\sigma_C^2$
AC_{ik}	$\sigma_\varepsilon^2 + nb\sigma_{ACD}^2 + nbd\sigma_{AC}^2$
BC_{jk}	$\sigma_\varepsilon^2 + na\sigma_{BCD}^2 + nad\sigma_{BC}^2$
ABC_{ijk}	$\sigma_\varepsilon^2 + n\sigma_{ABCD}^2 + nd\sigma_{ABC}^2$
D_m	$\sigma_\varepsilon^2 + nab\sigma_{CD}^2 + nabc\sigma_D^2$
AD_{im}	$\sigma_\varepsilon^2 + nb\sigma_{ACD}^2 + nbc\sigma_{AD}^2$
BD_{jm}	$\sigma_\varepsilon^2 + na\sigma_{BCD}^2 + nac\sigma_{BD}^2$
ABD_{ijm}	$\sigma_\varepsilon^2 + n\sigma_{ABCD}^2 + nc\sigma_{ABD}^2$
CD_{km}	$\sigma_\varepsilon^2 + nab\sigma_{CD}^2$
ACD_{ikm}	$\sigma_\varepsilon^2 + nb\sigma_{ACD}^2$
BCD_{jkm}	$\sigma_\varepsilon^2 + na\sigma_{BCD}^2$
$ABCD_{ijkm}$	$\sigma_\varepsilon^2 + n\sigma_{ABCD}^2$
$\varepsilon_{q(ijkm)}$	σ_ε^2

No direct test on A, B, or AB.

6.7 a. $Y_{ijk} = \mu + T_i + S_j + TS_{ij} + \varepsilon_{k(ij)}$.

b.

Source	df	EMS
T_i	3	$\sigma_\varepsilon^2 + 2\sigma_{TS}^2 + 10\phi_T$
S_j	4	$\sigma_\varepsilon^2 + 8\sigma_S^2$
TS_{ij}	12	$\sigma_\varepsilon^2 + 2\sigma_{TS}^2$
$\varepsilon_{k(ij)}$	20	σ_ε^2

c. See arrows above.

6.9

Source	EMS
O_i	$\sigma_\varepsilon^2 + 2\sigma_{OI}^2 + 8\sigma_O^2$
I_j	$\sigma_\varepsilon^2 + 2\sigma_{OI}^2 + 8\sigma_I^2$
OI_{ij}	$\sigma_\varepsilon^2 + 2\sigma_{OI}^2$
$\varepsilon_{k(ij)}$	σ_ε^2

Instruments:	40 percent
Operators:	32 percent
Operator/	
instrument interaction:	15 percent
Error:	13 percent

6.11 $n_o = 4.44$, $s_\varepsilon^2 = 1.16$, $s_A^2 = 5.32$.

6.15   To test A, $MS = MS_{AB} + MS_{AC} - MS_{ABC}$

B, $MS = MS_{AB} + MS_{BC} - MS_{ABC}$

C, $MS = MS_{AC} + MS_{BC} - MS_{ABC}$

Chapter 7

7.1

Source	df	SS	MS	EMS
L_i	2	27.42	13.71	$\sigma_\varepsilon^2 + 3\sigma_R^2 + 12\phi_L$
$R_{j(i)}$	9	36.38	4.04	$\sigma_\varepsilon^2 + 3\sigma_R^2$
$\varepsilon_{k(ij)}$	24	21.80	0.91	σ_ε^2
Totals	35	85.60		

Difference between rolls within lots is significant at the 1 percent level.

7.3

Source	EMS
A_i	$\sigma_\varepsilon^2 + 2\sigma_C^2 + 6\sigma_B^2 + 24\phi_A$
$B_{j(i)}$	$\sigma_\varepsilon^2 + 2\sigma_C^2 + 6\sigma_B^2$
$C_{k(ij)}$	$\sigma_\varepsilon^2 + 2\sigma_C^2$
$\varepsilon_{m(ijk)}$	σ_ε^2

Tests are obvious.

7.5

Source	df	SS	MS	EMS
T_i	1	5,489,354	5,489,354	$\sigma_\varepsilon^2 + 6\sigma_M^2 + 12\phi_T$
$M_{j(i)}$	2	408,250	204,125	$\sigma_\varepsilon^2 + 6\sigma_M^2$
S_k	1	8971	8971	$\sigma_\varepsilon^2 + 3\sigma_{MS}^2 + 12\phi_S$
TS_{ik}	1	37,445	37,445	$\sigma_\varepsilon^2 + 3\sigma_{MS}^2 + 6\phi_{TS}$
$MS_{kj(i)}$	2	31,520	15,760	$\sigma_\varepsilon^2 + 3\sigma_{MS}^2$
$\varepsilon_{m(ijk)}$	16	316,930	19,808	σ_ε^2

7.7

Source	EMS
T_i	$\sigma_\varepsilon^2 + 18\sigma_m^2 + 36\phi_t$
$M_{j(i)}$	$\sigma_\varepsilon^2 + 18\sigma_m^2$
S_k	$\sigma_\varepsilon^2 + 9\sigma_{ms}^2 + 36\phi_s$
TS_{ik}	$\sigma_\varepsilon^2 + 9\sigma_{ms}^2 + 18\phi_{LS}$
$MS_{kj(i)}$	$\sigma_\varepsilon^2 + 9\sigma_{ms}^2$
P_m	$\sigma_\varepsilon^2 + 6\sigma_{mp}^2 + 24\phi_p$
TP_{im}	$\sigma_\varepsilon^2 + 6\sigma_{mp}^2 + 12\phi_{tp}$
$MP_{mj(i)}$	$\sigma_\varepsilon^2 + 6\sigma_{mp}^2$
SP_{km}	$\sigma_\varepsilon^2 + 3\sigma_{msp}^2 + 12\phi_{sp}$
TSP_{ikm}	$\sigma_\varepsilon^2 + 3\sigma_{msp}^2 + 6\phi_{tsp}$
$MSP_{mkj(i)}$	$\sigma_\varepsilon^2 + 3\sigma_{msp}^2$
$\varepsilon_{q(ijkm)}$	σ_ε^2

7.9 a, b.

Source	df	EMS
C_i	1	$\sigma_\varepsilon^2 + 2\sigma_F^2 + 8\sigma_C^2$
$F_{j(i)}$	6	$\sigma_\varepsilon^2 + 2\sigma_F^2$
V_k	1	$\sigma_\varepsilon^2 + \sigma_{FV}^2 + 4\sigma_{CV}^2 + 8\phi_V$
CV_{ik}	1	$\sigma_\varepsilon^2 + \sigma_{FV}^2 + 4\sigma_{CV}^2$
$FV_{kj(i)}$	6	$\sigma_\varepsilon^2 + \sigma_{FV}^2$

c. 60 percent.

7.11 Newman–Keuls shows experimental posttest group means better than control and one experimental group better than the other.

7.13 Newman–Keuls gives: <u>54.66</u> 54.96 57.00 <u>65.66</u>
So posttest experimental is best.

7.15 Discuss based on EMS and results.

7.17 $Y_{ijkm} = \mu + G_i + C_j + GC_{ij} + S_{k(ij)} + T_m + GT_{im} + CT_{jm}$
$\qquad\qquad + GCT_{ijm} + TS_{mk(ij)}$

7.19

Source	df	EMS
D_i	2	$2\sigma_S^2 + 22\sigma_D^2$
G_j	1	$\sigma_{SG}^2 + 11\sigma_{DG}^2 + 33\phi_G$
DG_{ij}	2	$\sigma_{SG}^2 + 11\sigma_{DG}^2$
$S_{k(i)}$	30	$2\sigma_S^2$
$SG_{jk(i)}$	30	σ_{SG}^2

7.21

Source	df	EMS
R_i	1	$\sigma_\varepsilon^2 + 6\sigma_H^2 + 36\sigma_R^2$
$H_{j(i)}$	10	$\sigma_\varepsilon^2 + 6\sigma_H^2$
P_k	2	$\sigma_\varepsilon^2 + 2\sigma_{PH}^2 + 12\sigma_{RP}^2 + 24\phi_P$
RP_{ik}	2	$\sigma_\varepsilon^2 + 2\sigma_{PH}^2 + 12\sigma_{RP}^2$
$PH_{kj(i)}$	20	$\sigma_\varepsilon^2 + 2\sigma_{PH}^2$
$\varepsilon_{m(ijk)}$	36	σ_ε^2

7.23

Source	df	EMS
F_i	4	$\sigma_\varepsilon^2 + 9\sigma_S^2 + 54\phi_F$
C_j	1	$\sigma_\varepsilon^2 + 9\sigma_S^2 + 135\phi_C$
FC_{ij}	4	$\sigma_\varepsilon^2 + 9\sigma_S^2 + 27\phi_{FC}$
$S_{k(ij)}$	20	$\sigma_\varepsilon^2 + 9\sigma_S^2$
P_m	2	$\sigma_\varepsilon^2 + 3\sigma_{SP}^2 + 90\phi_P$
FP_{im}	8	$\sigma_\varepsilon^2 + 3\sigma_{SP}^2 + 18\phi_{FP}$
CP_{jm}	2	$\sigma_\varepsilon^2 + 3\sigma_{SP}^2 + 45\phi_{CP}$
FCP_{ijm}	8	$\sigma_\varepsilon^2 + 3\sigma_{SP}^2 + 9\phi_{FCP}$
$SP_{mk(ij)}$	40	$\sigma_\varepsilon^2 + 3\sigma_{SP}^2$
$\varepsilon_{q(ijkm)}$	180	σ_ε^2

7.25 Recommend 1:1 concentration and any of the fillers—iron filings, iron oxide, or copper.

7.27

Source	df	EMS
S_i	9	$\sigma_\varepsilon^2 + 2\sigma_{T_2}^2 + 4\phi_S$
$T_{j(i)}$	10	$\sigma_\varepsilon^2 + 2\sigma_T^2$
P_k	1	$\sigma_\varepsilon^2 + \sigma_{PT}^2 + 20\phi_P$
SP_{ik}	9	$\sigma_\varepsilon^2 + \sigma_{PT}^2 + 2\phi_{SP}$
$PT_{kj(i)}$	10	$\sigma_\varepsilon^2 + \sigma_{PT}^2$
$\varepsilon_{m(ijk)}$	0	σ_ε^2

Test S vs. T and test P and SP vs. PT. No test on T.

Chapter 8

8.1

Source	df	SS	MS	EMS
R_i	2	337.15	168.58	$\sigma_\varepsilon^2 + 18\sigma_R^2$
S_j	1	16.66	16.66	$\sigma_\varepsilon^2 + 9\sigma_{RS}^2 + 27\phi_S$
RS_{ij}	2	126.78	63.39	$\sigma_\varepsilon^2 + 9\sigma_{RS}^2$
H_k	2	15.59	7.80	$\sigma_\varepsilon^2 + 6\sigma_{RH}^2 + 18\phi_H$
HS_{jk}	2	80.12	40.06	$\sigma_\varepsilon^2 + 3\sigma_{SRH}^2 + 9\phi_{HS}$
RH_{ik}	4	62.85	15.71	$\sigma_\varepsilon^2 + 6\sigma_{RH}^2$
SRH_{ijk}	4	229.44	57.36	$\sigma_\varepsilon^2 + 3\sigma_{SRH}^2$
$\varepsilon_{m(ijk)}$	36	200.00	5.56	σ_ε^2
Totals	53	1068.59		

Replications (blocks) and replications by all other factors are significant at the 5 percent level; however, these are not the factors of chief interest.

8.3

Source	df	SS	EMS
R_i	1	As before	$\sigma_\varepsilon^2 + 3\sigma_{RM}^2 + 12\sigma_R^2$
T_j	1	with R in	$\sigma_\varepsilon^2 + 3\sigma_{RM}^2 + 6\sigma_M^2 + 6\sigma_{RT}^2 + 12\phi_T$
RT_{ij}	1	place of S	$\sigma_\varepsilon^2 + 3\sigma_{RM}^2 + 6\sigma_{RT}^2$
$M_{k(j)}$	2		$\sigma_\varepsilon^2 + 3\sigma_{RM}^2 + 6\sigma_M^2$
$RM_{ik(j)}$	2		$\sigma_\varepsilon^2 + 3\sigma_{RM}^2$
$\varepsilon_{m(ijk)}$	16		σ_ε^2
Total	23		

No direct test on types, but results look highly significant. No other significant effects.

8.5

Source	df	EMS	
R_i	2	$10\sigma_R^2$	No test
M_j	4	$2\sigma_{RM}^2 + 6\phi_M$	
RM_{ij}	8	$2\sigma_{RM}^2$	No test
S_k	1	$5\sigma_{RS}^2 + 15\phi_S$	Poor test
RS_{ik}	2	$5\sigma_{RS}^2$	No test
MS_{jk}	4	$\sigma_{RMS}^2 + 3\phi_{MS}$	
RMS_{ijk}	8	σ_{RMS}^2	No test

8.7

Source	df	EMS
R_i	2	$30\sigma_R^2$
V_j	4	$6\sigma_{RV}^2 + 18\phi_V$
RV_{ij}	8	$6\sigma_{RV}^2$
S_k	5	$5\sigma_{RS}^2 + 15\phi_S$
RS_{ik}	10	$5\sigma_{RS}^2$
VS_{jk}	20	$\sigma_{RVS}^2 + 3\phi_{VS}$
RVS_{ijk}	40	σ_{RVS}^2

8.9

Source	df	EMS
R_i	1	$\sigma_C^2 + 36\sigma_R^2$
P_j	2	$\sigma_C^2 + 12\sigma_{RP}^2 + 24\phi_P$
RP_{ij}	2	$\sigma_C^2 + 12\sigma_{RP}^2$
H_k	5	$\sigma_C^2 + 6\sigma_{RH}^2 + 12\phi_H$
RH_{ik}	5	$\sigma_C^2 + 6\sigma_{RH}^2$
PH_{jk}	10	$\sigma_C^2 + 2\sigma_{RPH}^2 + 4\phi_{PH}$
RPH_{ijk}	10	$\sigma_C^2 + 2\sigma_{RPH}^2$
$C_{m(ijk)}$	36	σ_C^2

Poor test

8.11 Compare models.

8.13 Complete: $Y_{ijkm} = \mu + R_k + M_i + RM_{ik} + H_{j(i)} + RH_{kj(i)} + \varepsilon_{m(ijk)}$.
Reduced: $Y_{ijkm} = \mu + R_k + M_i + H_{j(i)} + \varepsilon_{m(ijk)}$.
Take fewer readings per head.

8.15 Complete: $Y_{ijkmq} = \mu + R_q + M_i + RM_{iq} + G_j + RG_{jq} + MG_{ij} + RMG_{ijq}$
$$+ T_{k(j)} + RT_{qk(j)} + MT_{ik(j)} + RMT_{iqk(j)} + \varepsilon_{m(ijkq)}$$
Reduced: Omit all interactions with R.

Chapter 9

9.1

Source	df	SS	MS
Factor A	1	0.08	0.08
Factor B	1	70.08	70.08**
$A \times B$ interaction	1	24.08	24.08
Error	8	36.68	4.58
Totals	11	130.92	

9.3

Source	df	SS	MS
A	1	2704.00	2704.00
B	1	26,732.25	26,732.25
AB	1	7744.00	7744.00
C	1	24,025.00	24,025.00
AC	1	16,256.25	16,256.25
BC	1	64,516.00	64,516.00
ABC	1	420.25	420.25
Error	8	246,284.00	30,785.50
Totals	15	388,681.75	

Nothing significant at the 5 percent level.

9.5   No plots, since there are no significant effects at the 5 percent level.

9.7

Source	df	SS	MS
A	1	13,736.5	
B	1	6188.3	
AB	1	22,102.5	Same
C	1	22.7	as
AC	1	22,525.0	SS
BC	1	12,051.3	
ABC	1	20,757.0	
D	1	81,103.8	
AD	1	145,665.0*	
BD	1	9214.3	
ABD	1	126,630.3*	
CD	1	148.8	
ACD	1	6757.0	
BCD	1	294.0	
ABCD	1	19,453.8	
Error	16	431,599.4	26,975.0
Totals	31	918,249.7	

9.9   a. (1), a, b, ab, c, ac, bc, abc, d, ad, bd, abd, cd, acd, bcd, $abcd$.

b.

Source	df
Main effects	1 each for 4
Two-way interaction	1 each for 6
Three-way interaction	1 each for 4
Four-way interaction	1 each for 1
Total	15

Three-way interaction and Four-way interaction: } Use as error

c. Find A effect by plus signs where a is present in treatment combination and minus signs where absent. Same for D and then multiply signs to get AD signs.

9.11 a. 32

b.

Source	df
Main effects	1 each for 5
Two-way interaction	1 each for 10
Three-way interaction	1 each for 10
Four-way interaction	1 each for 5
Five-way interaction	1 each for 1
Total	31

Four-way interaction and Five-way interaction: } Use as error

c. Get signs for A, C, and E and multiply.

d. True if no interactions are present. Otherwise use Newman–Keuls on treatment means.

9.13 Feed rate, 0.015; tool condition, new; tool type, precision.

9.15 2^6 factorial. Probably use four-, five-, and six-way interactions as error term with 22 df.

9.17

Source	df	SS and MS
Dropout temperature (A)	1	3.00
Back-zone temperature (B)	1	4.06*
$A \times B$ interaction	1	1.20
Atmosphere (C)	1	3.00
$A \times C$ interaction	1	1.71
$B \times C$ interaction	1	4.65*
$A \times B \times C$ interaction	1	15.96**

Significant based on MS for error of 0.37 with 3 df.

9.19

Source	df	SS and MS
A	1	45.60*
B	1	1.36
AB	1	53.56*
C	1	0.01
AC	1	0.00
BC	1	0.01
ABC	1	0.04

A and AB significant based on MS for error of 4.41 with 3 df.
Newman–Keuls on A, B means: <u>0.00   4.20   4.35   9.95.</u>
No significant gaps as error is too large.

9.21 Plot the following: $(1) = 0$, $a = -3$, $b = -2$, $ab = 5$, $c = -7$, $ac = 0$, $bc = 1$, and $abc = -2$.

9.23 No. Results of cold test are better on B. No need for a cold test on lot A.

9.25

Source	df	MS	F
Type (T)	1	2468	7.01*
Position (P)	1	14,921	42.39**
T * P	1	1582	4.49*
Edge (E)	1	4255	12.09**
T * E	1	195	< 1
P * E	1	5645	16.03**
T * P * E	1	570	1.62
Error	24	352	

Because the $P * E$ is significant at the 5 percent level, we examine the four means and find <u>0   3.5   20.12   69.75</u>.

Chapter 10

10.1

Source	df	SS	MS
Temperature	2	348.44	174.22
Humidity	2	333.77	166.88
$T \times H$ interaction	4	358.23	89.56
Error	9	680.00	75.66
Totals	17	1720.44	

None significant at the 5 percent level.

10.3 $AB = 320.11$
$\left.\begin{array}{l}\\ \end{array}\right\}$ 358.22
$AB^2 = \ 38.11$

10.5

Source	df	SS	MS
V	2	2.79	1.40*
T	2	9.01	4.50**
$V \times T$	4	1.99	0.50
Error	9	2.15	

10.7 $VT = 0.94$ and $VT^2 = 1.04$, each with 2 df.

10.9   Any combination except $T_H V_H$.

10.11 After coding data by subtracting 2.6 and multiplying by 10, we have

Source	df	SS	MS
Surface thickness A	2	2544.71	1272.35
Base thickness B	2	4787.37	2393.68
$A \times B$ interaction	4	185.18	46.29
Subbase thickness C	2	4165.15	2082.57
$A \times C$ interaction	4	189.72	47.48
$B \times C$ interaction	4	27.74	6.93
$A \times B \times C$ interaction	8	100.06	12.51
Error	27	71.50	2.65

All except $B \times C$ are significant at the 5 percent level, but the main effects predominate.

10.13

Source	df	SS
AB	2	66.93
AB^2	2	118.26
AC	2	58.93
AC^2	2	130.82
BC	2	5.82
BC^2	2	21.93
ABC	2	17.60
ABC^2	2	8.26
AB^2C	2	53.48
AB^2C^2	2	20.71

$AB, AB^2 \Rightarrow$ 185.19
$AC, AC^2 \Rightarrow$ 189.75
$BC, BC^2 \Rightarrow$ 27.75
$ABC, ABC^2, AB^2C, AB^2C^2 \Rightarrow$ 100.05

10.15 Plots show strong linear effects of A, B, and C. They show AB and AC interaction— but it is slight compared with the main effects. A slight quadratic trend in A is also noted.

Chapter 11

11.1

	Source	df	EMS
Whole	R_i	1	$\sigma_\varepsilon^2 + 6\sigma_{RS}^2 + 18\sigma_R^2$
plot	H_j	2	$\sigma_\varepsilon^2 + 2\sigma_{RHS}^2 + 4\sigma_{HS}^2 + 6\sigma_{RH}^2 + 12\phi_H$
	RH_{ij}	2	$\sigma_\varepsilon^2 + 2\sigma_{RHS}^2 + 6\sigma_{RH}^2$
Split	S_k	2	$\sigma_\varepsilon^2 + 6\sigma_{RS}^2 + 12\sigma_S^2$
plot	RS_{ik}	2	$\sigma_\varepsilon^2 + 6\sigma_{RS}^2$
	HS_{jk}	4	$\sigma_\varepsilon^2 + 2\sigma_{RHS}^2 + 4\sigma_{HS}^2$
	RHS_{ijk}	4	$\sigma_\varepsilon^2 + 2\sigma_{RHS}^2$
	$\varepsilon_{m(ijk)}$	18	σ_ε^2
	Total	35	

11.3   a. Fill nine spaces in random order.

b. Choose one of the orifice sizes at random, randomize flow rate through this orifice, then move to a second orifice size, and so on.

c.

Source	df
R_i	2
O_j	2
RO_{ij}	4
F_k	2
RF_{ik}	4
OF_{jk}	4
ROF_{ijk}	8

d. Take more replications.

11.5

	Source	df	EMS
Whole	R_i	3	$\sigma_\varepsilon^2 + 8\sigma_{RA}^2 + 40\sigma_R^2$
plot	S_j	1	$\sigma_\varepsilon^2 + 4\sigma_{RSA}^2 + 16\sigma_{SA}^2 + 20\sigma_{RS}^2 + 80\phi_S$
	RS_{ij}	3	$\sigma_\varepsilon^2 + 4\sigma_{RSA}^2 + 20\sigma_{RS}^2$
	J_k	1	$\sigma_\varepsilon^2 + 4\sigma_{RJA}^2 + 16\sigma_{JA}^2 + 20\sigma_{RJ}^2 + 80\phi_J$
	RJ_{ik}	3	$\sigma_\varepsilon^2 + 4\sigma_{RJA}^2 + 20\sigma_{RJ}^2$
	SJ_{jk}	1	$\sigma_\varepsilon^2 + 2\sigma_{RSJA}^2 + 10\sigma_{RSJ}^2 + 8\sigma_{SJA}^2 + 40\phi_{SJ}$
	RSJ_{ijk}	3	$\sigma_\varepsilon^2 + 2\sigma_{RSJA}^2 + 10\sigma_{RSJ}^2$

	Source	df	EMS
Split	A_m	4	$\sigma_\varepsilon^2 + 8\sigma_{RA}^2 + 32\sigma_A^2$
plot	RA_{im}	12	$\sigma_\varepsilon^2 + 8\sigma_{RA}^2$
	SA_{jm}	4	$\sigma_\varepsilon^2 + 4\sigma_{RSA}^2 + 16\sigma_{SA}^2$
	RSA_{ijm}	12	$\sigma_\varepsilon^2 + 4\sigma_{RSA}^2$
	JA_{km}	4	$\sigma_\varepsilon^2 + 4\sigma_{RJA}^2 + 16\sigma_{JA}^2$
	RJA_{ikm}	12	$\sigma_\varepsilon^2 + 4\sigma_{RJA}^2$
	SJA_{jkm}	4	$\sigma_\varepsilon^2 + 2\sigma_{RSJA}^2 + 8\sigma_{SJA}^2$
	$RSJA_{ijkm}$	12	$\sigma_\varepsilon^2 + 2\sigma_{RSJA}^2$
	$\varepsilon_{q(ijkm)}$	80	σ_ε^2

11.7

	Source	df	MS	EMS
Whole	R_i	2	60.46	$\sigma_\varepsilon^2 + 5\sigma_{RO}^2 + 20\sigma_R^2$
plot	D_j	4	276.19	$\sigma_\varepsilon^2 + \sigma_{RDO}^2 + 3\sigma_{DO}^2 + 4\sigma_{RD}^2 + 12\phi_D$
	RD_{ij}	8	40.63	$\sigma_\varepsilon^2 + \sigma_{RDO}^2 + 4\sigma_{RD}^2$
Split	O_k	3	16.73	$\sigma_\varepsilon^2 + 5\sigma_{RO}^2 + 15\sigma_O^2$
plot	RO_{ik}	6	57.44	$\sigma_\varepsilon^2 + 5\sigma_{RO}^2$
	DO_{jk}	12	12.78	$\sigma_\varepsilon^2 + 2\sigma_{RDO}^2 + 3\sigma_{DO}^2$
	RDO_{ijk}	24	46.83	$\sigma_\varepsilon^2 + \sigma_{RDO}^2$
	Total	59		

Days are significant by an F' test.

11.9 In a nested experiment the levels of B, say, are different within each level of A. In a split-plot experiment the same levels of B (in the split) are used at each level of A.

11.11

Laboratory means	1	3	2
	11.22	11.25	12.53
Mix means:	B	A	C
	10.29	11.58	13.13

Laboratory × mix means all different in decreasing order: 165, 155, 145 with B, A, and C.

11.13   To test T, MS = 0.93 and $F' = 1660$

To test M, MS = 0.40 and $F' = 180$

To test $T * M$, MS = 0.09 and $F' = 121$

All highly significant.

11.15

Source	df	EMS
R_i	$r-1$	$\sigma_\varepsilon^2 + 15\sigma_R^2$
T_j	4	$\sigma_\varepsilon^2 + 3\sigma_{RT}^2 + 3r\phi_T$
RT_{ij}	$4(r-1)$	$\sigma_\varepsilon^2 + 3\sigma_{RT}^2$
M_k	2	$\sigma_\varepsilon^2 + 5\sigma_{RM}^2 + 5r\phi_M$
RM_{ik}	$2(r-1)$	$\sigma_\varepsilon^2 + 5\sigma_{RM}^2$
TM_{jk}	8	$\sigma_\varepsilon^2 + \sigma_{RMT}^2 + r\phi_{MT}$
RTM_{ijk}	$8(r-1)$	$\sigma_\varepsilon^2 + \sigma_{RMT}^2$

Preferable design.

11.17 ANOVA if $r = 2$.

Source	df	MS
R_i	1	3230
M_j	2	7
RM_{ij}	2	78
T_k	5	99
RT_{ik}	5	47
MT_{jk}	10	23
RMT_{ijk}	10	36

No effects significant except replications.

Chapter 12

12.1 Show. $SS_{blocks} = SS_{AC} = 10.125$.

12.3

replication I confound AB		replication II confound AC		replication III confound BC	
(1)	a	a	(1)	(1)	b
ab	b	c	ac	bc	c
c	ac	ab	b	a	ab
abc	bc	bc	abc	abc	ac

12.5 BC.

12.7 $SS_{blocks} = 29,153.5 = SS_{AB} + SS_{ACD} + SS_{BCD}$.

12.9 The principal block contains: (1), ab, ad, bd, c, abc, acd, bcd.

12.11 One scheme is to confound *ABC*, *CDE*, and *ABDE*. The principal block contains (1), *ab*, *de*, *abde*, *ace*, *bce*, *acd*, *bcd*.

Source	df
Main effects	5
Two-way interaction	10
Three-way interaction	8
Four-way interaction	4
Five-way interaction	1
Blocks (or *ABC*, *CDE*, *ABDE*)	3
Total	31

12.13 One scheme is to confound $ABCD^2$. The principal block contains 0000, 1110, 1102, 1012, 2111, 2122, 0221, 0210, 0120, 0101, 1211, 1200, 2220, 0022, 0011, 0202, 0112, 1121, 1020, 1001, 1222, 2010, 2201, 2100, 2002, 2021, 2212.

12.15 Verify.

12.17 Confounded *ABC*, *CDE*, *ADF*, *BEF*, *ABDE*, *BCDF*, *ACEF* gives a principal block of (1), *acd*, *bce*, *abf*, *abde*, *bcdf*, *acef*, *def*.

Source	df
Main effects	1 each for 6
Two-way interaction	1 each for 15
Three-way interaction (except confounds)	1 each for 16
Four-way interaction (except confounds)	1 each for 12
Five-way interaction	1 each for 6
Six-way interaction	1 each for 1
Blocks (or confounds)	7
Total	63

12.19 Yates with seven columns—pool to get temperature and its interactions.

12.21 Confound *TABC* with blocks (mod 2), $L = X_1 + X_2 + X_3 + X_4$.

12.23 Principal blocks are

Replication I	(1), *ab*, *de*, *abde*, *bcd*, *acd*, *bce*, *ace*
Replication II	(1), *be*, *cd*, *bcde*, *ace*, *abc*, *ade*, *abd*
Replication III	(1), *bc*, *abe*, *ace*, *cde*, *bde*, *abcd*, *ad*

Source	df	
Replication	2	
Blocks in replication	9	11
Main effects	1 each for 5	
Two-way interaction	1 each for 10	
Three-way interaction	1 each for 10	
Four-way interaction	1 each for 5	
Five-way interaction	1 each for 1	
Replications × all	53	84
Total	95	

Chapter 13

13.1   Aliases:

Source	df	SS	
$A = C$	1	30.25	
$B = ABC$	1	2.25	No tests.
$AB = BC$	1	0.25	

13.3   Aliases are $A = BC = AB^2C^2$

$$B = AB^2C = AC$$
$$C = ABC^2 = AB$$
$$AB^2 = AC^2 = BC^2$$

13.5 $A = BD, B = AD, C = ABCD, D = AB, AC = BCD, BC = ACD, CD = ABC.$

13.7

$$A = AB^2C^2D = BCD^2$$
$$B = AB^2CD^2 = ACD^2$$
$$C = ABC^2D^2 = ABD^2$$
$$D = ABC = ABCD$$
$$AB = ABC^2D = CD^2$$
$$AB^2 = AC^2D = BC^2D$$
$$AC = AB^2CD = BD^2$$
$$AC^2 = AB^2D = BC^2D^2$$
$$AD = AB^2C^2 = BCD$$
$$BC = AB^2C^2D^2 = AD^2$$
$$BC^2 = AB^2D^2 = AC^2D^2$$
$$BD = AB^2C = ACD$$
$$CD = ABC^2 = ABD$$

13.9

Source	df
Main effects with 5- or 6-way	7
2-way interaction with 4- or 6-way	21
3-way interaction, and so on as error	35
Total	63

13.11

Source	df	SS		MS
Heats (or 2-way)	6	51.92		8.65*
Treatments (or part of H)	1	12.04		12.04*
Positions (or 3-way)	2	0.58		0.29
PH (or 3-way)	12 ⎫	29.83 ⎫	33.42	2.39
PT (or 2-way)	2 ⎭	3.59 ⎭		
Totals	23	97.96		

13.13 Combining three- and four-way interactions as error with 5 df gives

Source	MS
A	12.44
B	53.40*
AB	0.00
C	33.44
AC	7.06
BC	0.00
D	31.22
AD	42.48*
BD	1.62
CD	0.07
Error	5.73

B and AD significant at 5 percent level.

13.15 a. Replication I: (1), ab, de, $abde$, bcd, acd, bce, ace.
b. $BC = DE = ABD = ACE$
c. Replications $\times$ three-way = 14 df.

13.17 a. ACD

b.

Source	df	MS
A (or CD)	1	5729.85*
B (or ABCD)	1	91.80
C (or AD)	1	4199.86*
D (or AC)	1	1217.71
AB, BC, BD, and their aliases	3	281.88

Some bad aliases, better to confound ABCD.

13.19 BCDEF confounded.

Source	df
Main (or four-way)	1 each for 5
Two-way or three-way	1each for 10

No good error term unless one assumes no interactions.

13.21 2_{III}^{5-2}

13.23 2_{III}^{6-3}

13.25 Results are indecisive as there are large interactions in the example and Plackett–Burman only screens main effects.

Chapter 14

14.1 Taguchi on these first observations gives the same order to main effects as in Example 1.2. The percentages are tool type 6 percent, bevel angle 15 percent, and cut 79 percent.

14.3 Results agree with Section 10.2 if the SS are divided by 2 as each number is the sum of two observations.

14.5 Results agree exactly with Table 13.13.

Chapter 15

15.1 $Y_X' = 6.57 + 0.50X_j$.

15.3 $\eta^2 = 0.5981$, $r^2 = 0.5023$, $S_{Y.X} = 0.756$.

15.5 Scattergram indicates quadratic.

15.7 $Y_x' = -437.98 + 120.02x - 9.309x^2 + 0.320x^3 - 0.00409x^4$.

15.9 a. Second degree.
b. $Y_x' = 98.26 - 4.09x + 1.13x^2$

15.11 $r^2 = 0.22$, $\eta^2 = 0.27$.

15.13

Source	df	SS		
Between Ages	4	82.28		
Linear	1		72.25^{**}	
Quadratic	1		9.78^{*}	$MS_{error} = 0.39$
Cubic	1		0.25	
Quartic	1		0.00	
Error	45	17.50		

Curve is $Y'_x = -30.99 + 6.13x - 0.264x^2$ and $R^2 = 0.82$.

15.15

Source	df	SS	MS
Gate	2	10.58	5.29^{**}
Resin	1	0.01	0.01
$G \times R$ interaction	2	0.00	0.00
Error	6	0.07	0.01

15.17 Linear only on gate setting giving: $Y'_u = 2.517 + 1.15u$ or $Y'_X = 0.217 + 0.575X_j$.

15.19

Source	df		Source	df	
R	1		$G \times W$	4	
G	2		$G_L \times W_L$		1
G_L		1	$G_L \times W_Q$		1
G_Q		1	$G_Q \times W_L$		1
$R \times G$	2		$G_Q \times W_Q$		1
$R \times G_L$		1	$R \times G \times W$	4	
$R \times G_Q$		1	$R \times G_L \times W_L$		1
W	2		$R \times G_L \times W_Q$		1
W_L		1	$R \times G_Q \times W_L$		1
W_Q		1	$R \times G_Q \times W_Q$		1
$R \times W$	2		Error	18	
$R \times W_L$		1	Total	35	
$R \times W_Q$		1			

15.21 Graph.

15.23

Source	df		Source	df
Radius (R)	3		$R_L \times F_L$	1
R_L	1		$R_L \times F_Q$	1
R_Q	1		$R_L \times F_C$	1
R_C	1		$R_Q \times F_L$	1
Force (F)	3		$R_Q \times F_Q$	1
F_L	1		$R_Q \times F_C$	1
F_Q	1		$R_C \times F_L$	1
F_C	1		$R_C \times F_Q$	1
$R \times F$ interaction	9		$R_C \times F_C$	1
			Error	32
			Total	47

15.25   Plot.

15.27 $Y' = 68.78 + 1.92X_1$ adequate. $r_{Y1}^2 = 0.34$.
For both x's: $Y' = 64.69 + 1.92X_1 + 0.0127X_2$.
Note: $r_{12} = 0$ as X_1 and X_2 are taken orthogonal.

15.29   a. One with largest correlation with Y.
b. One with largest partial correlation with Y after first X is removed.
c. When no more partials are significant.
d. $F_{2, 20} = 3.33 < 3.49$ at 5 percent, so stop here.

15.31   a. $R_{Y.1234} = 0.71$.
b. X_2 and X_3 only.
c. ± 12.89 (two standard errors).

15.33   Adequate equation: $Y' = -0.66 + 0.29X_1 + 0.45X_3$, where X_1 = ethnic group
and X_3 = interview score. $R_{Y.13}^2 = 0.65$.

15.35   a. Scattergram looks like quadratic.
b. $Y' = 52.11 + 9.11X - 0.22X^2$ and $R_{Y.XX^2}^2 = 0.94$.
c. Linear: $Y' = 132.92 + 0.40X$ and $r_{YX}^2 = 0.11$.

Chapter 16

16.1   Before covariance: $F_{6, 100} = 50$ and significant.
After covariance: $F_{6, 99} = 16.5$ and still significant.

16.3

Source	df	$\sum x^2$	$\sum xy$	$\sum y^2$	Adjusted $\sum y^2$	df	MS
Between lots	3	500	−620	790			
Error	16	1182	428	406	251.02	15	16.73
Totals	19	1682	−192	1196	1174.08	18	
					923.06	3	307.69*

16.5

End of cycle	Estimate of s	Significance
2, block 1	0.60	None
2, block 2	0.64	All except AB
3, block 1	0.60	
3, block 2	0.61	All

16.7

Source	df	MS
Machines M	1	0.0136
Lumber grade L	1	0.0003
ML	1	0.0066
Replications R	1	0.0001
MR	1	0.0028
LR	1	0.0021
MLR	1	0.0015

16.9

Source	df	SS	MS
Days	3	34,292	11,431
Treatments (adjusted)	3	508	169
Error	5	5025	1005

Treatments not significant at 5 percent.

16.11 If seven were used,

Source	df
Treatments	1
Blocks (men)	6
Error	6
Total	13

Index

α probability error, 21
Aliases, 304–8, 330
Analysis, 3, 6
Analysis of attribute data, 423–26
  factorial chi square, 423
  philosophy, 423
Analysis of variance (ANOVA), 12,
    209–10
  assumptions, 69–70
  factorial experiments rationale, 131–36
  four-way, 138
  with interactions, 138
  nested experiments rationale, 179–80
  one-way, 51–56, 128–29, 138
  randomized block design rationale,
    99–101
  restrictions on randomization, 202–6
  SNK Test and, 75–88
  three-way, 12–14, 138
Anderson, V. L., 264
Autocorrection coefficient, 74

β probability error, 21, 22–24
BACKWARD (B) program, 368, 370–72
Balanced incomplete block design, 427
Barnett, E. H., 413
Bartlett-Box F test, 395
Batson, H. C., 423, 425
Before-after experiment, 43
Behrens-Fisher problem, 30
Bennett, C. A., 156, 168–69

Binomial variable, 7
Blocks, 95. *See also* Randomized block
    design
Box, G. E. P., 412, 413
Burr, I. W., 247

Canonical form, 413
Central composite design, 413
Change in mean effect (CIM), 414, 417
Chi square, 423–26
Coefficient of determination r^2, 353
Coefficient of variation, inverse of, 325
Completely randomized design, 8, 11,
    49–86, 126, 138, 209
Components of variance, 64–69, 85–86,
    135
Computer programs. *See* Statistical
    Analysis System (SAS)
Concomitant variable X, 387
Confidence interval, 19–20
Confidence limits, 20, 64, 135
Confounded design, 94, 136
Confounding in blocks, 278–99
  complete, 287–88
  confounding systems, 279–82
  in fixed model, 170
  partial, 288–91
  SAS program, 297–99
  in split-plot experiments, 261, 263
  with replication, 287–96
  without replication, 282–87

Confounding in blocks *(continued)*
  in 2^f factorials, 282–86
  in 3^f factorials, 292–96
Consistent statistic, 19
Contrasts, 58–61
Correlation, 2
Correlation ratio R, 357
Covariance, 390–97
Covariance analysis, 387–97
  philosophy, 387–90
  problems, 394–97
  SAS program, 397–401
Critical region, 20
Crossed factors, 173
Cycle, in EVOP, 414

Data collection, 6
Data reduction, 6
Davies, O. L., 26, 294, 423
Defining contrast, 280
Degrees of freedom, 13
  in fractional replication, 303, 310, 316
  in nested experiment, 176
  in one observation per treatment, 137
  with restriction on randomizaton, 203,
      206, 208, 429–30
  in single-factor experiments, 54
Dependent variable, 3
Design resolution, 310–12
Designs
  balanced incomplete block, 427
  central composite, 413
  completely confounded, 287–88
  completely randomized, 8, 11, 49–86,
      126, 138, 209
  confounded, 94, 136
  first-order, 412
  general, 4–6
  Graeco-Latin square, 106–7
  incomplete block, 107
  incomplete randomized block, 427–33
  Latin square, 103–6
  partially confounded, 287, 288–92
  randomized block, 95–103, 138
  repeated-measures, 184–89
  rotatable, 412
  second-order, 412–13

Designs *(continued)*
  split plot, 258–74
  split-split plot, 261–62
  Taguchi, 325–46
  Youden square, 107, 434–36
Developmental research, 2
Durbin-Watson statistic, 74

Effect of a factor, defined, 219
Equispacing, 353–54, 376
Error
  restriction, 264
  split plot, 261–62
  type I, 21
  type II, 21, 22–23
  whole-plot, 261–62
Error mean square, 164
Error sum of squares, 13, 176, 392
Estimation, 18–20
Eta squared, 353
Evolutionary operation (EVOP)
  cycle, 414, 413–22
  form rationale, 418–22
  phase, 414
  philosophy, 413–14
Expected mean square (EMS), 65–69,
      158–70
  derivations, 66–69, 165–67
  in nested experiments, 175
  in nested-factorial experiments, 180–
      83, 189–90
  with one observation per cell, 169–70
  pseudo-F tests and, 169
  in randomized block design, 202–3
  in repeated-measures design, 185,
      187–89
  rules, 160–65
  in two-factor experiments, 160–65
Expected value, 18, 65–69
Experimental research, 1–3
Experimental unit, 1
Experiments
  defined, 1
  factorial, 4, 121–48, 209, 217–30,
      235–55, 258–74, 278–99
  general, 3–4
  nested, 4, 174–79, 209

Experiments *(continued)*
  nested-factorial, 179–89, 209–10
  single-factor, 49–86, 93–114, 157–58,
    165, 229
  three-factor, 14, 133–36, 185–88
  two-factor, 131–33, 158–60, 165–67,
    201–12, 230, 254
Ex-post-facto research, 2, 365–67, 376

F distribution, 27–28
F ratios, 135
F test
  in covariance analysis, 396–97
  EMS values and, 163
  in incomplete block design, 427
  pseudo-F, 167–69
  restrictions on randomization and,
    206–7
  in split-split-plot design, 267
  in Taguchi method, 329
Factorial chi square, 423–26
Factorial experiments, 4, 121–48
  advantages, 123
  ANOVA rationale, 131–36
  confounding in blocks, 278–99
  defined, 123–24
  interactions, 124
  interpretations, 129–31
  in Latin square design, 207–12
  with one operation per cell, 136–37
  in randomized block design, 201–7,
    208–12
  SAS programs, 138–48
  in split-plot design, 258–74
  summarized, 138
  2^f, 217–30
  3^f, 235–55, 282–86, 312–16
  2^2, 217–25, 278–79, 405, 414
  2^3, 10, 11, 225–30, 303–6, 406, 412
  2^4, 284
  3^2, 236–46, 254
  3^3, 346–54, 255
Factors, levels of
  qualitative and quantitative, 3–4, 7
  random and fixed, 7
  Taguchi method, 325–28, 339–40
Feedback, 6

First-order designs, 412
First-order equations, 403
Fisher, R. A., 104
Fixed effects, 3
Fixed model, 49, 170
  defined, 156
  single-factor, 157–58
  two-factor, 158–60, 166
FORWARD (F) program, 367–68, 369
Fractional factorial, 304, 312–16
Fractional replication, 303–20
  aliases, 304–8
  Plackett-Burman designs, 317–18
  resolution in, 310–12
  SAS program, 319–20
  2^f, 306–10
  3^f, 312–16
Franklin, N. L., 156, 168–69
Fundamental equation of analysis of
    variance, 53

General Linear Model (GLM), 79–85
Graeco-Latin square design, 106–7, 138,
    208

Hierarchical experiments, 4, 173, 209
High-order interactions, 303–8
Historical research, 2
Homogeneous variances, 69
Hypotheses
  in experimental design, 5
  statistical, 20
  testing, 20–22, 26–28

I and J components of interaction, 242,
    247, 251, 253
Incomplete block design, 107, 427–33
Independent variables
  defining, 3
  ways of handling, 5–6
Interaction
  confounded, 279, 281–84, 287–89,
    292–94
  factorial experiments, 124, 137, 138–
    39, 148
  high-order, 303–8

Interaction *(continued)*
I and *J* components, 242, 247, 251, 253
involving fixed factors, 164
L_4 design, 326–30
L_8 design, 330–36
L_9 design, 339–42
L_{16} design, 337–38
L_{18} design, 345
in randomized block design, 205–7
split-split plot design, 267
two-way, 294, 310
2^2 factorial, 218–122
2^3 factorial, 226–271
3^2 factorial, 238–45, 246, 249, 251–55
Interference. *See* Statistical interference
Interval estimate, 18–19

Kempthorne, O., 280, 292
Keuls. *See* Newman-Keuls range test

Latin square design, 103–6, 138
complete, 209, 210
factorial experiments in, 207–10
fractional replication and, 315
nested or nested-factorial experiments in, 208
SAS program, 108–14, 210–12
Youden square, 107, 434–36
Least squares method, 350, 355
Least squares normal equations, 350, 406
Linear contrast, 235–42
Linear graphs, 325, 326, 331, 334–35, 337–39, 342
Linear regression, 347–54

McLean, 264
Main effects, 324, 330–32, 345
Manipulation, 2
Mathematical model, 5, 6
Matrix algebra, 406
Maximum yield, 124–26
Max-min-con principle, 6
MAXR program, 368, 373–74
Mean square (MS), 13, 19, 54–64
Means
confidence limits on, 64
tests on, 28–32, 56

Minimum variance, 19
Minimum-variance estimate, 19
Missing values, 101–3
Mixed model, 163
defined, 157
EMS for, 175
test, 170
two-factor, 159–60, 167
Modulus, 3, 242–43
Multiple regression, 404, 365–67, 377
SAS program, 367–76

National Bureau of Standards, 312, 316
Nelson, L. S., 317, 318
Nested experiments, 4, 174–79, 209
ANOVA rationale, 178–79
defined, 173
randomized block or Latin square design, 208
repeated-measures designs as, 184
SAS programs, 189–92
Nested-factorial experiments, 179–80, 209–10
randomized block or Latin square design, 208
repeated-measures designs as, 184
SAS program, 189–92
Nesting notation, 189
Newman-Keuls range test, 61–63
ANOVA and, 75, 111–14
in factorial experiment interpretation, 129–31, 143, 145
in nested experiment, 177
in nested-factorial experiment, 184
of 3^2 factorial, 245–46
Normally and independently distributed (NID) random effects, 50
Null hypothesis form, 5

One-eighth replications, 312
One-fourth replications, 308, 311
One-half replications, 304–8, 310–11
One-third replications, 313–17
One-way analysis of variance, 51
Operating characteristic (OC) curve, 22–24
Operator, 43

Orthogonal arrays, 325–46
Orthogonal contrasts, 57–61
  method, 57
  SAS program, 79–80, 235–36, 357–58
Orthogonal polynomials, 357–58
Ostle, B., 388, 398
Owen, D. B., 26

Parameters, 17
Path of steepest ascent method, 402–3
Pearson product-moment correlation
  coefficient, 353
p-factor effect, 311
Phase, in EVOP, 414
Plackett-Burman designs, 317–18, 346
Plots, 259
Point estimate, 18
Poisson variable, 70
Polynomial regression, 357–58, 376–77
Pooling, described, 105, 111
Population, 1, 17
  standard deviation, 17
  variance, 17
Power curves, 24
Principal block, 281, 284, 292
Probability, 16
PROC ANOVA, 347, 362–63
  for completely randomized designs,
    74–79
  in factorial experiments, 138–48
  for randomized blocks and Latin
    squares, 108–14
PROC GLM, 79–85, 358–64, 347,
  399–401
PROC MEANS, 44, 139
PROC NESTED, 85–86
PROC PLOT, 139
PROC REG, 347
PROC TTEST, 40–44
Proportion tests, 32–33
Pseudo-*F* test, 167–69

Quadratic contrasts, 235–42
Quadratic regression, 354–57
Qualitative and quantitative factors,
  3–4, 7
Quasi-experiments, 2

Random effects, 3
Random model
  defined, 49, 156
  single-factor, 157–58
  two-factor, 158–60, 166
  variance and, 64–66
Random order, 4–5
Random samples, 17
Random variable, 17
Randomization, 2
  of order of experimentation, 4–5
  restrictions, 210–12, 278–79
Randomized block design
  ANOVA rationale, 99–101
  complete, 95–103, 138, 209
  factorial experiments, 201–7,
    208–10
  incomplete, 107, 427–33
  method for balanced blocks, 427
  missing values, 101–3
  nested or nested-factorial, 208, 209
  SAS program, 108–14, 210–12
  single-factor experiments, 49–86
Range tests, 61–64
Regression, 2, 347–77
  covariance analysis and, 388,
    390–97
  curvilinear, 354–57
  linear, 347–54
  multiple, 365–67, 377
  orthogonal polynomial, 357–58,
    376–77
  SAS programs, 347, 358–64, 367–76
Repeated-measures designs, 184–89
Replication
  block confounding with, 287–96
  fractional, 303–20
Replication effect, 262, 265, 267
Research
  defined, 1
  types, 2
Research hypothesis, 5
Residuals analysis, 70–74
Resolution, design, 310–12
Resolution III, 310
Resolution IV, 310–11
Resolution V, 311

Response-surface experimentation, 402–13
    more complex surfaces, 412–13
    philosophy, 402
    two-fold problem, 402–12
Response variable Y, 3
Restriction error, 264
Rotatable design, 412

Sample, random, 17
Sample size, determining, 4, 25–26
Sample statistics, 17
Satterthwaite, 168
SAS. *See* Statistical Analysis System (SAS)
Scattergrams, 377
Scheffé's test, 61, 63–64
Second-degree (quadratic) model, 354–57
Second-order design, 412–13
Second-order equation, 403–4
Signal-to-noise ratio, 325
Single-factor experiments, 49–86, 157–58
    completely randomized, 49–86
    covariance analysis and, 388
    defined, 49
    Graeco-Latin square design, 106–7, 138
    Latin square design, 103–6, 138
    randomized block design, 95–103
    summarized, 229
Single-variance tests, 26–27
Snedecor, G. W., 388
SNK test. *See* Newman-Keuls range test
Solid analytic geometry methods, 404, 413
Split-plot design, 258–74
Split-plot error, 261–62
Split-split-plot design, 266–70
Standard deviation, in EVOP, 415–22
Standard error of estimate, 353
Statistical Analysis System (SAS), 33–45
    analysis of variance, 74–86
    data set, 34–35
    single-factor experiments, 74–87
Statistical hypothesis, 20

Statistical interference, 16–45
    defined, 17
    estimation, 18–20
    operating characteristic curve, 22–24
    sample size determination, 25–26
    in SAS analysis of experiments, 33–45
    test of hypotheses, 20–22
    tests on means and, 28–32
    tests on proportions and, 32–33
    tests on variances and, 26–28
Statistics, defined, 16
STEPWISE program, 368, 372
Straight-line fit, 348, 350
Student-Newman-Keuls range test. *See* Newman-Keuls range test
Sum of squares (SS), 13
    in covariance analysis, 391–93
    described, 18–19
    in linear regression, 352–54
    in multiple regression, 365–66
    in nested experiment, 175–76, 178, 179
    orthogonal, 60–61
    in quadratic regression, 356
    in randomized complete block design, 100–101
    in response-surface experiments, 405, 407
    in single-factor experiments, 53–56, 66
    in Taguchi method, 333, 343
    in three-factor problem, 133–35
    in two-factor problem, 131–33, 165–66
    in 2^2 factorial experiment, 219, 221, 222, 223–25
    in 2^3 factorial experiment, 228
    in 3^2 factorial experiment, 239, 242
    in 3^3 factorial experiment, 252
Survey research, 2
Symmetrical balanced incomplete randomized block design, 430
Systematic research, 1

Taguchi, Genichi, 325
Taguchi method, 325–46
    L_4 design, 325–30
    L_8 design, 330–36
    L_9 design,, 339–44

Taguchi method *(continued)*
  L_{16} design, 336–39
  L_{18} design, 344–45
Test statistics, 6, 20
Treatment effect, 49, 157, 170
True experiment, 1
Type-I error, 21
Type-II error, 21, 22–23

Unbiased statistic, 18
Universe, 1

Variance
  components, 64–69
  of population, 17
Variance ratio, 27
Variance tests, 26–28
  homogeneity, 69
  of two independent populations, 27–28
  on single population, 26–27

Whole-plot error, 261–62
Winer, B. J., 188

Yates, 104, 224
Yates method
  in covariance analysis, 397–401
  in factorial chi squares, 423–24
  for factorial experiments, 138–48,
    210–12, 297–99
  for fractional factorial, 309
  in fractional replication, 319–20
  for nested and nested-factorial experi-
    ments, 189–92
  in orthogonal polynomials, 357–58
  in polynomial regression, 358–64
  for randomized blocks and Latin
    squares, 108–14
  for randomized incomplete blocks,
    432–33
  for split-plot designs, 270–74
  for 2^2 factorial, 224–25
  for 2^3 factorial, 227, 228
  for 2^4 factorial, 285–86
  for 3^f factorial analysis, 255
Yield, 124–25
Youden square, 107, 434–36